C0-ASQ-958

INTRODUCTION TO SYSTEMS ENGINEERING, DETERMINISTIC MODELS

TUNG AU
Carnegie-Mellon University

INTRODUCTION TO SYSTEMS ENGINEERING, DETERMINISTIC MODELS

ADDISON-WESLEY PUBLISHING COMPANY
Reading, Massachusetts
Menlo Park, California · London · Don Mills, Ontario

This book is in the
ADDISON-WESLEY SERIES IN CIVIL ENGINEERING

Consulting Editor
THOMAS E. STELSON

Second printing, August 1973

 Printed in the United States of America. Published simultaneously in Canada. Library of Congress Catalog Card No. 69-13155.

ISBN 0-201-00363-5
CDEFGHIJKL-CO-79876543

PREFACE

This book is written primarily for an introductory course in systems engineering for undergraduate students. The development of such a course in the undergraduate curriculum has received its impetus on two fronts. First, there is the need for treating the process of decision making on a more objective and rational basis as the problems of engineering design and operation grow in size and complexity; and second, the development of mathematical methods in operations research and the expanded use of electronic computation have made it practical to recast engineering problems in a broader and more analytical context. I believe that the application of the methods of systems engineering will multiply rapidly once engineers in practice are familiar with the basic concepts of this approach.

In order to provide an adequate foundation of systems engineering methods for possible application to various areas of engineering, the introductory course is designed to emphasize the fundamental principles underlying different mathematical methods of analysis, with only simplified examples for illustration. To appreciate the advantages of systems methods, one must realize that much larger real-life problems can be solved by the same methods with the aid of computers. The importance of this introductory course in systems engineering may be likened to the first course in rigid-body mechanics in which the emphasis is also placed on the understanding of basic principles, with only illustrative elementary examples for application to different fields.

In selecting suitable materials for a first course, I have decided to include only deterministic models, which do not require knowledge of probability and statistics. A second course, which deals with probabilistic models as presented in a companion volume,* is a logical sequel. To carry further the analogy of mechanics, in which rigid-body mechanics is treated as a special case of deformable-body mechanics, the deterministic models may be regarded as a special case of the probabilistic models.

I have taught portions of the material in this book to undergraduates since 1962, and have varied the contents from class to class. I have also used the materials for a continuing education course for practicing engineers.

---

*Tung Au, Richard M. Shane, and Lester A. Hoel, *Fundamentals of Systems Engineering, Probabilistic Models*, Addison-Wesley, 1972.

The contents that are finalized in this book are so organized that related subjects are treated coherently although they have been presented as isolated topics in their developments; the materials included are more than sufficient for a one-semester course. The order of presentation has been arranged in such a manner that the later sections in most chapters (denoted by asterisks in the Table of Contents) may be omitted without disrupting the continuity and chapter sequence.

The prerequisites in mathematics for this book have been kept at the level of undergraduate engineering students. When additional mathematical techniques are required, they are introduced through an intuitive approach. Hence in each chapter elementary examples precede the presentation of the general method of solution. This approach is found to be effective in motivating students, and is also well received by practicing engineers.

In the preparation of this book, the published works of others who have made major contributions to various problems in this field have been most helpful, and these references are listed at the end of the chapters. However, the bibliography is selective rather than exhaustive. As a rule, references are made only to those to which the readers may want to turn for additional information. Problems for assignment are also included at the end of each chapter.

I wish to acknowledge my appreciation to many present and former colleagues in the Department of Civil Engineering at Carnegie-Mellon University, particularly to Thomas E. Stelson for his cooperation. Others who have either read portions of the manuscript or have furnished a problem for solution include George Bugliarello, Harold J. Day, Joseph S. Drake, William G. Fleck, Lester A. Hoel, Edward M. Krokosky, Ernest W. Parti, Richard M. Shane, Joseph F. Werner, and Andrew K. C. Wong. The manuscript has been painstakingly reviewed by Prof. Robert M. Stark of the University of Delaware and Prof. Gus Mavrigian of Youngstown State University to whom I am most grateful.

Pittsburgh, Pennsylvania T. A.
October 1973

CONTENTS

CHAPTER 1

INTRODUCTION

1-1 BASIC CONCEPTS

The role of engineers in the design and operation of modern complex engineering systems often entails technological and managerial decisions with far-reaching consequences. In general, an engineering system consists of a large number of inter-connected components, each of which may serve a different function but all of which are intended for a common purpose. The degree of success in achieving the common goal is a measure of the effectiveness of the system. In order to optimize successfully the effectiveness of either the design or operation of a highly complex engineering system, the process of decision making must be placed on an objective and rational basis. The development of operations research in the past two decades has provided much of the impetus for seeking the conceptual clarification of rational decision making as well as a high level of mathematical rigor in the methods of analysis. The term *systems engineering* is used in its broadest context to cover the comprehensive aspects of engineering practice, and the application of the modern rational approach to the formulation and solution of technological problems.

An urban area or a city can be looked upon as a complex system. Like many systems, it is a conglomeration of interacting subsystems. We speak of transportation systems, water supply systems, lighting and electric power systems, sanitary systems, school systems, distribution systems, heating systems, flood warning and control systems, communication systems, and many others. Each of these systems has many subsystems; for example, a transportation system may include bus systems, train systems, street-highway systems, traffic signal systems, pedestrian-sidewalk systems, parking systems, airport systems, and so on.

No system is really independent of others. For example, when parking is restricted or when parking fees are increased, a shift of modal split in transportation may occur. People who formerly drove may ride mass transit. Such a shift may cure traffic congestion which could not be solved by the most sophisticated applications of traffic signal and control theory. Basically, the ultimate goal of all decisions in engineering is either to minimize the effort or to maximize the desired benefit. Since many real decision situations are too complex to be treated mathematically, it is often necessary to introduce simplifying assumptions, i.e., by considering only the variables, relationships, and objectives that are deemed to be important or relevant by the decision maker. Thus the two key activities in systems engineering studies are (1) synthesis of a system, and (2) analysis and optimization of the system. The former refers to the conception of an acceptable system reasonably responsive to the real world situation and yet amenable to quantitative analysis; the latter infers the use of mathematical methods for analyzing and optimizing the effectiveness of the system.

1-2 SYSTEM SYNTHESIS

System synthesis is the initial conception of an engineering system for the purpose of satisfying certain human needs. In order to make an intelligent selection, engineers are often required to specify design and operation criteria, study system properties, propose alternative schemes, and assess technological and economical feasibility.

Let us consider the selection of a new mass transit system for a metropolitan area. We must first decide what the true objectives are in constructing this new facility, and what social and economical values are attached to different objectives. A system which can best satisfy the most important objective may fail completely to satisfy other objectives; conversely, a system which appears to be only mildly attractive in fulfilling the most important objective may turn out to fulfill the overall requirements most satisfactorily. For example, a newly developed transit system which provides the fastest speed for moving people but which also requires infrequent stops may prove to be less satisfactory than a system with slower moving vehicles that can make more frequent stops. This is clearly a trade-off between moving more people and moving people faster. If we do not recognize the real needs at the initial stage, we may end up with a system which is not only useless, but beyond rectification. We must also consider many other properties of the system, such as the speed and noise level of the vehicles, and the appearance and right-of-way requirements of the supporting structure. After examining these factors, we begin to consider alternative schemes that are in existence or anticipated, exploring paths that are relatively promising and ignoring paths that are relatively sterile. Finally, we must select a system which can be constructed—and done so economically.

An engineering system may be so enormous that, because of practical considerations of time or other difficulties, the decision maker is forced to deal with the subsystems making up the whole system instead of with the system itself. Thus the objectives of each subsystem, rather than those of the whole system, are optimized. While the best solution of the total system is not attainable, this approach at least provides a rational method of obtaining a reasonable approximation, if no conflicting factors are introduced in setting up the conceptual models for the subsystems. Actually, every system is a subsystem of a yet larger system, and every system also consists of subsystems or component systems. The systems approach aims to make the subsystems complement rather than hinder each other.

Consider, for instance, the design of a river crossing in a metropolitan area. First the river crossing is isolated as a separate engineering system from a larger system, i.e., the entire highway project of which this river crossing is a subsystem or component system. The topography, soil conditions, and traffic patterns near the proposed crossing site are a few of the more relevant factors affecting the design. The proposed crossing may be a bridge or an underwater tunnel, depending on which will lead to more desirable results. Thus the basic decision is to choose between these two alternatives. If the cost of construction is the principal criterion for making the choice, the optimal solution is one which requires a minimum cost. A wise decision can be made only after extensive investigation of the interactions of various phases of the engineering system encompassing the entire project. After a preliminary study, however, it may become obvious that the construction of a tunnel is either infeasible or uneconomical, and further studies may be confined to the optimization of the bridge design. In this more limited problem, again a preliminary study is made of economics of various bridge types, such as a truss, an arch, or a suspension bridge, for the proposed crossing site. If one bridge type is clearly far superior to the others, the remaining problem is to seek the most economical arrangements of members in the selected bridge type. Hence the optimal design of a river crossing is reduced in this example to the optimal design of a particular bridge type.

We cannot overemphasize the importance of independent judgment in assessing the values and objectives of a proposed engineering project. If an incorrect premise is assumed, no amount of mathematical rigor or elegance can save the resulting system from the error of misjudgment. As an example, we consider the planning of a network of highways connecting three towns which are equidistant from each other. If the primary objective of the network is to provide a minimum distance of travel between any two towns, the road system in Fig. 1–1(a) will result. On the other hand, if the primary objective is to provide a minimum total mileage of roads in order to reduce cost, the road system in part (b) of the figure will prevail. Thus we find that the conceptual model must reflect the characteristics of the real system which are to be investigated.

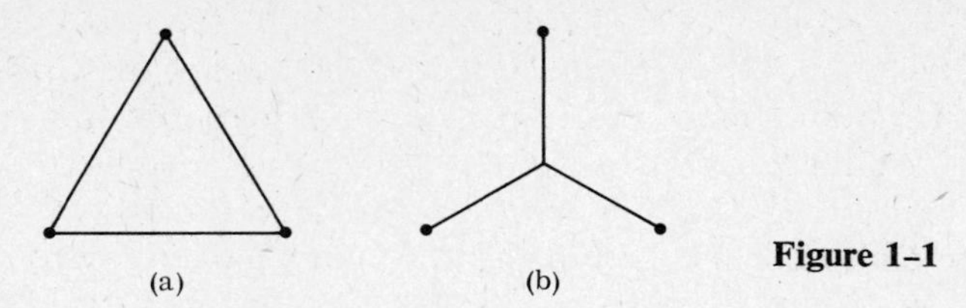

Figure 1-1

It is evident that creative technological innovation and sound engineering judgment hold the key to imaginative solution of engineering problems. However, the design or operation process can be recast in an expanded context of a large engineering system so that the decision aspect of the problem can also be formulated quantitatively. It is also clear that the larger the selected engineering system, the better the opportunities for a trade-off of undesirable characteristics in order to establish the best solution. On the other hand, the whole system may be unmanageable because of its magnitude and one must be satisfied with the optimization of the subsystems. Therefore the process of formalizing a problem involves the selection of a conceptual model which expresses the proper relationships of relevant variables and parameters, and the optimization of the desired outcome on the basis of the conceptual model. The task of identifying proper variables in forming the conceptual model and that of evaluating the results obtained from the mathematical solution are essential parts of the entire systems engineering process.

1-3 SYSTEM ANALYSIS AND OPTIMIZATION

The process of system analysis and optimization includes essentially the following steps: (1) the outcomes for each possible action or feasible solution are predicted on the basis of analysis; (2) these outcomes are then evaluated according to some scale of value or desirability; and (3) a criterion for decision, based on the objectives of the system, is used to determine the most desirable action or optimal solution. This process is illustrated schematically in Fig. 1-2.

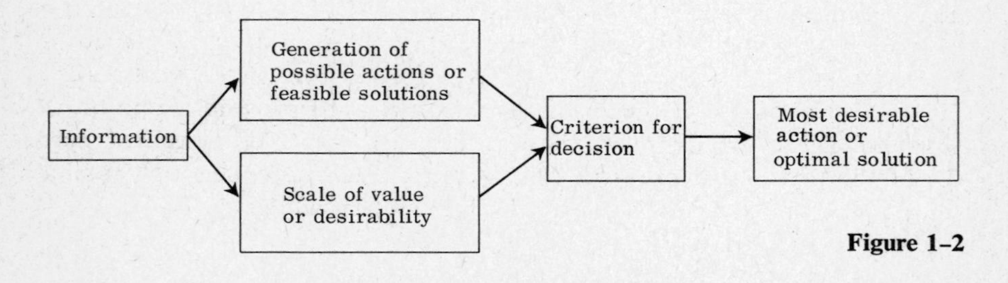

Figure 1-2

The methods of quantitative decision making are useful tools for system analysis and optimization. Although many engineering systems are inherently ill-structured, it is reasonable to expect that at least in some phase of design and operation, the process of decision making can be formulated quantitatively, particularly in view of the great variety of mathematical methods of analysis and optimization that are currently available.

From the viewpoint of methods of analysis, we shall distinguish two major types of decision problems. A decision problem in which the choice of a set of values for the decision variables can be assumed to lead invariably to a specific value of the function is called *deterministic*, and one which does not satisfy this requirement is broadly classified as *nondeterministic*. In other words, each decision in a deterministic problem is known to lead invariably to a specific outcome under conditions of certainty; but each decision in a nondeterministic problem leads to a set of possible outcomes under conditions of uncertainty. In the latter case, a decision implies a gamble with respect to future conditions, and the probabilities of occurrence of the possible outcomes have important effects on the choice of a course of action as the best bet. In reality, the demarcation of certainty and uncertainty is not always clear-cut. Furthermore, in each case all decision variables may be determined either simultaneously or sequentially.

The mathematical methods for the solution of *deterministic problems* may be placed in the broad category of *mathematical programming*, which provides mathematical frameworks and computational methods for choosing the most desirable programs or schedules of action among many possible alternatives. Although problems in the real world may be very complex and involve many variables, very often only a small number of variables can be freely controlled by the decision maker. Other variables and parameters which are determined either by the state of nature or by relevant physical principles are uncontrollable. In general, an objective function relating various decision variables is introduced for the purpose of ranking the desirability of the outcome of different decisions. After a decision problem is formulated, the best decision or optimal solution is sought by means of available mathematical methods. The final decision may be a set of values or a policy to be followed in the course of action.

The decision variables in a problem may involve cost, material, space, time, activity, or any combination of these quantities. The funtional relationships of a model may or may not be independent or consistent. There may appear nonnegative constraints, equality constraints, inequality constraints, or no constraints at all. The objective function may be linear or nonlinear with respect to decision variables. Hence an optimal solution may not even exist in some cases. However, many significant engineering problems can be tackled by mathematical programming methods such as *linear*, *nonlinear*, *and dynamic programming*, *network flow*, *and critical path scheduling*.

The consideration of *probability* and *statistical inference* has opened a new vista in the solution of *nondeterministic problems* in which the state of nature is probabilistic or uncertain. The weights by which a decision-maker evaluates the possible states of nature are expressed by the probabilities of occurrence; and his preference for possible consequences is expressed by the measure of value, or utility. When the state of nature is uncertain, further information is obtained from experimentation. The objective of the statistical methods is to choose, with or without experimentation, a course of action cosistent with the probabilities of the occurrence of the states and with utilities attached to the consequences.

The application of probability and statistical methods to the solution of engineering problems is rising rapidly. Many problems can be treated in this context by the use of *stochastic process*, *queueing theory*, *information theory*, and *statistical decision theory*.

1-4 CHARACTERISTICS OF DETERMINISTIC MODELS

The formulation of a quantitative decision problem represents a process of abstraction to transform the external environment into a conceptual or mathematical model which encompasses the functional relationships of pertinent variables. A feasible solution to the problem represents the selection of a set of variables which satisfy the required functional relationships. A criterion must be established to measure the value or the effectiveness of a solution in comparison with other feasible solutions. The set of variables which produces an optimal solution represents the best decision.

Very often the probabilistic nature of the decision variables can be ignored, at least in the first model of investigation. Then, the decision problems are treated as deterministic, and such problems are the primary concern of this book. The general characteristics of deterministic models are as follows:

1. A set of decision variables. Although the physical problems may be very complex and involve many variables, only the variables that can be freely controlled by the decision maker are regarded as decision variables in the mathematical model. The variables which respond to the decisions inside the system are called *endogenous* variables, and the variables which represent influences outside the system are called *exogenous* variables.

2. A set of constraints. The conditions of external environment impose a number of constraints on the decison variables. For example, many physical variables cannot have a negative value (although they can be zero); such conditions are referred to as nonnegative constraints. Technical feasibility and economical consideration may also be the constraints of a system. The constraint conditions are usually given as equations and/or inequalities.

3. An objective function. The task of identifying proper objectives is an essential step in the decision process. An objective function relating various decision variables is introduced for the purpose of ranking the desirability of the outcome of different decisions. In general, we either maximize the effectiveness or minimize the cost of the system as represented by the objective function.

4. An optimal solution. After a decision problem is formulated quantitatively, the optimal solution or the best decision is sought by means of available mathematical methods. The optimal solution may be expressed by a set of values or by a policy for the decision variables leading to the desired objective.

At the risk of oversimplification, several elementary examples of deterministic decision problems in engineering are used for illustration, despite the fact that the real value of the systems engineering approach lies in the solution of problems of extended complexity.

Example 1–1. A single wheel load P moves along a crane runway girder of span length L. Determine the maximum bending moment in the girder if (a) the wheel can move freely, and (b) there is a stopper at a fixed distance b ($<L/2$) from the left support as the wheel moves from the left to the right.

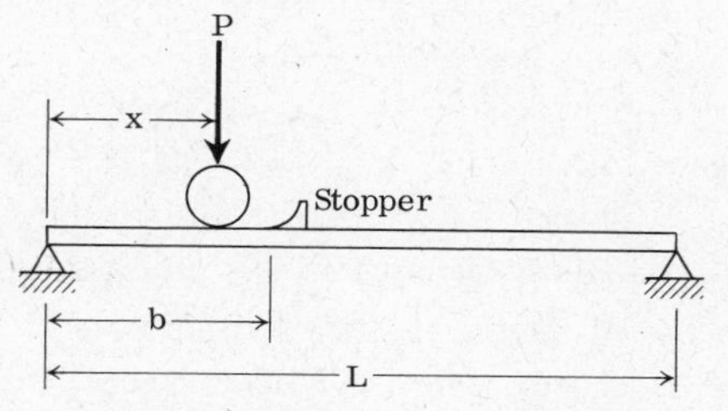

Figure 1–3

This is an elementary problem in mechanics which is familiar to all engineering students. Yet it has the basic characteristics of a deterministic decision problem. The location of the moving load P is the only decision variable in this maximization problem; and we shall denote by x the distance between the left support and the moving load P. The conceptual model can be derived from Fig. 1–3 by noting the functional relationship between the decision variable and other parameters. From static equilibrium, the bending moment M under the load is found to be

$$M = Px(L - x)/L\,.$$

Furthermore, according to the physical constraint that the wheel cannot move to the left of the left support, it is seen that $x \geq 0$. Also, for case (a) $x \leq L$; and for case (b) $x \leq b$. The objective function is the maximization of the bending moment M with respect to the distance x. Thus the problem can be stated as follows:

a) Maximize $M = Px(L - x)/L\,,$
subject to $0 \leq x \leq L\,.$

b) Maximize $M = Px(L - x)/L\,,$
subject to $0 \leq x \leq b < L/2\,.$

The conditions $0 \leq x \leq L$ in case (a) and $0 \leq x \leq b$ in case (b) are constraints which play an important part in determining the optimal value of the objective function. Physically, we can visualize that the maximum bending moment occurs at $x = L/2$ in case (a), but that it will occur at $x = b$, where $b < L/2$, in case (b). Thus the optimal solution for case (a) is found to be $M = PL/4$ and that for case (b) is $M = Pb(L - b)/L$.

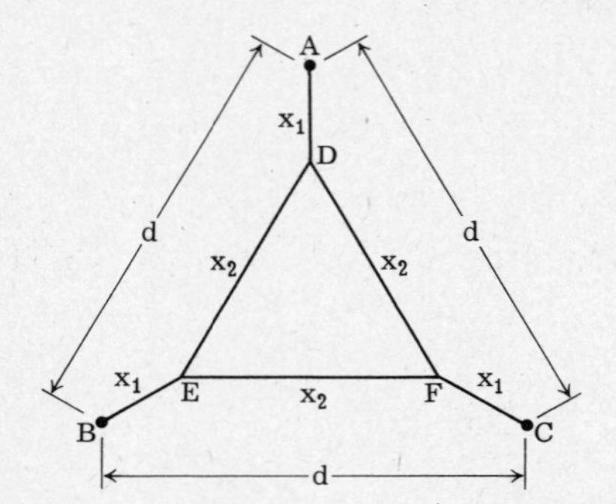

Figure 1-4

Example 1-2. Three towns, A, B, and C, are equidistant (at d mi) from each other. A network of roads in the form shown in Fig. 1-4 has been proposed for linking the three towns. The residents of the towns want a network which will minimize the distance of travel between any two towns, but the road commissioner wants a network which will minimize the total mileage of roads connecting all three towns. Find the optimal solution if

a) the wishes of the residents are followed, or
b) the opinion of the road commissioner prevails.

Let x_1 be the distance from a town to a junction in the network, and x_2 be the distance between two junctions, as shown in Fig. 1-4. Therefore for the equilateral triangles ABC and DEF,

$$d = 2x_1 \cos 30^\circ + x_2 = \sqrt{3}x_1 + x_2 .$$

For case (a), the problem can be stated as follows:

$$\begin{aligned} &\text{Minimize} && z = 2x_1 + x_2 , \\ &\text{subject to} && x_1 \geq 0 , \qquad x_2 \geq 0 , \\ &&& \sqrt{3}x_1 + x_2 = d . \end{aligned}$$

For case (b), the problem becomes:

$$\begin{aligned} &\text{Minimize} && z = 3x_1 + 3x_2 , \\ &\text{subject to} && x_1 \geq 0 , \qquad x_2 \geq 0 , \\ &&& \sqrt{3}x_1 + x_2 = d . \end{aligned}$$

The solution for each case can easily be formed by substituting the constraint condition into the objective function

a) In the first case, let $x_2 = d - \sqrt{3}x_1$, or

$$\text{Minimize} \qquad z = 2x_1 + (d - \sqrt{3}x_1) = d + (2 - \sqrt{3})x_1 .$$

Thus minimum z occurs when $x_1 = 0$, since $(2 - \sqrt{3})$ is positive; the corresponding $x_2 = d$ and $\min z = d$. The total mileage of roads is therefore $3d$.

b) In the second case, let $x_1 = (d - x_2)/\sqrt{3}$, or

$$\text{Minimize} \qquad z = \frac{3}{\sqrt{3}}(d - x_2) + 3x_2 = \frac{3}{\sqrt{3}}[d + (\sqrt{3} - 1)x_2] .$$

Then, minimum z occurs when $x_2 = 0$, since $(\sqrt{3} - 1)$ is positive; the corresponding $x_1 = d/\sqrt{3}$ and $\min z = 3d/\sqrt{3} = \sqrt{3}d$.

Example 1-3. A contractor has three units of heavy construction equipment each available in Cleveland and New York, and he has construction jobs in Philadelphia, Chicago, and Pittsburgh that require two, three, and one units of such equipment, respectively. The unit shipping costs c_{ij} between cities i and j (for $i = 1, 2$ and $j = 1, 2, 3$) are noted on the lines linking respective origins and destinations in Fig. 1-5. Formulate the problem of finding the shipping pattern which minimizes the cost.

Referring to Fig. 1-5, let the quantity to be shipped from origin 1 to destination 1 be denoted by x_{11}, origin 1 to destination 2 be denoted by x_{12}, etc. Since the total amount in supply $(3 + 3 = 6)$ is equal to the

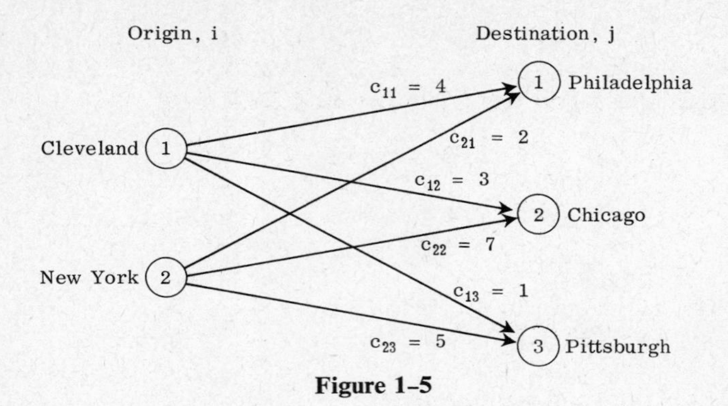

Figure 1-5

total amount in demand $(2 + 3 + 1 = 6)$, the total shipment from each origin to all destinations is the amount available at the origin. Thus

$$\text{at origin 1,} \qquad x_{11} + x_{12} + x_{13} = 3\,;$$
$$\text{at origin 2,} \qquad x_{21} + x_{22} + x_{23} = 3\,.$$

Similarly, the total shipment received by each destination from all origins is the amount required at the destination. Thus

$$\text{at destination 1,} \qquad x_{11} + x_{21} = 2\,;$$
$$\text{at destination 2,} \qquad x_{12} + x_{22} = 3\,;$$
$$\text{at destination 3,} \qquad x_{13} + x_{23} = 1\,.$$

Hence this transportation problem can be formulated by minimizing the total shipping cost z, that is:

$$\text{Minimize} \qquad z = 4x_{11} + 3x_{12} + x_{13} + 2x_{21} + 7x_{22} + 5x_{23}\,,$$
$$\text{subject to} \qquad x_{11} \geq 0\,, \quad x_{12} \geq 0\,, \quad x_{13} \geq 0\,,$$
$$x_{21} \geq 0\,, \quad x_{22} \geq 0\,, \quad x_{23} \geq 0\,.$$
$$\sum_{j=1}^{3} x_{1j} = 3\,, \qquad \sum_{j=1}^{3} x_{2j} = 3\,,$$
$$\sum_{i=1}^{2} x_{i1} = 2\,, \qquad \sum_{i=1}^{2} x_{i2} = 3\,, \qquad \sum_{i=1}^{2} x_{i3} = 1\,.$$

Example 1-4. In the highway network shown in Fig. 1-6, the nodes represent cities of (1) Cleveland, (2) New York, (3) Philadelphia, (4) Chicago, and (5) Pittsburgh. Some of these cities are linked directly and others only indirectly by limited access routes. The unit shipping costs between directly linked cities for transporting the construction equipment units described in the previous example are given at the links in Fig. 1-6. For the same supply and demand cited in the previous example, formulate the problem of finding the shipping pattern which minimizes the cost.

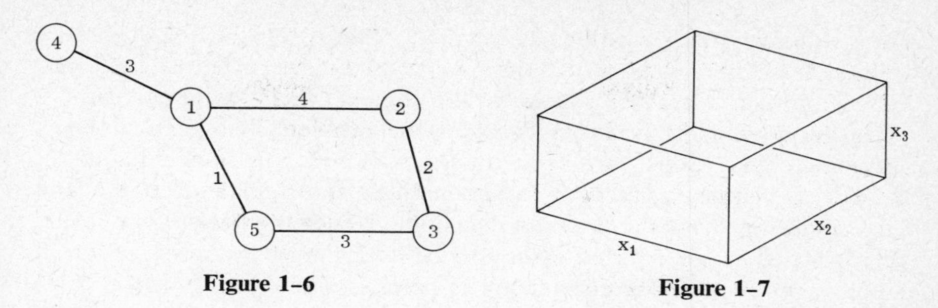

Figure 1-6

Figure 1-7

Let us first allow that the unit shipping costs between cities not directly linked by the network are obtained by using the minimum cost routes. For example, a shipment from New York to Pittsburgh may go through Cleveland or Philadelphia, since the unit cost on each route is the same ($4 + 1 = 5$ vs. $2 + 3 = 5$); on the other hand, a shipment from Cleveland to Philadelphia should go through Pittsburgh instead of New York because the former route costs less ($1 + 3 = 4$ vs. $4 + 2 = 6$). If all shipping unit costs for all indirect routes are computed in this manner, the resulting problem is identical to Example 1-3.

When the many cities are included in a network, the computation of unit shipping costs for indirect routes becomes rather tedious if minimum cost routes must be selected. Another approach which treats shipment between cities as "flows" in a network may be used. Then the problem becomes the determination of the flow pattern for which the cost is a minimum.

Example 1-5. A manufacturer of large crates has in stock a large quantity of steel bars which are used for the frames of crates. In order to make use of these bars which are b ft long, each bar is to be cut into three sections x_1 x_2, and x_3, corresponding to length, width and depth of the crate respectively, as shown in Fig. 1-7. (Note that four sets of cut bars are used for each frame.) Determine the lengths x_1, x_2, and x_3 so that the volume of the crates will be maximum. Formulate the problem only.

In this case, the decision variables are x_1, x_2, and x_3, and the volume V of each crate is seen to be the product of x_1, x_2, and x_3. Then the problem can be formulated as follows:

$$\begin{aligned} &\text{Maximize} && V = x_1 x_2 x_3 , \\ &\text{subject to} && x_1 \geq 0 , \quad x_2 \geq 0 , \quad x_3 \geq 0 , \\ &&& x_1 + x_2 + x_3 = b . \end{aligned}$$

We may proceed to solve this problem by substituting the constraint condition $x_1 + x_2 + x_3 = b$ into the objective function, thus eliminating one of the variables. Then the maximization can be carried out by the differentiation of the modified objective function. We shall not discuss the details of calculation at this time; nevertheless we recognize the problem as one which may be solved by the classical methods of calculus.

On the other hand, we may consider the problem as one involving a sequence of decisions, one after another. After x_3 is selected, the remaining decisions, x_2 and x_1, must constitute an optimal solution at the state resulting from the previous decision. Hence the decisions x_1, x_2, x_3 are regarded as an optimal policy to be determined by optimization in sequence with the same constraints as before. That is,

$$\operatorname{Max} V = \operatorname*{Max}_{x_3} \left[x_3 \operatorname*{Max}_{x_2} (x_2 \operatorname*{Max}_{x_1} x_1) \right],$$

in which the maximization at each stage is confined to a single variable.

It is sufficient to point out at this time that the first approach to the solution of this problem is a single-stage or simultaneous optimization process, while the second approach is essentially a multistage or sequential optimization procedure. Usually, a physical problem fits more naturally into one category or the other although it can be set up artificially, at the sacrifice of computational efficiency, to suit the less logical process if necessary.

Example 1-6. Two general contractors are invited to participate jointly in a huge construction project which, the owner thought, would require the resources of both. On the other hand, each contractor feels that he can handle the whole project, and wants to get as big a share of profit from the project as he can. Therefore, the owner decides that one of the contractors will divide the project into two parts but the other contractors will then be permitted to select his part first. How should the project be divided, on the basis of percentage of the total construction cost, so that each can get as much profit as possible? Discuss the problem qualitatively.

This is a problem of optimization with competition. Each contractor must consider not only his own preference but also those of his competitors. If each of the contractors, A and B, knows that his competitor is as competent as himself, there is no reason to believe that his competitor will select the part which is less profitable. In the final analysis, the rational strategy is to divide it in such a way that each part will yield the same profit regardless who makes the division. This approach might seem pessimistic, but it is the only logical solution if each contractor wants to get as much profit as possible, which is half of the total profit for the project in this case.

If both contractors are interested in getting the largest amount of work to take up the slack period in their respective organizations instead of getting the biggest share of profit, the strategies of the competitors would have been different. In this case, the rational decision would be to divide the amount of work equally for both contractors.

1-5 GENERAL APPROACH TO DETERMINISTIC DECISION PROBLEMS

After examining several simple examples, it is possible to generalize the formulation of deterministic decision problems. The decision variables in such problems are usually nonnegative real numbers. A given deterministic decision problem can be formulated by introducing a suitable mathematical model describing the functional relationship of the decision variables. In general, the objective is to optimize a function of nonnegative real variables subject to certain arbitrary constraints. The optimizing process may either be maximization or minimization, according to the physical requirements of the problem. We can select a mathematical model involving only maximization or minimization without losing generality.

A deterministic decision problem which involves the optimization of n nonnegative real variables subject to m arbitrary constraints is called a *general mathematical programming problem*. It may be expressed in the minimization form:

$$\begin{aligned} \text{Min} \quad & z = f(x_1, x_2, \ldots, x_n)\,, \\ & x_j \geq 0\,, \qquad j = 1, 2, \ldots, n\,, \\ \text{subject to} \quad & g_i(x_1, x_2, \ldots, x_n) \geq 0\,, \qquad i = 1, 2, \ldots, m\,. \end{aligned}$$

in which $j = 1, 2, \ldots, n$ refers to n *separate* nonnegative constraints, and $i = 1, 2, \ldots, m$ refers to m *separate* inequality constraints. This minimization problem may be changed to the maximization form if we multiply the function z and the inequality constraints by -1, that is,

$$-z = -f(x_1, x_2, \ldots, x_n)$$

and

$$-g_i(x_1, x_2, \ldots, x_n) \leq 0\,.$$

Letting $A = -z$ and noting that minimization of z is equivalent to the maximization of A, we find that the equivalent maximization form becomes

$$\begin{aligned} \text{Max} \quad & A = F(x_1, x_2, \ldots, x_n)\,, \\ & x_j \geq 0\,, \qquad j = 1, 2, \ldots, n\,, \\ \text{subject to} \quad & G_i(x_1, x_2, \ldots, x_n) \leq 0\,, \qquad i = 1, 2, \ldots, m\,, \end{aligned}$$

in which

$$F(x_1, x_2, \ldots, x_n) = -f(x_1, x_2, \ldots, x_n),$$
$$G_i(x_1, x_2, \ldots, x_n) = -g_i(x_1, x_2, \ldots, x_n).$$

Note that in all discussions in this book, Min z is an abbreviation for "Minimize z" while min z is an abbreviation for the "minimum value of z"; similar abbreviations "Max z" and "max z," respectively, are used in maximization. The inequality sign $\geq$ (greater than or equal to), is used instead of the strict inequality $>$ (greater than) for the nonnegative constraints, since zero is an admissible value; similarly, the inequality sign $\leq$ is used instead of the strict inequality $<$ for nonpositive constraints. The nonnegative constraints are, in fact, lower bounds of the decision variables, the upper bounds being infinity unless otherwise specified. The constraint conditions generally represent the boundary of the feasible region for the objective function, and they may either be equality, strict inequality, or mixed inequality.

The change of sign in the objective function, when a minimization problem is transformed into the maximization form, can be illustrated graphically by a simple example of the minimization of a function of one variable without constraint, i.e.,

$$\text{Min } z = f(x).$$

Let $A = -z = -f(x)$. Then we have

$$\text{Max } A = -f(x).$$

If $z = f(x)$ is represented by the curve in Fig. 1-8(a), then $A = -f(x)$ is represented by the curve shown in part (b). Conversely, for a given maximization problem,

$$\text{Max } A = F(x),$$

an equivalent minimization form is obtained by letting $z = -A = -F(x)$.

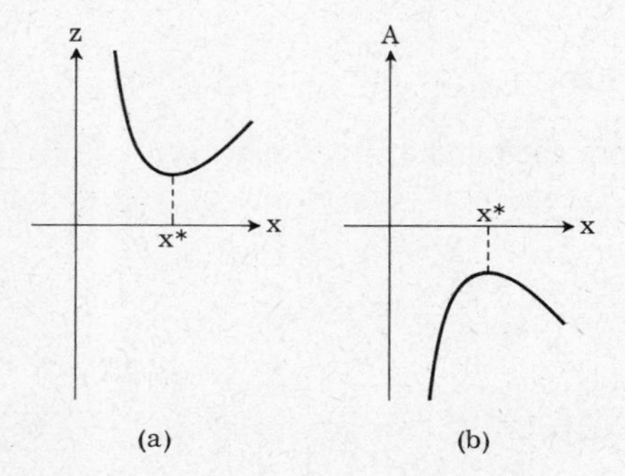

Figure 1-8

Hence we get

$$\text{Min } z = -F(x) .$$

The inequality constraints in a minimization problem actually remain unchanged in the equivalent maximization form, and conversely. However, their appearance in one form differs from that in the other by the direction of the inequality sign because of the multiplication by -1 to both sides of each inequality during the transformation. This arrangement is a matter of convenience since the direction of the inequality sign for the inequality constraints corresponds to the relation of z to min z, or A to max A, as the case may be. For example, in minimization,

$$f(x_1, x_2, \ldots, x_n) \geq \min z ,$$
$$g_i(x_1, x_2, \ldots, x_n) \geq 0 , \qquad i = 1, 2, \ldots, m ;$$

and in maximization,

$$F(x_1, x_2, \ldots, x_n) \leq \max A ,$$
$$G_i(x_1, x_2, \ldots, x_n) \leq 0 , \qquad i = 1, 2, \ldots, m .$$

There is no single method for the solution of the general mathematical programming problem. Instead, a number of optimization methods have been developed for various special cases of the objective functions and/or constraints. However, the optimal solution of a mathematical programming problem, if it exists, can in general be expressed by the set

$$(x_1^*, x_2^*, \ldots, x_n^*, z^*) \qquad \text{or} \qquad (x_1^*, x_2^*, \ldots, x_n^*, A^*) ,$$

in which $(x_1^*, x_2^*, \ldots, x_n^*)$ is the set of values of the variables leading to the optimal solution, and $z^* = \min z$ or $A^* = \max A$ is the value of the optimal solution itself. It is possible to have more than one set of values of the variables which lead to the same value for the optimal solution. In other words, although $z^* = \min z$ or $A^* = \max A$ is unique, the set $(x_1^*, x_2^*, \ldots, x_n^*)$ is not necessarily so. For some problems, it is difficult or even impossible to obtain an optimal solution; we may try to find a near-optimal solution. However, it is usually difficult to evaluate how good a near-optimal solution is.

The mathematical methods of optimization have long been developed in geometry and calculus. For differentiable functions with continuous variables, the methods of calculus can still be used to a limited extent. However, for problems involving large number of variables and constraints, iterative search methods or optimization techniques are most powerful. When the objective function, as well as the constraints, are linear functions of decision variables, the methods of linear programming are introduced for the solution of such problems. Special forms of linear programming are applicable to a large class of deterministic decision problems in transportation and allocation. Further extensions of linear programming lead to the development of methods for network analysis and critical-path scheduling. Certain nonlinear problems can also be solved by direct-search methods. The consideration of sequential optimization instead of simultaneous optimization leads to dynamic programming. Finally, the method of optimization under competition in which the decision variables are subject to the control of more than one decision maker provides a vital link between linear programming and theory of games.

The objective of this book is to present a broad view of the important prototype models and the mathematical methods pertinent to the solution of deterministic decision problems. It will serve as a critical introduction to the formulation and solution of such problems without an extensive presentation of the theories underlying the mathematical methods. Emphasis is placed in the interrelations of different prototype models and the alternatives that may be used for a given class of problems, calling attention to the unity of purpose of various methods.

1-6 FUTURE CHALLENGE

The rapid progress in the development and application of systems engineering in recent years has been phenomenal. Thanks to interdisciplinary efforts, great strides have been made in the rational approach to system synthesis as well as system optimization. Even the processes of creativity and human judgment have been, to some degree, formalized and simulated on the computer by psychologists and computer scientists. Pioneering work in management science and mathematics has also added a great wealth of knowledge to system modeling and opimization methods. Impressive though this progress may appear, we have barely scratched the surface of this new field. Engineers today face a great challenge and opportunity in furthering the development and application of systems engineering to produce a real impact on engineering practice.

The mathematical methods for the solution of prototype models are growing rapidly, and many of them are applicable to both the design and the operation of engineering systems. However, the conceptual models for engineering operation are usually simpler than those for engineering design, since engineering designs are usually further complicated by factors affected by the physical laws governing such systems. Consequently, the application of these methods to engineering operation and industrial management is widely known as the methods of *operations research*. On the other hand, similar methods have been applied extensively to the design of automatic control, communication, and computer systems. Thus the term *systems engineering* conveys a restricted connotation to those system specialists. However, the current trend of development takes on a broader view in regarding systems engineering as an integrated approach to decision problems in engineering operation and design.

The role of systems engineering in decision making promises to be much enlarged as the mathematical methods are supplemented by new heuristic approaches in information gathering, problem solving, optimum searching, and evaluating processes. Examples of such applications are *system simulations* and *heuristic games*. Since system effectiveness can best be achieved by considering the largest overall system instead of its subsystems, there is a tendency to treat decision problems in larger and larger context. Thus the systems approach has an interdisciplinary outlook, and often requires a team of specialists in behavioral sciences as well as in engineering. As can be expected, the systems approach can be carried to the extreme by considering the largest possible system, which is the whole universe, and a *general systems theory* is the fond speculation of philosophers in a technological society. It is our modest hope that the reader will be able to master the elementary methods of quantitative decision making in this book without necessarily being a philosopher.

REFERENCES

1-1. GOSLING, W., *The Design of Engineering Systems*, Heywood, London, 1962.

1-2. SAATY, THOMAS L., *Mathematical Methods of Operations Research*, McGraw-Hill, New York, 1959.

1-3. CHURCHMAN, C. W., R. L. ACKOFF, and E. L. ARNOFF, *Introduction to Operations Research*, Wiley, New York, 1957.

1-4. SASIENI, M., A. YASPAN, and L. FRIEDMAN, *Operations Research Methods and Problems*, Wiley, New York, 1959.

1-5. CARR, C. R., and C. W. HOWE, *Quantitative Decision Procedures in Management and Economics*, McGraw-Hill, New York, 1964.

1-6. BROSS, I. D. J., *Design for Decision*, MacMillan, New York, 1953.

1-7. AU, T., "Heuristic Games for Structural Design," *Journal of Structural Division, ASCE*, **92,** No. ST6 (1966), 499-509.

PROBLEMS

P1-1. In Example 1-1, the functional relationship between the bending moment M under load P and the distance x of load P from the left support is expressed by

$$M = Px(L - x)/L\ .$$

Plot this relationship of M versus x for $x = 0$, $L/4$, $L/2$, $3L/4$, and L. Determine graphically the maximum moment M in the beam given that the constraint condition is $0 \leq x \leq L$.

P1–2. Represent both cases (a) and (b) in Example 1–2 graphically by using in each case the optimal value of z obtained analytically in the example.

P1–3. In Example 1–3, the constraint conditions have been found to be as folllows:

$$\begin{aligned} x_{11} + x_{12} + x_{13} &= 3 , \\ x_{21} + x_{22} + x_{23} &= 3 , \\ x_{11} + x_{21} &= 2 , \\ x_{12} + x_{22} &= 3 , \\ x_{13} + x_{23} &= 1 . \end{aligned}$$

Show that this set of equations is not independent. In other words, any equation in the set of equations may be derived from the remaining four equations.

P1–4. In traveling from town A to town B by the intertown bus, one must go through a web of arterial roads of 20 mi or more. In order to cut down the distance of traveling, a bypass is to be constructed from A to C which is some intermediate point between A and B (see Fig. P1–4). Because only limited construction funds are available, the length of the new bypass (which is expected to be one-half of the length of existing route AC) will be shorter than or equal to the distance between C and B through the arterial roads. What is the minimum possible distance of travel between A and B if the new bypass AC is to be constructed under such constraints? Formulate the problem.

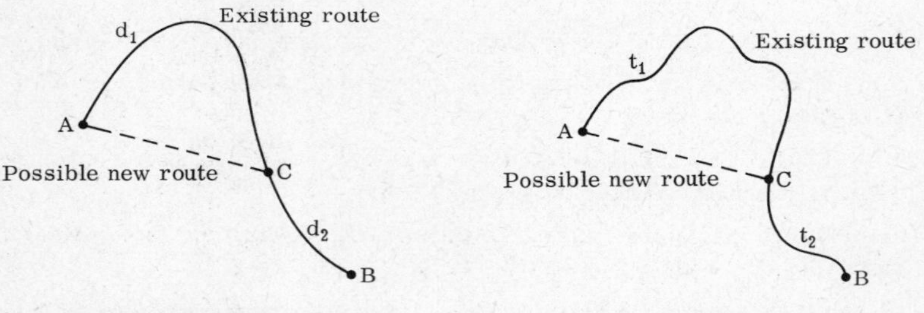

Figure P1–4 **Figure P1–5**

P1–5. The travel time by car through the existing arterial roads between the two towns A and B takes 1 hr or more. In order to reduce the total travel time, a freeway is to be constructed from A to C, where C is an intermediate point between A and B. The travel time on the new freeway between A and C is expected to be one-half the original time. Because only limited funds are available for construction, the length AC of the freeway must be so restricted that the new travel time on the freeway should be less than or equal to travel time in arterial roads between C and B (see Fig. P1–5). What is the minimum possible travel time between A and B if the freeway is constructed under such constraints? Formulate the problem.

P1–6. Determine graphically the optimal solution of the following problem (for $x \geq 0$, $y \geq 0$):

$$\begin{aligned} \text{Max} \quad & A = x^2 + y^2 , \\ \text{subject to} \quad & \sqrt{3}x + y = 8 . \end{aligned}$$

P1–7. Consider the following optimization problem with equality constraints:

$$\begin{aligned} \text{Maximize} \quad & A = kx_1x_2 , \\ \text{subject to} \quad & 0 \leq x_1 , \qquad 0 \leq x_2 , \\ & x_1 + x_2 = L , \end{aligned}$$

in which x_1 and x_2 are variables, and k and L are constants. By use of the equality constraint, eliminate one of the variables from the objective function and determine the maximum value of A.

P1–8. Sealed bids are solicited for the sales of two groups of surplus construction equipment evaluated at $60,000 and $40,000 for groups I and II, respectively. Two bidders, A and B, each planning to spend no more than a total of $70,000 for both groups of equipment, are bidding for both groups of equipment simultaneously. In other words, when a bidder places a bid of x dollars for group I, he has at the most $(70{,}000 - x)$ dollars for group II. If both bidders happen to submit identical bids for both groups, each will be awarded one group (I or II) by the owner. What should be the bids of A and B if each bidder is interested in a maximum gain (in dollar values)?

CHAPTER 2

METHODS OF CALCULUS

2–1 ELEMENTARY EXAMPLES

The methods of differential calculus for determining maxima and minima (generally speaking, extrema) of functions of real variables are applicable to a class of decision problems in which the objective function has a maximum and/or a minimum within the bounded region. In the absence of constraint equations, the optimization process involves nothing more than the application of derivatives in investigating the extrema of the objective function; or if constraint equations can be solved first explicitly in terms of some variables, these variables can be eliminated by direct substitution into the objective function before differentiation. A few simple examples involving functions of one or two variables are given as illustrations.

It will soon become obvious that the optimization of functions of more than one variable generally leads to great complexities, particularly when equality constraints, nonnegative constraints, and/or inequality constraints are present. In order to appreciate these complexities, certain mathematical concepts must be introduced and explained in this chapter. The primary purpose of this exposition is to illustrate the *limitations* rather than the capabilities of the classical methods of calculus in the solution of the general mathematical programming problem, thus leading logically to the treatment of only linear problems in subsequent chapters. The mathematical rigor in this chapter can be glossed over or omitted entirely if so desired.

Example 2–1. A cement manufacturer has an agreement to supply a builder at a constant rate of R bags of cement per day. If the manufacturer starts a run, it can produce at a constant rate of K bags per day (with $K > R$). Usually, a total volume of Q bags will be produced before a run is stopped, because a known fixed cost C_0 will be involved in setting up each production run. On the other hand, if the product is not shipped out on the same day, the known storage cost will be C_1 per bag per day. What is the optimal size of each run so that the average production cost per day will be minimum? How frequently should the runs be made?

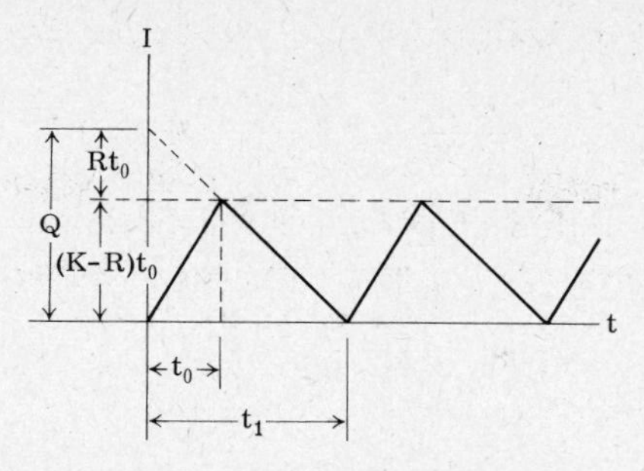

Figure 2–1

The fluctuation of inventory I with respect to time t can be represented by a series of triangles, as shown by solid lines in Fig. 2–1. Let t_0 be the number of days for each run, and t_1 be the number of days between production runs. Since the volume Q produced in t_0 days is exhausted in t_1 days, we have $t_0 = Q/K$ and $t_1 = Q/R$. During the period of production t_0, the inventory is accumulated at a rate of $(K - R)$ bags per day, since K bags are produced while R bags are depleted per day. The inventory reaches $(K - R)t_0$ bags when the production stops. Thus

$$(K - R)t_0 = (1 - R/K)Q \,.$$

During the nonproduction period $(t_1 - t_0)$, this amount will further be depleted at a rate of R bags per day until it is completely exhausted. The area of each triangle in the figure, that is, $(K - R)t_0t_1/2$, represents the number of bag-days for storage in each cycle of period t_1.

Let C be the average production cost per day. Then for each production run the total production cost is

$$Ct_1 = C_0 + \tfrac{1}{2}C_1(K - R)t_0t_1 \,.$$

Thus the objective is to minimize C, or

$$\text{Min } C = \frac{1}{t_1}\left[C_0 + \frac{1}{2}C_1\left(1 - \frac{R}{K}\right)Qt_1\right],$$

or

$$\text{Min } C(Q) = \frac{C_0R}{Q} + \frac{C_1Q}{2}\left(1 - \frac{R}{K}\right).$$

In order to obtain the optimal size for each production run, it is necessary that the first derivative of C vanish. Thus

$$\frac{dC}{dQ} = -\frac{C_0R}{Q^2} + \frac{C_1}{2}\left(1 - \frac{R}{K}\right) = 0\,,$$

from which

$$Q = \sqrt{\frac{2C_0R}{C_1(1 - R/K)}}\,.$$

Note that the second derivative of C is (for $C_0 > 0$)

$$\frac{d^2C}{dQ^2} = \frac{2C_0R}{Q^3} > 0 .$$

Hence the optimal value of Q leads to a minimum cost. The optimal time interval between production runs can be obtained by substituting the optimum value of Q into the relation $t_1 = Q/R$.

If the production time for each run is negligible, that is, $R/K = 0$ compared to 1, the optimal value of Q becomes

$$Q = \sqrt{2C_0R/C_1} .$$

This is generally known as the economic lot-size formula for holding inventory of merchandise.

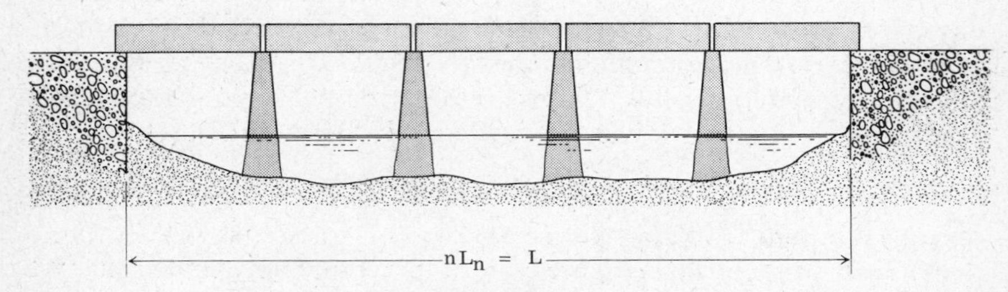

Figure 2-2

Example 2-2. A bridge is to be constructed for a long crossing over a river (see Fig. 2-2). From the preliminary study it is concluded that a series of equal-span simple trusses or girders is most suitable and that piers can be constructed anywhere at the crossing at about the same cost per pier. Determine the optimal span length such that the total cost of the bridge is to be a minimum.

Let C be the total cost of the bridge, which will be expressed in terms of the following notation:

L = total length of the crossing, which is assumed fixed.
L_n = span length of individual truss.
n = number of spans = L/L_n.
A = cost of the two abutments.
P = cost of one pier.
w = weight of steel in pounds, per foot of span for trusses or girders, including bracing.
c = cost per pound of steel for trusses or girders, including bracing.
m = cost per foot of span for floor system, roadway and miscellaneous items, such as railings and sidewalks.

The weight of steel per foot of span is a function of span length L_n, and is assumed to be proportional to L_n, that is, $w = kL_n$, where k is a constant for a given type of truss or girder.

Thus the total cost can be given as the sum of the cost of abutments, the cost of piers, the cost of steel superstructures, and the cost of floor system:

$$C = A + (n - 1)P + cwL + mL\,.$$

Since $n = L/L_n$ and $w = kL_n$, the objective function may be expressed in the form

$$\text{Min } C = A + \left(\frac{L}{L_n} - 1\right)P + ckL_nL + mL\,.$$

In order to optimize the span length, let $dC/dL_n = 0$:

$$\frac{dC}{dL_n} = -\frac{LP}{L_n^2} + ckL = 0\,,$$

from which

$$L_n^2 = P/ck \qquad \text{or} \qquad L_n = \sqrt{P/ck}\,.$$

The second derivative of C with respect to L_n gives

$$\frac{d^2C}{dL_n^2} = \frac{2LP}{L_n^3} > 0\,.$$

The total cost C is therefore a minimum for the span length L_n thus obtained.

Another way of expressing the result of optimization is to note that $L_n^2 = P/ck$ can be written as

$$P = ckL_n^2 = cwL_n\,.$$

This expression implies that the cost of one pier is equal to the cost of one span of trusses or girders. This is a simple rule for determining the optimal span length at a long crossing.

It should be emphasized again that the mathematical analysis is only as good as the validity of the assumption that piers can be constructed anywhere at the crossing at about the same cost per pier. For example, if part of the crossing is very shallow and part of it is over a deep valley, the above formulation is no longer correct and the simple conclusion thus derived is not applicable to such a case.

Example 2-3. A developer for a housing project plans to erect a fence along the perimeter of a triangular area for recreational facilities. In order to provide convenient access, one side of the triangular area is fixed in length. If the total length of the fence available is also predetermined, find the lengths of the remaining two sides which will maximize the enclosed area.

Let the fixed side of the triangle be a and the other two sides be b and c. If the perimeter is denoted by $2s$ and the enclosed area by A, then, from geometry,

$$2s = a + b + c\,,$$
$$A = \sqrt{s(s-a)(s-b)(s-c)}\,.$$

Hence the problem may be formulated as follows:

$$\text{Max} \quad A = \sqrt{s(s-a)(s-b)(s-c)}\,, \qquad b, c > 0\,,$$
$$\text{subject to} \quad 2s = a + b + c\,.$$

Since s and a are constants, the area A is a function of b and c. Solving the constraint equation for c, that is,

$$c = 2s - a - b\,,$$

and substituting it into the original objective function, we see that A is reduced to a function of variable b only. Thus

$$A = \sqrt{s(s-a)(s-b)(a+b-s)}\,.$$

Letting $dA/db = 0$, we have

$$s(s-a)(2s-a-2b) = 0\,.$$

Since $s \neq 0$ and $s - a \neq 0$, it is necessary that

$$2s - a - 2b = 0,$$

or

$$b = s - \frac{a}{2} \quad \text{and} \quad c = s - \frac{a}{2}\,.$$

Consequently,

$$\max A = \frac{a}{2}\sqrt{s(s-a)}\,.$$

We should verify that this set of values for b and c indeed leads to a maximum A. It may be noted further that, for the sake of simplicity in differentiation, we can optimize A^2 instead of A, since both objectives lead to the same result.

Example 2-4. In the previous example, given that the triangular area A is kept constant but the perimeter is varied as required while one side of the triangle remains fixed in length, determine the lengths of the remaining two sides which will minimize the length of the perimeter.

Again, from geometry, this problem can be expressed in the form:

$$\text{Min} \quad 2s = a + b + c\,, \qquad b, c > 0\,,$$
$$\text{subject to} \quad s(s-a)(s-b)(s-c) = A^2\,.$$

By eliminating s from the constraint equation, it becomes

$$\tfrac{1}{16}(a+b+c)(b+c-a)(c+a-b)(a+b-c) - A^2 = 0\,.$$

Solving for $(a+b+c)$ from the above expression and substituting into the objective function, we have

$$\text{Min } 2s = \frac{16A^2}{(b+c-a)(c+a-b)(a+b-c)}\,.$$

Since a and A are constant, the necessary conditions for a minimum $2s$ are that $\partial s/\partial b = 0$ and $\partial s/\partial c = 0$. For $\partial s/\partial b = 0$, we obtain

$$(b+c-a)(c+a-b) - (b+c-a)(a+b-c) + (c+a-b)(a+b-c) = 0\,.$$

Upon simplification, we have

$$-3b^2 + 2bc + 2ab + c^2 - 2ac + a^2 = 0\,.$$

Similarly, for $\partial s/\partial c = 0$, we obtain

$$-3c^2 + 2bc + 2ac + b^2 - 2ab + a^2 = 0\,.$$

From these two equations, we get

$$-4b^2 + 4ab = -4c^2 + 4ac \qquad \text{or} \qquad (b-c)(b+c-a) = 0\,.$$

Since $b + c - a \neq 0$ in a triangle, $b = c$ must hold. Then, by letting $b = c$ in the constraint equation, we note that

$$b = c = \frac{1}{2a}\sqrt{a^4 + 16A^2}\,.$$

Substituting the values of b and c into the objective function, we get

$$\min 2s = a + \frac{1}{a}\sqrt{a^4 + 16A^2}\,.$$

Again, we must verify that the value $2s$ thus obtained is indeed a minimum,

2–2 OPTIMIZATION WITHOUT CONSTRAINTS

Although the examples cited in the previous section are extremely simple, they clearly illustrate the point that an optimal solution exists only if the objective function possesses certain mathematical properties. In order

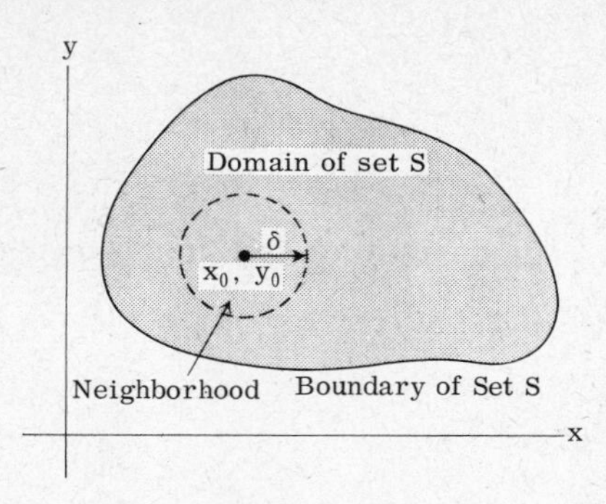

Figure 2-3

to apply the methods of calculus to more general problems, it is desirable to review some basic concepts in the differentiation of functions of one or more variables and their application to the determination of maxima and minima of such functions. We shall begin with optimization of functions without constraints.

For functions of one variable, the behavior of a function is given in terms of the function defined over an interval. In an extension of this concept, the behavior of a function of several variables must be specified in terms of the function defined within a "domain." For a function of two variables, the domain is seen to be an area. Although the concept of domain can easily be extended to many dimensions, the graphical representation becomes either cumbersome or impossible for more dimensions. For the sake of clarity, we shall formulate the concept of domain in terms of functions of two variables without losing its generality.

Let us define a *set of points*, S, in the xy-plane as any sort of collection of points, finite or infinite in number. The *neighborhood* of a point (x_0, y_0) is defined as a set of points inside a circle having (x_0, y_0) as center and δ as radius (see Fig. 2-3). Each point (x, y) in the neighborhood satisfies the restriction represented by the inequality

$$[(x - x_0)^2 + (y - y_0)^2] < \delta^2 .$$

The *complement* of a set of points S is defined as the set containing all points on the plane which do not belong to the set S. A point (x_0, y_0) is said to be an *interior point* of a set S if there exists a neighborhood of (x_0, y_0) which consists entirely of points of S. A point (x_0, y_0), every neighborhood of which contains both points of S and points which do not belong to S, is called a *boundary point*. For example, a point in the interior of a circle is an interior point of the circle; a point on the circumference of a circle is a boundary point of the circle. A set is said to be *bounded* if the entire set can be enclosed in a circle of sufficiently large radius; otherwise, it is said to be *unbounded*.

A set of points is said to be *open* if every point in the set is an interior point. A set is said to be *closed* if every point (interior point or boundary point) of the set belongs to the set. Thus the interior of a circle is an open set, but the interior of a circle plus the circumference forms a closed set. It may be observed that the complement of a closed set is open. For example, the points on and outside the circumference of a circle form a closed set; the points inside of the circumference of a circle then form an open set. An open set is said to be *connected* if any two of its points can be joined by a broken line lying entirely within the set. Two non-overlapping open sets cannot be connected. A *closed connected* set is defined as a closed set which cannot be separated into two nonempty closed sets that have no point in common.

A set which is both open and connected is called a *domain*. The term *region* is sometimes used to denote a set of points consisting of a domain plus none, some, or all of its boundary points, If all boundary points are included, the region is called a *closed* region. In most practical problems, a domain is defined by one or more equations and the boundary of a domain is defined by one or more strict inequalities. A closed region is specified by a combination of the two.

In seeking the optimal solution of a problem, we must distinguish between *absolute* optimum and *relative* optimum, which are sometimes referred to as *global* optimum and *local* optimum, respectively. The most general conditions for the existence of an absolute extremum have been stated by Weierstrauss as follows: "A function, continuous in a bounded closed set, has a greatest and a smallest value." If a function has only two extrema in a closed region, it is obvious that one must correspond to the maximum and the other to the minimum. However, a function may have several relative extrema, some representing maxima and some minima in the region. It may have relative minima which are greater than relative maxima. Thus the absolute maximum or minimum of a function in general can be ascertained only by determining all the relative extrema within the region, and by comparing the values of the functions at these points and those at the boundary points. The largest of these values then is the absolute maximum and the smallest is the absolute minimum.

Optimization without constraints refers to the determination of relative extrema at the interior points of a function. The necessary conditions for finding relative maxima and minima of functions are generally not sufficient. Except for functions of one variable, the criteria for specifying the sufficient conditions, even if they can be established, meet with much difficulty in application to practical problems. We shall examine some of these difficulties later.

A special class of functions which are either convex or concave for the entire range of the variables in a closed region is of great practical importance because there exists at the most one relative minimum or maximum of the function in the region. Hence the extremum obtained by the methods of calculus can easily be recognized as a maximum or a minimum without using the sufficient conditions for determining relative extremum. We shall examine the properties of convex and concave functions in detail.

After discussing the optimization of functions of one, two, or more variables without constraints, we shall then examine the effects of constraints which define the boundary of the closed region.

2-3 FUNCTIONS OF ONE VARIABLE

The simplest case of optimization without constraints is the determination of extrema of functions of one variable. Let $y = f(x)$ be defined in a closed interval $a \leq x \leq b$. The basic problem can be represented either by the minimization of the function,

$$\text{Min } y = f(x) ,$$

or by the maximization

$$\text{Max } y = f(x) .$$

Let us examine first the general properties concerning maxima and minima of the function $f(x)$. The function $f(x)$ is said to have a *relative minimum* at x_0 if $f(x) \geq f(x_0)$ for x sufficiently close to x_0, and $f(x)$ is said to be *convex* in this neighborhood. Similarly, the function $f(x)$ is said to have a *relative maximum* at x_0 if $f(x) \leq f(x_0)$ for x sufficiently close to x_0, and $f(x)$ is said to be *concave* in the neighborhood. If the equality sign does not enter into these expressions, i.e., only strict inequalities prevail, the function $f(x)$ is said to be *strictly convex* or *strictly concave* in the neighborhood according to the direction of the inequality.

A necessary condition for a relative extremum of a function $f(x)$ at $x = x_0$ is that either $f(x)$ has a derivative $f'(x_0) = 0$ or $f'(x_0)$ does not exist. For example, in Fig. 2-4, $f'(x) = 0$ at both points A and B, but $f'(x)$ does not exist at either point C or D. A point of discontinuity, such as point C or D in the figure, is called a *node*. It is noted that the one-sided derivatives on the right and on the left of such a point are different. If both one-sided derivatives equal infinity but are of opposite sign, the point is called a *cusp*.

The sufficient condition for a relative extremum of a function $f(x)$ may be determined by its second derivative. The function $f(x)$ has a relative minimum at x_0 if $f''(x_0) > 0$ in addition to $f'(x_0) = 0$; similarly, $f(x)$ has a

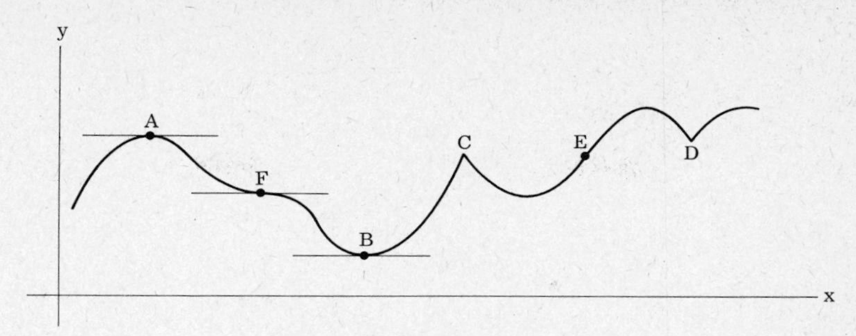

Figure 2–4

relative maximum at x_0 if $f''(x_0) < 0$ in addition to $f'(x_0) = 0$. If $f''(x_0) = 0$ as well as $f'(x_0) = 0$, higher derivatives must be considered. Suppose that $f'(x_0) = 0$, $f''(x_0) = 0, \ldots, f^{(n)}(x_0) = 0$, but $f^{(n+1)}(x_0) \neq 0$. Then $f(x)$ has a relative minimum at x_0 if n is odd and $f^{(n+1)}(x_0) > 0$; $f(x)$ has a relative maximum at x_0 if n is odd and $f^{(n+1)}(x_0) < 0$; $f(x)$ has neither relative maximum nor relative minimum at x_0 if n is even.

If a function $f(x)$ is differentiable and has a derivative $f'(x_0) = 0$ in the interval $a < x_0 < b$, the point x_0 is called the *critical point* of $f(x)$. Hence points of relative minimum and relative maximum are critical points, but the converse is not necessarily true. A critical point representing neither relative maximum nor relative minimum is called a *horizontal point of inflection*, as distinguished from ordinary *point of inflection*, which represents the transition from a convex to a concave interval or the reverse. In Fig. 2–4, for example, E is an ordinary point of inflection, while F is the horizontal point of inflection.

If a function $f(x)$ increases monotonically in the interval $a \leq x \leq b$, it has its smallest value at a and its greatest value at b; if $f(x)$ decreases monotonically, it has its greatest value at a, and its smallest value at b. If a function has only one relative extremum in the interval, then the function has its smallest value at the point of a relative minimum, or its largest value for a relative maximum. A function $f(x)$ is said to be convex if it has at most one relative minimum and no point of inflection in the interval. Furthermore, a convex function is termed *strictly convex* if $f(x) > f(x_0)$ for all x in the interval. Concave function and *strictly concave* functions are similarly defined by making reference to the relative maximum. Examples of convex and concave functions are shown in Figs. 2–5 and 2–6, respectively. Note that any straight line segment satisfies both the definitions of convex and concave functions but not those of strictly convex or strictly concave functions.

If the function $f(x)$ has a first derivative $f'(x)$ at all points of the interval, the necessary and sufficient condition for $f(x)$ to be convex is that $f'(x)$ be a nondecreasing function, and the condition for $f(x)$ to be concave

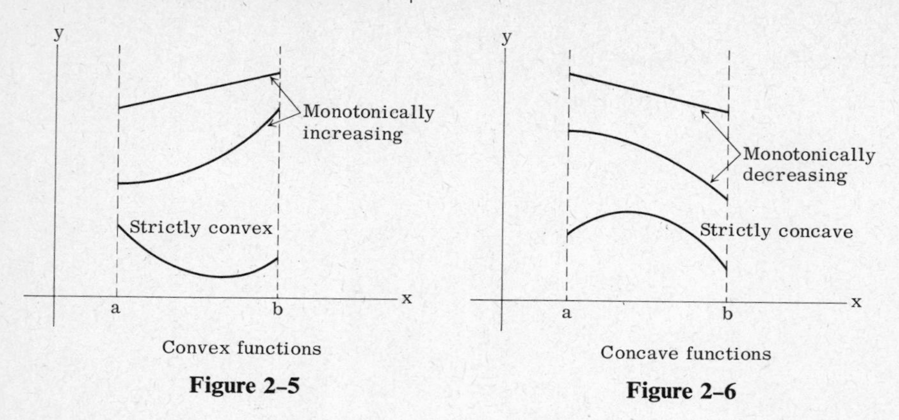

Convex functions

Figure 2-5

Concave functions

Figure 2-6

is that $f'(x)$ be a nonincreasing function. Furthermore, if the function $f(x)$ has a second derivative $f''(x)$ at all points of the interval, the necessary and sufficient condition for $f(x)$ to be convex is that $f''(x) \geq 0$, and the condition for $f(x)$ to be concave is that $f''(x) \leq 0$, for any x in the interval.

Example 2-5. A rigid ball of weight W is placed successively in three different positions A, B, and C, as shown in Fig. 2-7. Determine the state of potential energy associated with each position of equilibrium.

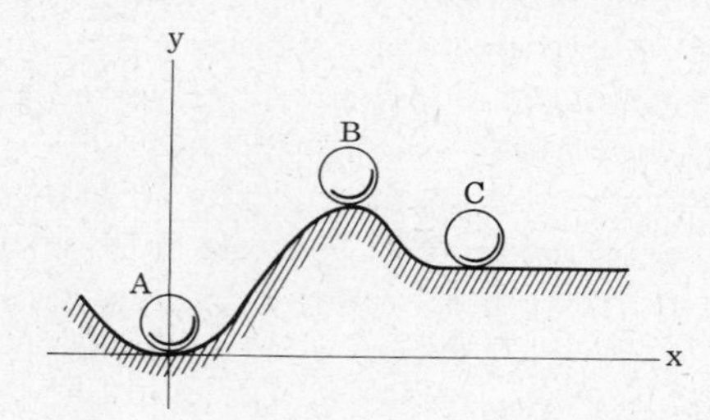

Figure 2-7

In mechanics, it is well known that minimum potential energy is associated with stable equilibrium and maximum potential energy is associated with unstable equilibrium. This can be demonstrated from the consideration of the variation of potential energy from the position of equilibrium.

The potential energy U of a rigid ball of weight W placed at a point on the contour located at a distance y above the origin is

$$U = Wy .$$

When the system is in equilibrium, its potential energy is said to be stationary, that is $\delta U = W\delta y = 0$. Hence from the definition of the differential δy we have

$$\delta U = W[y(x_0 + \delta x) - y(x_0)] = 0\,.$$

Expanding $y(x_0 + \delta x)$ by Taylor's series, we get

$$\delta y = y'(x_0)\delta x + \frac{y''(x_0)}{2!}(\delta x)^2 + \frac{y'''(x_0)}{3!}(\delta x)^3 + \cdots\,.$$

Taking only the first term, when δx is infinitely small, we have

$$\delta U = Wy'(x_0)\delta x = 0,$$

which implies that $y'(x_0) = 0$, or that the tangent of the contour is horizontal at any point where the ball is in equilibrium.

When δx is finite but small, the change of potential energy is given by

$$\delta U = \frac{Wy''(x_0)}{2!}(\delta x)^2 + \frac{Wy'''(x_0)}{3!}(\delta x)^3 + \cdots\,.$$

If δx is sufficiently small, the sign of δU depends on the sign of the first term on the right-hand side of the above equation. Hence δU has the same sign as $y''(x_0)$, since δx is squared. In Fig. 2–7, a positive rate of change of slope, i.e., positive $y''(x_0)$, corresponds to a valley and a negative rate of change of slope, i.e., negative $y''(x_0)$, corresponds to a summit. Thus the departure from valley A increases the potential energy of the system since δU is positive, whereas the departure from the summit B decreases the potential energy of the system since δU is negative. If $y'(x_0) = y''(x_0) = 0$ but $y'''(x_0) \neq 0$, then the ball must be located at a point such as C on a horizontal contour. This position of neutral equilibrium can easily be disturbed since $(+\delta x)^3$ and $(-\delta x)^3$ will cause different signs for δU. On the other hand, if $y'''(x_0)$ also equals zero but $y''''(x_0) \neq 0$, then the sign of δU follows that of $y''''(x_0)$.

2-4 FUNCTIONS OF TWO VARIABLES

A more general optimization problem without constraints is the case involving functions of two variables. While the characteristics of functions of two variables are typical of those of functions of many variables, they can be illustrated more easily by graphical representations and algebraic operations.

Let the function $z = f(x, y)$ be defined in a closed region. Without losing generality, the basic problem in optimization may be represented by the minimization of the function

$$\text{Min } z = f(x, y)\,.$$

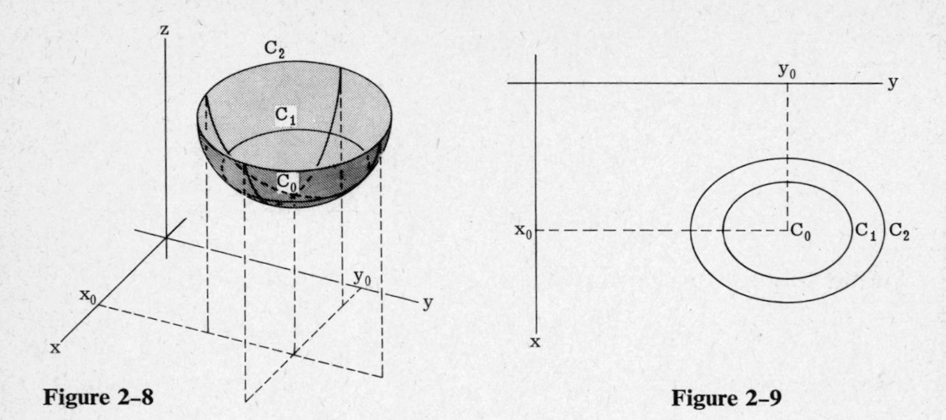

Figure 2–8

Figure 2–9

Although $z = f(x, y)$ can be represented graphically in three-dimensional space, it is often useful to represent the function in the xy-plane by its contour lines which are plots of the loci

$$f(x, y) = C_1, \qquad f(x, y) = C_2, \ldots, \qquad f(x, y) = C_n,$$

where $C_1, C_2, \ldots, C_n$ are constants, usually chosen in sequence. For example, the surface in Fig. 2–8 has been represented by the contours in Fig. 2–9.

A function $z = f(x, y)$ which is defined and continuous in a domain D is said to have a *relative maximum* at (x_0, y_0) if $f(x, y) \leq f(x_0, y_0)$ for (x, y) sufficiently close to (x_0, y_0), and a *relative minimum* at (x_0, y_0) if $f(x, y) \geq f(x_0, y_0)$ for (x, y) sufficiently close to (x_0, y_0). If both partial derivatives $\partial z/\partial x$ and $\partial z/\partial y$ exist and are zero at (x_0, y_0), the point (x_0, y_0) is called a *critical point*. Hence points of relative minimum and relative maximum are critical points. The nature of a critical point may be determined by examining the functions $f(x, y_0)$ and $f(x_0, y)$ in the planes $y = y_0$ and $x = x_0$, respectively. If one of these functions has a minimum and the other has a maximum at (x_0, y_0), the critical point is said to be a *saddle point*. Both of these functions must have a relative minimum at (x_0, y_0) if the critical point is a relative minimum for the function $z = f(x, y)$. These conditions are not sufficient to ensure a relative minimum for z. Even though the surface may have a relative minimum for $f(x, y_0)$ and $f(x_0, y)$ at (x_0, y_0), the critical point is a saddle point if a relative maximum exists in the curve on the vertical plane at an angle α_1 from the x-axis and if a relative minimum exists in the curve on the vertical plane at an angle α_2 from the x-axis (see Fig. 2–10).

If the partial derivatives of second order of $z = f(x, y)$ exist and are continuous in the neighborhood of (x_0, y_0), then from Taylor's theorem

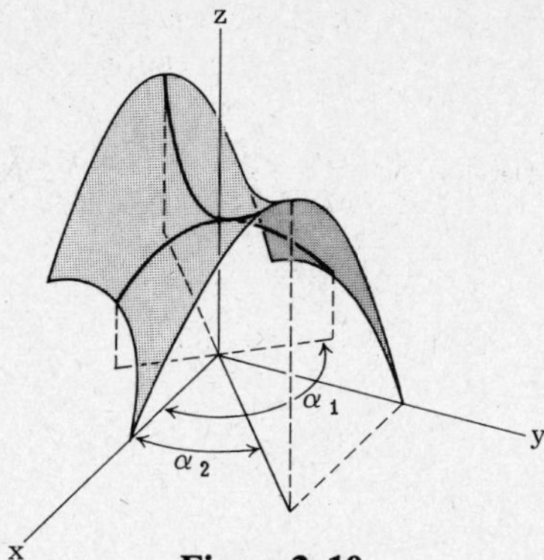

Figure 2–10

we have for increments of h and k in x- and y-directions, respectively,

$$\begin{aligned}\delta z &= f(x_0 + h, y_0 + k) - f(x_0, y_0)\\ &= \left(\frac{\partial z}{\partial x_0} h + \frac{\partial z}{\partial y_0} k\right) + \frac{1}{2!}\left(h^2 \frac{\partial^2 z}{\partial x_0^2} + 2hk \frac{\partial^2 z}{\partial x_0\, \partial y_0} + k^2 \frac{\partial^2 z}{\partial y_0^2}\right) + \cdots,\end{aligned}$$

in which all partial derivatives of $z = f(x, y)$ are evaluated at $x = x_0$ and $y = y_0$. If both partial derivatives of the first order are zero but not all the partial derivatives of the second order are zero, the sign of δz depends on the term involving partial derivatives of the second order in the series. Let

$$A = \frac{\partial^2 z}{\partial x_0^2}, \qquad B = \frac{\partial^2 z}{\partial x_0\, \partial y_0}, \qquad C = \frac{\partial^2 z}{\partial y_0^2}. \tag{2–1}$$

Then this term inside the parenthesis may be written in the following quadratic form for h and k:

$$I = h^2 A + 2hkB + k^2 C. \tag{2–2}$$

In order that the function which is continuously differentiable twice has an extremum at (x_0, y_0), it is necessary and sufficient that $\partial z/\partial x_0 = 0$ and $\partial z/\partial y_0 = 0$, and that δz does not change in sign. In other words, if the function has an extremum at (x_0, y_0), the quadratic form in (2–2) should retain a constant sign in the neighborhood of (x_0, y_0). Let the increments h and k in x- and y-directions be expressed in polar coordinates (r, θ) such that

$$h = r\cos\theta, \qquad k = r\sin\theta.$$

Then, the quadratic form for h and k may be given as follows:

$$Ar^2\cos^2\theta + 2Br^2\sin\theta\cos\theta + Cr^2\sin^2\theta, \tag{2–3}$$

in which $0 \leq \theta \leq 2\pi$. Thus the investigation of critical points for an extremum is reduced to the study of possible variation in the sign of (2-3).

Consequently, a necessary condition for a relative extremum of a function $z = f(x, y)$ at (x_0, y_0) is that the first differential exists, and that it is equal to zero, i.e.,

$$dz = \frac{\partial z}{\partial x_0} dx + \frac{\partial z}{\partial y_0} dy = 0 , \tag{2-4}$$

which implies that

$$\frac{\partial z}{\partial x_0} = 0 \quad \text{and} \quad \frac{\partial z}{\partial y_0} = 0 .$$

However, a function may also have extrema at points where the first differential does not exist.

The sufficient conditions for a relative extremum of a function $z = f(x, y)$ which has continuous first and second partial derivatives can be derived from the quadratic form in (2-3). If the first partial derivatives of the function at (x_0, y_0) are zero, the nature of the critical point (x_0, y_0) may be determined by the following criteria:†

1. a relative minimum if $B^2 - AC < 0$ and $A, C > 0$;
2. a relative maximum if $B^2 - AC < 0$ and $A, C < 0$;
3. a saddle point if $B^2 - AC > 0$;
4. indeterminate if $B^2 - AC = 0$.

If all second partial derivatives are zero, then the nature of the critical point can be examined by determining the third partial derivatives, etc.

The special class of functions which are either convex or concave for the entire range of the variables in the closed region is of special interest. For functions of two variables, $z = f(x, y)$ is said to be convex if it has at most one relative minimum in the region but no saddle point, and a convex function is termed strictly convex if $f(x, y) > f(x_0, y_0)$. Concave function and strictly concave function are similarly defined. It may be noted that the linear function

$$z = f(x, y) = C_1 x + C_2 y ,$$

in which C_1 and C_2 are constant, satisfies both the definitions of convex and concave functions. It can be represented graphically as a plane in the three-dimensional space.

Example 2-6. A force P is applied to a system of elastic springs connected to frictionless rigid bodies, as shown in Fig. 2-11. The springs,

---

For derivation and detailed discussion, see References 2-1 and 2-2.

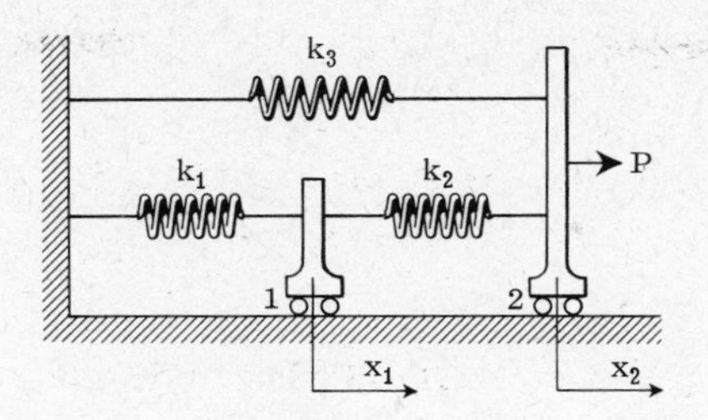

Figure 2-11

which have constants k_1, k_2, and k_3, respectively, are in their natural positions when $P = 0$. Determine the displacements x_1 and x_2 by the principle of minimum potential energy when the load P is applied.

The example illustrates an extremum problem in mechanics, in which the function has traditionally been defined in such a manner that a minimum corresponds to a state of stable equilibrium. A potential-energy function U of the given system consists of the strain energy U_i expressed in terms of the displacements and the potential energy U_e of the prescribed external force. Thus for the strain energy in the elastic springs,

$$U_i = \tfrac{1}{2}k_1x_1^2 + \tfrac{1}{2}k_2(x_2 - x_1)^2 + \tfrac{1}{2}k_3x_2^2\,,$$

and for the potential energy of the force P, $U_e = -Px_2$. For the entire system,

$$U = \tfrac{1}{2}k_1x_1^2 + \tfrac{1}{2}k_2(x_2 - x_1)^2 + \tfrac{1}{2}k_3x_2^2 - Px_2\,.$$

Note that the function $U = U(x_1, x_2)$ is a paraboloid surface and has only one extreme value which is a global minimum. The existence of a minimum is observed a priori from the principle of minimum potential energy which states that, of all geometrically compatible displacement configurations, the one corresponding to the state of stable equilibrium has a minimum potential energy. The conditions for determining the minimum of U are given by

$$\frac{\partial U}{\partial x_1} = k_1x_1 - k_2(x_2 - x_1) = 0\,,$$

$$\frac{\partial U}{\partial x_2} = k_2(x_2 - x_1) + k_3x_2 - P = 0\,.$$

These two equations correspond to the conditions of equilibrium if the rigid bodies 1 and 2 are isolated as free bodies. Hence x_1 and x_2 for the equilibrium state can be solved simultaneously from these two equations.

Let us examine this problem further by assigning the following numerical values to the given quantities: $k_1 = 2$ kips/in, $k_2 =$ kips/in, $k_3 =$ 6 kips/in, and $P = 11$ kips. By expanding and collecting terms,

$$U = 3x_1^2 + 5x_2^2 - 4x_1x_2 - 11x_2\,.$$

Hence

$$\frac{\partial U}{\partial x_1} = 6x_1 - 4x_2 = 0 ,$$

$$\frac{\partial U}{\partial x_2} = -4x_1 + 10x_2 - 11 = 0 ,$$

from which $x_1 = 1$ in and $x_2 = 1.5$ in. The minimum U corresponding to this set of displacements is -8.25 in-kips.

However, we can verify that the critical point is indeed a minimum by the use of criteria for sufficient conditions. At point (1, 1.5),

$$A = \frac{\partial^2 U}{\partial x_1^2} = 6 , \qquad B = \frac{\partial^2 U}{\partial x_1 \, \partial x_2} = -4 , \qquad C = \frac{\partial^2 U}{\partial x_2^2} = 10 ,$$

we have

$$B^2 - AC = (-4)^2 - (6)(10) = -44 \qquad (< 0) ,$$

$$A + C = 16 \qquad (> 0) .$$

Hence the point (1, 1.5) is a relative minimum.

The contour lines representing the projections of $U = -5$, 0, and 5 on the x_1x_2-plane are shown in Fig. 2-12.

Example 2-7. Show that the point (0, 0) is a critical point for the function $z = 4 - x^2 + 6xy - y^2$. Determine the nature of this critical point.

The first and second derivatives of the given functions are obtained as follows:

$$\frac{\partial z}{\partial x} = -2x + 6y , \qquad \frac{\partial z}{\partial y} = 6x - 2y ;$$

$$\frac{\partial^2 z}{\partial x^2} = -2 , \qquad \frac{\partial^2 z}{\partial x \, \partial y} = 6 , \qquad \frac{\partial^2 z}{\partial y^2} = -2 .$$

Since $\partial z/\partial x$ and $\partial z/\partial y$ are zero at $x = 0$ and $y = 0$, the point (0, 0) is a critical point. Furthermore, for point (0, 0), the second partial derivatives are $A = -2$, $B = 6$ and $C = -2$. Hence

$$B^2 - AC = 6^2 - (-2)(-2) = 32 , \qquad (> 0).$$

According to the criteria, the critical point is a saddle point. The contours of the surface for $z = 0$, 4, and 8 are plotted in Fig. 2-13.

2-5 FUNCTIONS OF MANY VARIABLES

In this section, we shall make further generalization on the optimization of functions of many variables. Let the function $z = f(x_1, x_2, \ldots, x_n)$ be defined in a closed region of the variables. For example, the basic

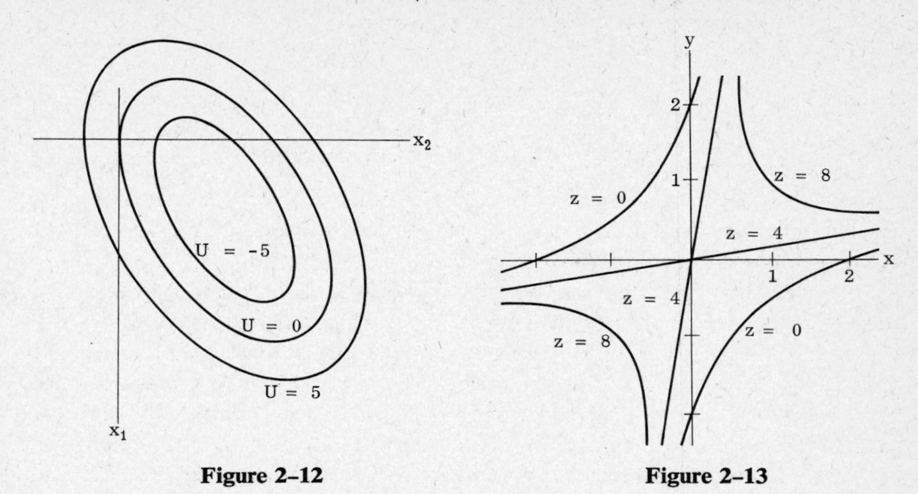

Figure 2-12

Figure 2-13

problem can be represented by the minimization of the function

$$\operatorname{Min} z = f(x_1, x_2, \ldots, x_n) .$$

For the function to have a relative extremum at $(x_1^*, x_2^*, \ldots, x_n^*)$, if the first differential exists, it is necessary that

$$dz = \frac{\partial z}{\partial x_1} dx_1 + \frac{\partial z}{\partial x_2} dx_2 + \cdots + \frac{\partial z}{\partial x_n} dx_n = 0 , \tag{2-5}$$

or

$$\frac{\partial z}{\partial x_i} = 0 , \qquad i = 1, 2, \ldots, n .$$

However, a function may also have extrema at points where the first differential does not exist.

If the function $z = f(x_1, x_2, \ldots, x_n)$ has continuous first and second partial derivatives in the neighborhood of $(x_1^*, x_2^*, \ldots, x_n^*)$, we have a polynomial for the term involving the second partial derivatives in the Taylor's series as follows:

$$\sum_{i=1}^{n} \sum_{j=1}^{n} \frac{\partial^2}{\partial x_i \, \partial x_j} f(x_1^*, x_2^*, \ldots, x_n^*) h_i h_j ,$$

in which h_i and h_j are increments of x_i and x_j, respectively, $(i = 1, 2, \ldots, n;$ $j = 1, 2, \ldots, n)$. This is called the *quadratic form* of the second partial derivatives, and is evaluated at the critical point $(x_1^*, x_2^*, \ldots, x_n^*)$ where the first partial derivatives equal zero. Let

$$f_{ij} = \frac{\partial^2}{\partial x_i \, \partial x_j} f(x_1^*, x_2^*, \ldots, x_n^*) . \tag{2-6}$$

Then the quadratic form may be written as follows:

$$I_n = \sum_{i=1}^{n} \sum_{j=1}^{n} f_{ij} h_i h_j \,. \tag{2-7}$$

The quadratic form is said to be *positive definite*, if $I_n > 0$, and *semipositive definite* if $I_n \geq 0$, for all possible combinations of h_i and h_j; similarly, it is termed *negative definite* if $I_n < 0$, and *seminegative definite* if $I_n \leq 0$, for all possible combinations of h_i and h_j.

In order that the function which is continuously differentiable twice has an extremum at the point $(x_1^*, x_2^*, \ldots, x_n^*)$, it is necessary and sufficient that the first partial derivatives equal zero at that point, and that the quadratic form obtained from the second partial derivatives does not change in sign. If the quadratic form is positive definite, the critical point is a relative minimum; if the quadratic form is negative definite, the critical point is a relative maximum.† For all other cases, the critical point is indeterminate and must be examined further by other means that may be available.

A quadratic form is positive definite if the determinant of the coefficients and all its principal minors are positive. Thus for the quadratic form in (2-7), we have

$$\begin{vmatrix} f_{11} & f_{12} & \cdots & f_{1n} \\ f_{21} & f_{22} & \cdots & f_{2n} \\ \vdots & & & \\ f_{n1} & f_{n2} & \cdots & f_{nn} \end{vmatrix} > 0 \,, \quad \ldots, \quad \begin{vmatrix} f_{11} & f_{12} \\ f_{21} & f_{22} \end{vmatrix} > 0 \,, \quad f_{11} > 0 \,. \tag{2-8}$$

A quadratic form is negative definite if the determinant formed from the negative of all coefficients and all principal minors of this determinant are positive.

Again, we pay special attention to the class of convex or concave functions which has at the most one extremum in the closed region. A function $z = f(x_1, x_2, \ldots, x_n)$ is said to be convex if, for the entire region in which $(x_1^*, x_2^*, \ldots, x_n^*)$ is a relative minimum, it has no saddle point but

$$f(x_1, x_2, \ldots, x_n) \geq f(x_1^*, x_2^*, \ldots, x_n^*) \,;$$

and it is said to be strictly convex if

$$f(x_1, x_2, \ldots, x_n) > f(x_1^*, x_2^*, \ldots, x_n^*) \,.$$

Concave and strictly concave functions are similarly defined.

In general, a function of n variables can be visualized as a *hypersurface* in the $(n + 1)$-dimensional space, and a linear function of n variables is called a *hyperplane*, because they are analogous respectively to the surface

---

† See References 2-2 and 2-3 for further discussion.

and the plane in three-dimensional space. Thus we may speak of contour lines of a hypersurface which represent the loci of $z = C_i$, where C_i $(i = 1, 2, \ldots, n)$ are constants in sequence. Hence the geometrical representation of functions of two variables can readily be extended, at least symbolically, to the interpretation of functions of several variables.

Example 2–8. Derive the criteria for the sufficient condition in finding a relative minimum or maximum of functions of one or two variables from the general criteria represented by (2–8).

In the case of functions of one variable, the sufficient condition depends only on f_{11} in (2–8). From (2–6)

$$f_{11} = \left. \frac{d^2 f(x_1)}{dx_1^2} \right|_{x_1 = x_1^*},$$

where x_1^* is a critical point. If $f_{11} > 0$, the critical point is a minimum; if $-f_{11} > 0$, or $f_{11} < 0$, the critical point is a maximum. These results are well known in elementary calculus.

For functions of two variables, the sufficient conditions depend on f_{22}, f_{12}, and f_{21} as well as f_{11}. From (2–6),

$$f_{11} = \frac{\partial^2 f(x_1^*, x_2^*)}{\partial x_1^2}, \qquad f_{22} = \frac{\partial^2 f(x_1^*, x_2^*)}{\partial x_2^2},$$

$$f_{12} = f_{21} = \frac{\partial^2 f(x_1^*, x_2^*)}{\partial x_1 \partial x_2}.$$

Hence the critical point is a minimum if

$$f_{11} > 0 \qquad \text{and} \qquad \begin{vmatrix} f_{11} & f_{12} \\ f_{21} & f_{22} \end{vmatrix} = f_{11} f_{22} - f_{12} f_{21} > 0 .$$

Similarly, the critical point is a maximum if

$$-f_{11} > 0 \qquad \text{and} \qquad \begin{vmatrix} -f_{11} & -f_{12} \\ -f_{21} & -f_{22} \end{vmatrix} = (-1)^2 \begin{vmatrix} f_{11} & f_{12} \\ f_{21} & f_{22} \end{vmatrix} > 0 ,$$

or

$$f_{11} < 0 \qquad \text{and} \qquad f_{11} f_{22} - f_{12} f_{21} > 0 .$$

Note that $f_{12} f_{21} = f_{12}^2 > 0$. Consequently,

$$f_{11} f_{22} > 0 .$$

Hence the sign of f_{22} must be the same as f_{11}, or

$$f_{22} > 0 \qquad \text{for} \qquad f_{11} > 0 ,$$

$$f_{22} < 0 \qquad \text{for} \qquad f_{11} < 0 .$$

These results are in effect identical to the criteria for relative minimum and maximum previously developed for functions of two variables.

2-6 OPTIMIZATION WITH EQUALITY CONSTRAINTS

In Examples 2-3 and 2-4, it has been shown that equality constraints can be substituted into the objective function so that such a problem is reduced to an optimization problem without constraints and with fewer variables. However, this procedure is not always convenient; hence a more general approach is introduced for the solution of this type of problem.

Let us consider first a special case of minimization of a function of three variables subject to only one equality constraint. For example,

$$\begin{aligned} \text{Min} \quad & z = f(x_1, x_2, x_3)\,, \\ \text{subject to} \quad & g(x_1, x_2\, x_3) = 0\,. \end{aligned}$$

If the constraint condition were expressed in terms of x_3 and substituted into the objective function, we would have obtained an optimization problem without constraints in the following form:

$$\text{Min } z = F(x_1, x_2)\,,$$

in which the function $F(x_1, x_2)$ is to be determined later. The necessary condition for the function z to have a relative minimum is that

$$dz = \frac{\partial F}{\partial x_1}\, dx_1 + \frac{\partial F}{\partial x_2}\, dx_2 = 0\,. \tag{2-9}$$

However, if we look at the original problem, we can obtain the first differentials for $f(x_1, x_2, x_3)$ and $g(x_1, x_2, x_3)$ as follows:

$$dz = \frac{\partial f}{\partial x_1}\, dx_1 + \frac{\partial f}{\partial x^2}\, dx_2 + \frac{\partial f}{\partial x_3}\, dx_3$$

and

$$\frac{\partial g}{\partial x_1}\, dx_1 + \frac{\partial g}{\partial x_2}\, dx_2 + \frac{\partial g}{dx_3}\, dx_3 = 0\,.$$

By introducing an undetermined parameter λ, multiplying it with the last equation, and adding the product from the first differential for f, we get

$$\begin{aligned} dz = {} & \left(\frac{\partial f}{\partial x_1} + \lambda\,\frac{\partial g}{\partial x_1}\right) dx_1 + \left(\frac{\partial f}{\partial x_2} + \lambda\,\frac{\partial g}{\partial x_2}\right) dx_2 \\ & + \left(\frac{\partial f}{\partial x_3} + \lambda\,\frac{\partial g}{\partial x_3}\right) dx_3\,. \end{aligned} \tag{2-10}$$

Comparing (2-10) with (2-9), we see that the coefficient for dx_3 must be zero:

$$\frac{\partial f}{\partial x_3} + \lambda \frac{\partial g}{\partial x_3} = 0 .$$

Furthermore, if z is to have a relative minimum, the first partial derivatives with respect to x_1 and x_2 must be zero. Thus

$$\frac{\partial f}{\partial x_1} + \lambda \frac{\partial g}{\partial x_1} = \frac{\partial F}{\partial x_1} = 0 , \qquad \frac{\partial f}{\partial x_2} + \lambda \frac{\partial g}{\partial x_2} = \frac{\partial F}{\partial x_2} = 0 .$$

If we introduce a function Φ such that

$$\Phi(x_1, x_2, x_3\ \lambda) = f(x_1, x_2, x_3) + \lambda g(x_1, x_2, x_3) , \tag{2-11}$$

we see that the direct optimization of Φ leads to

$$\frac{\partial \Phi}{\partial x_1} = 0 , \qquad \frac{\partial f}{\partial x_1} + \lambda \frac{\partial g}{\partial x_1} = 0 ;$$

$$\frac{\partial \Phi}{\partial x_2} = 0 , \qquad \frac{\partial f}{\partial x_2} + \lambda \frac{\partial g}{\partial x_2} = 0 ;$$

$$\frac{\partial \Phi}{\partial x_3} = 0 , \qquad \frac{\partial f}{\partial x_3} + \lambda \frac{\partial g}{\partial x_3} = 0 ;$$

$$\frac{\partial \Phi}{\partial \lambda} = 0 , \qquad g(x_1, x_2, x_3) = 0 .$$

The first three equations thus obtained are identical to the requirements of the original problem, provided that $\Phi(x_1, x_2, x_3\ \lambda)$ is reduced to $F(x_1, x_2)$ under the condition of the equality constraint which was restated in the last equation. Thus the problem of minimization with an equality constraint can be replaced by a direct minimization problem without constraint if the objective function of the new problem is constructed according to (2-11). The parameter λ thus introduced is called the *Lagrange multiplier*, and the function Φ is called the *Lagrange function*.

We can extend this concept to the general case of optimization of functions of many variables having many equality constraints. For example, consider,

$$\text{Min } z = f(x_1, x_2, \ldots, x_n) ,$$

subject to m constraints ($n > m$);

$$g_i(x_1, x_2, \ldots, x_n) = 0, \qquad i = 1, 2, \ldots, m ,$$

where all functions are differentiable. If m variables could be solved in terms of the other $(n - m)$ variables, the resulting objective function would have $(n - m)$ independent variables.

By introducing λ_i $(i = 1, 2, \ldots, m)$ as Lagrange multipliers, one for each equality, we can construct a new minimization problem without constraints, which is sometimes referred to as the *Lagrange problem*. Thus

$$\begin{aligned} &\text{Min}\ \Phi(x_1, x_2, \ldots, x_n, \lambda_1, \lambda_2, \ldots, \lambda_m) \\ &\quad = f(x_1, x_2, \ldots, x_n) + \sum_{i=1}^{m} \lambda_i g_i(x_1, x_2, \ldots, x_n)\,. \end{aligned} \tag{2-12}$$

It can be shown that the necessary conditions for minimizing the original problem are identical to those for minimizing the Lagrange problem. From the first partial derivatives of Φ in (2-12), we obtain a set of $n + m$ simultaneous equations for finding critical points:

$$\frac{\partial f}{\partial x_j} + \sum_{i=1}^{m} \lambda_i \frac{\partial g_i}{\partial x_j} = 0\,, \qquad j = 1, 2, \ldots, n\,;$$

$$g_i(x_1, x_2, \ldots, x_n) = 0\,, \qquad i = 1, 2, \ldots, m\,.$$

The Lagrange multipliers can be uniquely determined from this set of equations if certain conditions about $\partial g_i/\partial x_j$ are met. Furthermore, the criteria for determining whether the critical points are extrema under the conditions of equality constraints must also be established. All these details are rather complicated and therefore will not be considered here.†

Example 2-9. Construct a direct maximization problem using the Lagrange multiplier for the problem in Example 2-3, and determine b and c which satisfy the necessary condition for a relative maximum.

The original problem can be stated in the modified form as follows:

$$\begin{aligned} \text{Max} \qquad & f(b, c) = A^2 = s(s - a)(s - b)(s - c)\,, \\ \text{subject to} \qquad & g(b, c) = a + b + c - 2s = 0\,. \end{aligned}$$

Let λ be the Lagrange multiplier. The Lagrange problem can be expressed as the maximization of a function $\Phi = \Phi(b, c, \lambda)$:

$$\text{Max}\ \Phi = s(s - a)(s - b)(s - c) + \lambda(a + b + c - 2s)\,.$$

Then

$$\begin{aligned} \partial\Phi/\partial b &= -s(s - a)(s - c) + \lambda = 0\,, \\ \partial\Phi/\partial c &= -s(s - a)(s - b) + \lambda = 0\,, \\ \partial\Phi/\partial \lambda &= a + b + c - 2s = 0\,, \end{aligned}$$

† See References 2-2 and 2-3.

from which we obtain

$$s - c = \frac{\lambda}{s(s-a)}, \qquad c = s - \frac{\lambda}{s(s-a)},$$

$$s - b = \frac{\lambda}{s(s-a)}, \qquad b = s - \frac{\lambda}{s(s-a)},$$

$$a - \frac{2\lambda}{s(s-a)} = 0, \qquad \lambda = \frac{as(s-a)}{2},$$

Hence

$$b = c = s - a/2.$$

Example 2-10. Construct a direct minimization problem by using the Lagrange multiplier for the problem in Example 2-4, and determine b and c which satisfy the necessary conditions for a relative minimum.

The original problem can be stated in the modified form as follows:

$$\begin{aligned} \text{Min} \quad & f(b, c) = 2s = a + b + c, \\ \text{subject to} \quad & g(b, c) = \tfrac{1}{16}(a + b + c)(a + b - c) \\ & \qquad \times (a - b + c)(b + c - a) - A^2 \\ & \quad = -\tfrac{1}{16}[c^4 - 2(a^2 + b^2)c^2 + (a^2 - b^2)^2 + 16A^2] \\ & \quad = 0. \end{aligned}$$

The Lagrange problem can be expressed as the minimization of a function $\Phi = (b, c, \lambda)$:

$$\text{Min } \Phi = a + b + c - \frac{\lambda}{16}[c^4 - 2(a^2 + b^2)c^2 + (a^2 - b^2)^2 + 16A^2].$$

Then

$$\frac{\partial \Phi}{\partial b} = 1 - \frac{\lambda}{16}[-4bc^2 - 4b(a^2 - b^2)] = 0,$$

$$\frac{\partial \Phi}{\partial c} = 1 - \frac{\lambda}{16}[4c^3 - 4c(a^2 + b^2)] = 0,$$

$$\frac{\partial \Phi}{\partial \lambda} = -\frac{1}{16}[c^4 - 2(a^2 + b^2)c^2 + (a^2 - b^2)^2 + 16A^2] = 0.$$

From the first two equations $\partial\Phi/\partial b = 0$ and $\partial\Phi/\partial c = 0$, we get

$$-4bc^2 - 4ba^2 + 4b^3 \equiv 4c^3 - 4ca^2 - 4cb^2.$$

Hence $b = c$ must hold true. Then from the first two equations we have

$$\lambda = -\frac{4}{ba^2} = -\frac{4}{ca^2}.$$

In addition, from the third equation, $\partial\Phi/\partial\lambda = 0$, we get

$$a^4 - 4a^2b^2 + 16A^2 = 0$$

or

$$a^4 - 4a^2c^2 + 16A^2 = 0\,.$$

Hence

$$b = c = \frac{1}{2a}\sqrt{a^4 + 16A^2}$$

and

$$\lambda = -\frac{8}{a\sqrt{a^4 + 16A^2}}\,.$$

Example 2-11. Solve the following minimization problem by using the Lagrange multiplier:

$$\begin{aligned} &\text{Min} && U = 3x_1^2 + 5x_2^2 - 4x_1x_2 - 11x_2\,, \\ &\text{subject to} && x_1 + x_2 = 3\,. \end{aligned}$$

We can construct the Lagrange problem as follows:

$$\text{Min}\ \Phi = 3x_1^2 + 5x_2^2 - 4x_1x_2 - 11x_2 + \lambda(x_1 + x_2 - 3)\,.$$

Then

$$\begin{aligned} \partial\Phi/\partial x_1 &= 6x_1 - 4x_2 + \lambda = 0\,, \\ \partial\Phi/\partial x_2 &= 10x_2 - 4x_1 - 11 + \lambda = 0\,, \\ \partial\Phi/\partial\lambda &= x_1 + x_2 - 3 = 0\,. \end{aligned}$$

Solving these three equations simultaneously, we get

$$\lambda = -\tfrac{11}{12}\,, \qquad x_1 = \tfrac{31}{24}\,, \qquad x_2 = \tfrac{41}{24}\,.$$

Hence

$$\min U = 3(\tfrac{31}{24})^2 + 5(\tfrac{41}{24})^2 - 4(\tfrac{31}{24})(\tfrac{41}{24}) - 11(\tfrac{41}{24}) = -\tfrac{385}{48} \approx -8.021\,.$$

The solution of the problem is illustrated in Fig. 2-14 in which the objective function of the original problem is a paraboloid surface and the equality constraint is represented by a plane parallel to the U-axis. It is seen that the Lagrange problem is equivalent to seeking the extremum of the curve representing the intersection of the paraboloid surface and the plane. That is, the optimization process in this problem with an equality constraint may either be interpreted as seeking an extremum in the interior of an equivalent problem of reduced dimensions (one variable being eliminated), or as seeking an extremum at the boundary of the

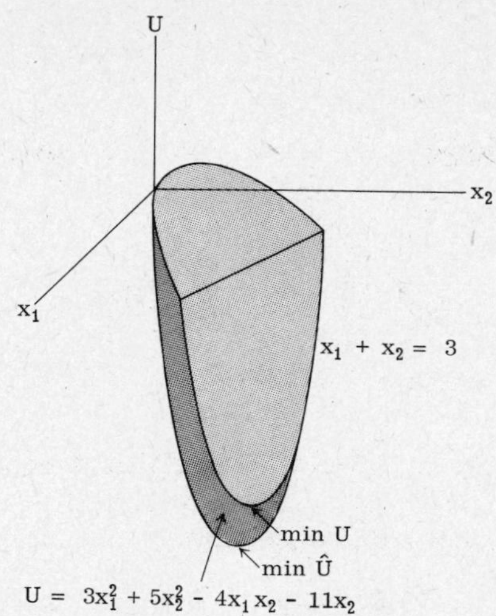

Figure 2–14

original problem (as represented by the intersecting curve). Hence the minimum U of this constraint problem is different from the minimum $\hat{U}$ for the function without constraints.

2–7 OPTIMIZATION WITH NONNEGATIVE CONSTRAINTS

The concept of using Lagrange multipliers for optimization problems with equality constraints may be extended to the solution of problems with inequality constraints. The equivalent optimization problem thus constructed from the constrained problem is called a *generalized Lagrange problem*. We may look upon the nonnegative constraints of variables as the simplest form of inequality constraints, and consider the necessary conditions for finding an optimum. Since we can visualize more readily convex or concave functions, we shall use them for illustration in the subsequent discussion, and shall bypass the discussion of sufficient conditions.

Let us consider first the minimization of the function of one variable. For example,

$$\begin{aligned} \text{Min} \quad & z = f(x)\,, \\ \text{subject to} \quad & x \geq 0\,. \end{aligned}$$

The complete solution of this problem is given by either one of the two cases represented by (a) and (b) in Fig. 2–15; that is, the point of minimum occurs either at an interior point $x^* > 0$ or at a boundary point $x^* = 0$. If the minimum occurs at $x^* > 0$, we can determine x^* by setting $df/dx = 0$ and find the minimum z corresponding to this value of x; and

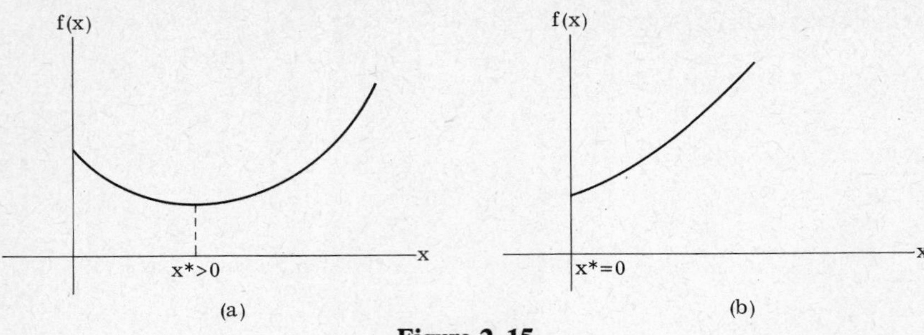

Figure 2-15

if the minimum occurs at $x^* = 0$, then $df/dx \geq 0$ and z is evaluated at $x = 0$.

Let us now construct a new problem which has similar characteristics:

$$\text{Min } \Phi(x, y) = f(x) + yx . \tag{2-13}$$

The necessary conditions for finding an optimum are

$$x \geq 0 , \qquad y \leq 0 , \tag{2-14}$$

$$yx = 0 . \tag{2-15}$$

In other words, the new variable y may be regarded as a Lagrange multiplier, and x may be treated as $g(x)$ if the constraint is $x \geq 0$. The combination of (2-14) and (2-15) gives the conditions that if $x > 0$, then $y = 0$; and if $y < 0$, then $x = 0$. By taking the partial derivative of Φ with respect to x in (2-13), we have

$$\frac{\partial \Phi}{\partial x} = \frac{df}{dx} + y = 0 .$$

Then

$$\frac{df}{dx} = 0 \qquad \text{for} \qquad x > 0 \qquad \text{and} \qquad y = 0 ;$$

$$\frac{df}{dx} \geq 0 \qquad \text{for} \qquad x = 0 \qquad \text{and} \qquad y \leq 0 .$$

Hence the generalized Lagrange problem represented by (2-13) leads to the conditions of the original minimization problem.

The same generalization may be applied to the minimization of nonlinear objective functions of several variables subject to nonnegative constraints. For example,

$$\begin{aligned} &\text{Min} && z = f(x_1, x_2, \ldots, x_n) , \\ &\text{subject to} && x_j \geq 0 , \quad j = 1, 2, \ldots, n . \end{aligned}$$

The corresponding Lagrange problem becomes

$$\text{Min } \Phi = f(x_1, x_2, \ldots, x_n) + \sum_{j=1}^{n} y_j x_j \,. \tag{2–16}$$

The necessary conditions for finding an optimum are

$$x_j \geq 0\,, \qquad y_j \leq 0\,, \qquad j = 1, 2, \ldots, n\,; \tag{2–17}$$

$$y_1 x_1 + y_2 x_2 + \cdots + y_n x_n = 0\,, \tag{2–18}$$

in which $\Phi = \Phi(x_1, x_2, \ldots, x_n, y_1, y_2, \ldots, y_n)$. Again y_j $(j = 1, 2, \ldots, n)$ are analogous to Lagrange multipliers and x_j $(j = 1, 2, \ldots, n)$ correspond to $g_i(x_1, x_2, \ldots, x_n) = 0$ for equality constraints. The equivalence of the original minimization problem and the generalized Lagrange problem can be noted by taking the partial derivatives of Φ with respect to x_j $(j = 1, 2, \ldots, n)$ in (2–16). Thus

$$\frac{\partial \Phi}{\partial x_j} = \frac{\partial f}{\partial x_j} + y_j\,, \qquad j = 1, 2, \ldots, n\,.$$

In view of (2–17) and (2–18), we have $y_j = 0$ if $x_j > 0$, and $x_j = 0$ if $y_j < 0$. Hence

$$\frac{\partial f}{\partial x_j} = 0\,, \qquad \text{for} \qquad x_j > 0\,,$$

$$\frac{\partial f}{\partial x_j} \geq 0\,, \qquad \text{for} \qquad x_j = 0\,,$$

The conditions for optimization are identical to those of the original minimization problem.

Example 2–12. Solve the following optimization problem in which the variables are constrained to be nonnegative:

$$\begin{aligned} \text{Max} \qquad & V = x_1 x_2 x_3\,, \\ \text{subject to} \qquad & x_1 + x_2 + x_3 = b\,, \\ & x_1, x_2, x_3 \geq 0\,. \end{aligned}$$

Let us solve for x_3 from the equality constraint and substitute it into the original objective function. Then

$$V = x_1 x_2 (b - x_1 - x_2)\,.$$

From the necessary condition for a relative maximum, $\partial V/\partial x_1 = 0$ and $\partial V/\partial x_2 = 0$, we obtain through the elimination of one of the two resulting equations:

$$(x_2 - x_1)(b - x_2 - x_1) = 0\,.$$

Since there are two critical points in the solution, we shall examine both of them. For $x_2 - x_1 = 0$ we have $x_1 = x_2$. Using similar arguments, we can obtain $x_2 = x_3$ or $x_3 = x_1$ Hence

$$x_1 = x_2 = x_3 = b/3 .$$

Furthermore, at this critical point

$$\frac{\partial^2 V}{\partial x_1^2} = -2x_2 = -\frac{2b}{3},$$

$$\frac{\partial^2 V}{\partial x_1 \partial x_2} = b - 2x_1 - 2x_2 = -\frac{b}{3},$$

$$\frac{\partial^2 V}{\partial x_2^2} = -2x_1 = -\frac{2b}{3}.$$

Hence the sufficient condition for a relative maximum at the point $(x_1 = b/3,\ x_2 = b/3,\ x_3 = b/3)$ is also satisfied. On the other hand, for $b - x_2 - x_1 = 0$, we have

$$x_1 + x_2 = b \qquad \text{and} \qquad x_3 = 0 .$$

So long as $x_3 = 0$, then $V = 0$ regardless of the values of x_1 and x_2. Hence $V = 0$ is a minimum at the boundary $x_3 = 0$, since V, as well as all variables, must be nonnegative.

Although this problem has been solved by direct substitution instead of by using Lagrange multipliers, the solution clearly indicates that a maximum occurs in the interior, and a minimum exists at the boundary of the region. Hence we must search for all optima of the function at the interior points and at the boundary points in the solution of such a problem.

2-8 OPTIMIZATION WITH INEQUALITY CONSTRAINTS

We shall now consider the general case of inequality constraints, which can also be treated by using the concept of the generalized Lagrange problem. Let us consider the general problem

$$\begin{aligned} \text{Max} \quad & A = f(x_1, x_2, \ldots, x_n), \\ \text{subject to} \quad & g_i(x_1, x_2, \ldots, x_n) \leq 0, \qquad i = 1, 2, \ldots, m; \\ & x_j \geq 0, \qquad j = 1, 2, \ldots, n. \end{aligned}$$

The special case analogous to that in the previous section can be constructed by introducing the multipliers $y_i \leq 0$ $(i = 1, 2, \ldots, m)$:

$$\text{Max } \Phi = f(x_1, x_2, \ldots, x_n) + \sum_{i=1}^{m} y_i g_i(x_1, x_2, \ldots, x_n). \tag{2-19}$$

The necessary conditions for a relative maximum have been developed by Kuhn and Tucker† as follows:

$$x_j \geq 0\,, \qquad \frac{\partial \Phi}{\partial x_j} \leq 0\,, \qquad x_j \frac{\partial \Phi}{\partial x_j} = 0\,;$$
$$j = 1, 2, \ldots, n \tag{2–20}$$

and

$$y_i \leq 0\,, \qquad \frac{\partial \Phi}{\partial y_i} \leq 0\,, \qquad y_i \frac{\partial \Phi}{\partial y_i} = 0\,;$$
$$i = 1, 2, \ldots, m\,. \tag{2–21}$$

The product terms in both (2–20) and (2–21) mean that $x_j = 0$ when $\partial\Phi/\partial x_j < 0$, and $\partial\Phi/\partial x_j = 0$ when $x_j > 0$; similarly $y_i = 0$ when $\partial\Phi/\partial y_i < 0$, and $\partial\Phi/\partial y_i = 0$ when $y_i < 0$. These conditions imply, as in the case of nonnegative constraints, that we must search for relative maxima of the function at the interior points and the boundary points in order to determine the absolute maximum. Except for concave or convex functions with inequality constraints satisfying certain conditions, the necessary conditions for extremum are not sufficient, and the solution of general optimization problems is usually met with great difficulties.

However, if the objective function of the problem is linear and the inequality constraints are also linear, the search for an optimum is confined only to the linear boundary segments, since the plane or hyperplane representing the objective function has no relative optimum at the interior points. In that case, an absolute or global maximum, if it exists, can be sought by searching the boundary only; and the methods of calculus for finding relative extrema are therefore not applicable.

Example 2–13. Solve the following optimization problem with an inequality constraint:

$$\text{Min} \qquad U = 3x_1^2 + 5x_2^2 - 4x_1x_2 - 11x_2\,,$$
$$\text{subject to} \qquad x_1 + x_2 \leq 3\,.$$

The minimization of U in the problem, but without the constraint, has been solved in Example 2–6, which has an optimal solution of $x_1 = 1$, $x_2 = 1.5$ and $U = -8.25$. The minimization of U with an equality constraint $x_1 + x_2 = 3$ has also been solved in Example 2–11, which has an optimal solution of $x_1 = \frac{31}{24}$, $x_2 = \frac{41}{24}$ and $U = -\frac{385}{48}$. As can be seen from Fig. 2–14, the solution for this problem with an inequality constraint is obtained by searching both the interior points on the paraboloid surface bounded by the inequality constraint $x_1 + x_2 \leq 3$, and the boundary points

† See Reference 2–4.

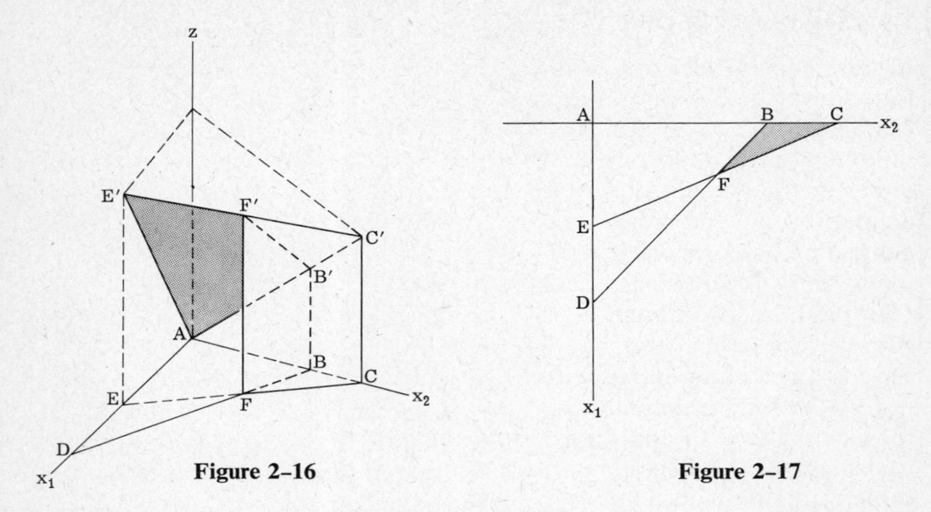

Figure 2-16

Figure 2-17

represented by $x_1 + x_2 = 3$. For this problem, it can be seen from the graphical representation that the optimum solution occurs at the interior point $x_1 = 1$ and $x_2 = 1.5$, as in Example 2-6.

If nonnegative constraints $x_1 > 0$ and $x_2 \geq 0$ are added to this problem, the minimum which occurs on each of the vertical planes $x_1 = 0$ and $x_2 = 0$ should be compared with the minimum at the boundary $x_1 + x_2 = 3$ as well as in the interior of the paraboloid surface bounded by

$$x_1 \geq 0, \quad x_2 \geq 0, \quad \text{and} \quad x_1 + x_2 \leq 3.$$

Example 2-14. Represent conceptually in a graph the solution of the following optimization problem:

$$\begin{aligned} \text{Min} \quad & z = 2x_1 + x_2, \\ \text{subject to} \quad & 5x_1 + 10x_2 \geq 8, \\ & x_1 + x_2 \leq 1, \\ & x_1 \geq 0, \quad x_2 \geq 0. \end{aligned}$$

The graphical representation of the solution of this problem is shown in Fig. 2-16, while its projection on the x_1x_2-plane is shown in Fig. 2-17. In this case, $x_2 \geq 0$ does not affect the solution at all, but the other three constraints form the three sides of a prismatic surface which intersects the x_1x_2-plane at the boundary of the triangle BCF. The objective function is a plane, represented by the area $AE'C'$, which intersects the prismatic surface at the boundary of the triangle $B'C'F'$. The absolute or global minimum occurs at the boundary point for which z is smallest. In this case, $z = BB'$ is seen by inspection to be the minimum of the objective function subject to the given constraints.

2–9 LIMITATIONS OF CALCULUS METHODS

In this chapter, we have examined the basic concepts of optimization of functions of real variables defined in a closed region. By using simple examples, we have illustrated that the global maximum (or global minimum) of an objective function subjected to various constraints occurs either in the domain or at the boundary. Since a domain is generally defined by one or more equations and the boundary of the domain is defined by one or more inequalities, we have considered methods of finding extrema of functions with or without constraints.

In the case of optimization without constraint or with equality constraints, the difficulty lies in the fact that the necessary conditions for the existence of an extremum are not sufficient, and that the criteria for testing the sufficient conditions, even if they can be established, are far too complicated to apply to real problems involving a large number of variables. Only when the functions are convex or concave can we remove some of the obstacles in obtaining an optimal solution.

The optimization of functions subject to inequality constraints generally compounds the difficulty in that the search for extrema is spread from the domain to the boundary. Thus the methods of calculus either become ineffective or are not applicable. When the objective function and all its constraints are linear, the search for a global optimum is confined to the boundary, and it can be carried out by use of linear programming.

In spite of these limitations, the discussion in this chapter has provided an overall view of the types of problems encountered in mathematical programming and their interrelationships. The reader is referred to the references at the end of this chapter for further study.

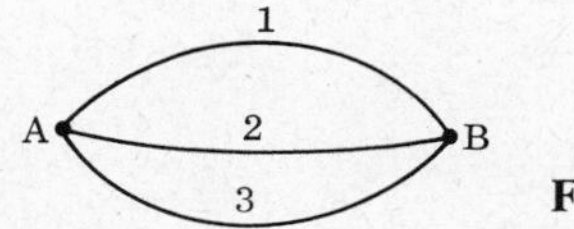

Figure 2–18

Example 2–15. In regulating traffic at rush hours, the cars at point A downtown heading for destination B are directed into three routes (see Fig. 2–18). In addition to some fixed factors, such as the distance and the topography along the route, the total traveling time t for all cars in each route depends of x, the number of cars in that route. Let the functional relations between t and x for routes 1, 2 and 3 be, respectively,

$$t_1 = 4x_1^2 - 5x_1 + 2\,, \qquad t_2 = 3x_2^2 + x_2 + 1\,, \qquad t_3 = x_3^3 + 2\,.$$

If 100 cars are going from A to B, how should they be directed to each route so that the sum of the total traveling time in all routes will be a minimum?

This is a simple problem which can be solved by using the Lagrange multiplier. However, even for such a simple problem, the determination of the multiplier involves the solution of a quadratic equation as shown below.

$$\text{Min} \quad f = t_1 + t_2 + t_3 \,,$$
$$\text{subject to} \quad x_1 + x_2 + x_3 = 100 \,.$$

In terms of variables x_1, x_2, and x_3, which are temporarily assumed to be continuous, we have

$$\text{Min}\, f = 4x_1^2 - 5x_1 + 3x_2^2 + x_2 + x_3^3 + 5 \,,$$
$$g = x_1 + x_2 + x_3 - 100 = 0 \,.$$

Let

$$\Phi(x_1, x_2, x_3, \lambda) = f(x_1, x_2, x_3) + \lambda g(x_1, x_2, x_3) \,.$$

Then the equivalent problem leads to

$$\frac{\partial \Phi}{\partial x_1} = 8x_1 - 5 + \lambda = 0 \,,$$
$$\frac{\partial \Phi}{\partial x_2} = 6x_2 - 1 + \lambda = 0 \,,$$
$$\frac{\partial \Phi}{\partial x_3} \, 3x_3^2 + \lambda = 0 \,,$$
$$\frac{\partial \Phi}{\partial \lambda} = x_1 + x_2 + x_3 - 100 = 0 \,,$$

from which

$$\lambda^2 + 682\lambda + 115200 = 0 \,.$$

For $\lambda = -306.6$, it is found that

$$x_1 = 39.0 \,, \qquad x_2 = 50.9 \,, \qquad \text{and} \qquad x_3 = 10.1 \,.$$

Since variables in this problem can only be integers and therefore are not continuous, we define $\hat{x}_i$ to be the integer nearest to the value of x_i for the integer function

$$\hat{x}_i = [x_i] \,, \qquad i = 1, 2, 3 \,.$$

Hence the solution for this problem becomes

$$\min \hat{z} = f(\hat{x}_1, \hat{x}_2, \hat{x}_3) \,,$$

in which

$$\hat{x}_1 = [x_1] = 39 \,, \qquad \hat{x}_2 = [x_2] = 51 \,, \qquad \hat{x}_3 = [x_3] = 10 \,.$$

Note that when the set of equations resulting from the necessary condition is nonlinear, as can be expected generally when the problem is nonlinear, even the solution of the algebraic equations may pose a problem. Furthermore, a solution based on the round-off values of the variables is not necessarily a good approximation of the truly optimal solution.

REFERENCES

2–1. KAPLAN, W., *Advanced Calculus*, Addison-Wesley, Reading, Mass., 1952.

2–2. HANCOCK, H., *Theory of Maxima and Minima*, Dover, New York, 1960.

2–3. ARAMONOVICH, I. G., R. S. GUTER, L. A. LYUSTERNIK, I. L. RAUKHVARGER, M. I. SKANAVI, A. R. YANPOL'SKII, *Mathematical Analysis*, Pergamon Press, Oxford, England, 1965.

2–4. KUHN, H. W., and A. W. TUCKER, "Nonlinear Programming," *Proceedings of the Second Berkeley Symposium on Mathematical Statistics and Probability*, University of California Press, Berkeley, Calif., 1951, 481–490.

2–5. SHEDD, T. C., *Structural Design in Steel*, Wiley, New York, 1934, 436–440.

2–6. AU, T., *Elementary Structural Mechanics*, Prentice-Hall, Englewood Cliffs, N. J., 1963, 106–110.

PROBLEMS

P2–1. Show that the necessary condition for maximum in Example 2–3 also satisfies the criteria for sufficient condition.

P2–2. Show that the necessary condition for minimum in Example 2–4 also satisfies the criteria for sufficient condition.

P2–3. Determine the maximum area of a triangle that can be inscribed in a semicircle of diameter d, as shown in Fig. P2–3.

P2–4. Find the maximum moment for the beam shown in Fig. P2–4 subject to a series of two wheels of equal weights at a constant distance b apart ($b < L$).

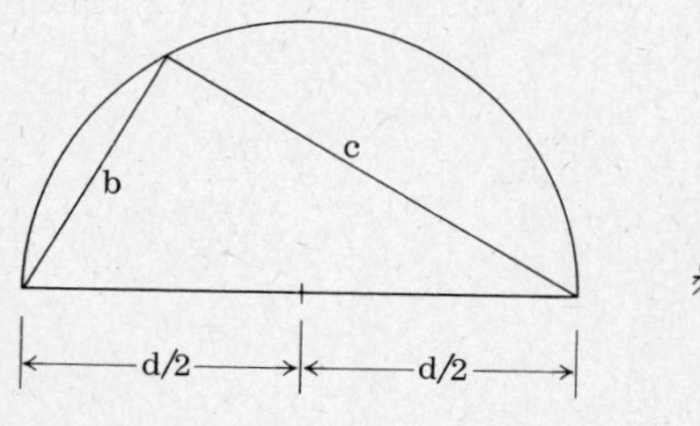

Figure P2–3

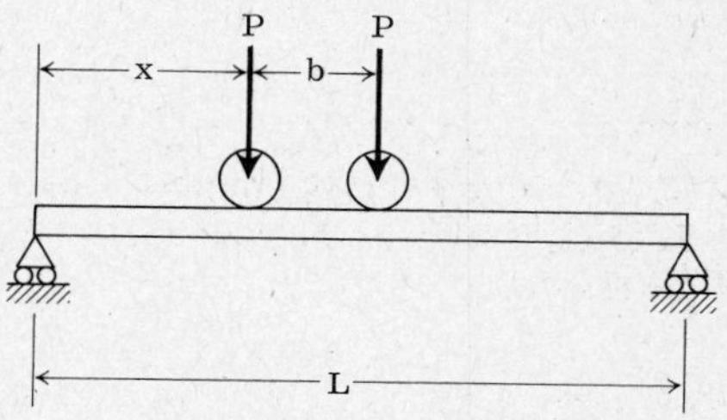

Figure P2–4

P2–5. Determine the location and nature of all critical points for the function $z = xy^2 + 2y^2 + x^2$.

P2–6. Determine the location and nature of all critical points for the function $z = \pm\sqrt{4 - x^2 - y^2}$.

P2–7. The length of the base line of a surveying network has been measured 12 times. For three times, the measured value is $x_1 = 2{,}100.25$ ft, for four times, the measured value is $x_2 = 2{,}100.42$ ft, and for five times, the measured value is $x_3 = 2{,}100.36$ ft. Let y_1, y_2, and y_3 be the corrections for x_1, x_2, and x_3, respectively, such that the adjusted length is given by

$$d = x_1 \pm y_1 = x_2 + y_2 = x_3 + y_3 .$$

It is desired that the sum of squares of the corrections to all measurements be a minimum, i.e.,

$$\text{Min } S = 3y_1^2 + 4y_2^2 + 5y_3^2 .$$

Determine the adjusted length of the base line. (This is known as the method of least squares.)

P2–8. In plotting a set of five points obtained from an experiment, it appears that a function

$$y = ax^2 + bx + c ,$$

may be used to represent the variation in x and y. Given that the coordinates (x, y) of these five points on the xy-plane are (1, 10), (2, 7), (3, 5), (4, 4), and (5, 3.5), determine the constants a, b, and c such that the sum of the squares of the errors in y at these five points will be a minimum.

P2–9. Solve the following maximization problem in which $p_1, p_2, \ldots, p_n$, and b are constants:

$$\begin{aligned} \text{Max} \quad & A = p_1 \ln x_1 + p_2 \ln x_2 + \cdots + p_n \ln x_n , \\ \text{subject to} \quad & x_1 + x_2 + \cdots + x_n = b , \\ & x_j \geq 0 , \qquad j = 1, 2, \ldots, n . \end{aligned}$$

P2–10. Solve the following minimization problem by using the Lagrange multiplier:

$$\begin{aligned} \text{Min} \quad & z = 3x_1^2 + 5x_2^2 - 2x_1x_2 - 11x_2 , \\ \text{subject to} \quad & 2x_1 + 2x_2 = 4 . \end{aligned}$$

P2–11. An ellipse whose axes are parallel to the coordinate axes is to be drawn through three points (x_1, y_1), (x_2, y_2), and (x_3, y_3). Formulate the problem for optimizing the area of such an ellipse.

P2–12. Determine the maximum radius r of a sphere, $x^2 + y^2 + z^2 = r^2$, which can be placed within an ellipsoid whose equation is $3x^2 + 2y^2 + z^2 = 6$. Formulate the problem only.

CHAPTER 3

LINEAR PROGRAMMING

3-1 ELEMENTARY EXAMPLES

Linear programming deals with the optimization of a linear objective function subject to a set of constraint conditions in the form of linear inequalities and/or equations. Since the constraint conditions represent the boundary of a feasible region, the basic purpose of linear programming is to seek among all feasible solutions an optimal solution satisfying the objective function. However, if the constraint conditions do not form a closed region, the problem may not have a finite optimal solution. Furthermore, if the constraint conditions are inconsistent or contradictory, the problem does not even have a feasible solution. We shall begin with several elementary examples.

Example 3–1. In order to produce a certain steel alloy with desired properties, it is required that for every ton of raw material, there should be no less than 1 lb of manganese, nor less than 3 lb of silicon, nor more than 6 lb of copper. The raw material is drawn from two iron ores with different compositions in these elements. Per ton of ore material, No. 1 ore contains 1 lb of manganese, 2 lb of silicon and 3 lb of copper; No. 2 ore contains 1 lb of manganese, 4 lb of silicon and 7 lb of copper. If the costs of material, including transportation, from ore Nos. 1 and 2 are $4 and $5 per ton, respectively, what amount should be shipped from each ore per ton of raw material used?

This is a typical problem in product mix. Let x_1 and x_2 be the quantities to be shipped from ore Nos. 1 and 2, respectively, per ton of raw material. A simple conceptual model can be formulated by observing the requirements of material balance and constraints. These conditions are

$$x_1 + x_2 \geq 1\,, \qquad 2x_1 + 4x_2 \geq 3\,, \qquad 3x_1 + 7x_2 \leq 6\,.$$

The objective function is a measure of the merit of various shipping combinations. Let the shipping cost z be the figure of merit, in dollars per ton. Then the objective function is given by

$$\text{Min } z = 4x_1 + 5x_2\,.$$

Since x_1 and x_2 can only be positive or zero, we can introduce the nonnegative constraint conditions $x_1 \geq 0$ and $x_2 \geq 0$ to limit the range of variables x_1 and x_2. The problem is thus completely stated and an optimal solution is sought. Since many sets of values of x_1 and x_2 may satisfy the constraint conditions (e.g., the set $x_1 = 0.3$ and $x_2 = 0.7$ constitutes a feasible solution, so is $x_1 = 0.5$ and $x_2 = 0.5$), we must determine the optimal solution, if one exists. At this point, it is sufficient to point out that for $x_1 = 0.3$ and $x_2 = 0.7$, $z = 4.7$; but for $x_1 = 0.5$ and $x_2 = 0.5$, $z = 4.5$. It is obvious that the former set does not lead to a minimum; whether the latter set yields a minimum cost or not remains to be shown. Our problem is to try to obtain a unique solution which is optimal, if it exists at all.

Example 3-2. A concrete mixing plant which supplies concrete in large quantity to a dam project uses sand and gravel mixtures of 30% sand and 70% gravel by weight. Natural deposits at five pits near the dam site are found to have different compositions and their costs, including transportation to site, also vary as shown in Table 3-1. For every pound mixture, how many pounds of deposits should be obtained from each source in order to minimize the total cost?

Table 3-1

SAND AND GRAVEL MIXTURE

	Pit number					Desired mixture
	1	2	3	4	5	
Sand	40%	20%	50%	80%	70%	30%
Gravel	60%	80%	50%	20%	30%	70%
Cost/10 lb	3 ¢	2 ¢	1 ¢	$1\frac{1}{2}$ ¢	$2\frac{1}{2}$ ¢	Min z

In this blending problem, let x_1, x_2, x_3, x_4, and x_5 (all ≥ 0) be the amounts of natural deposits to be taken from the five pits for every pound of mixture. Then the problem can be stated as follows:

$$\text{Min} \quad z = 3x_1 + 2x_2 + x_3 + 1.5x_4 + 2.5x_5 ,$$

subject to

$$0.4x_1 + 0.2x_2 + 0.5x_3 + 0.8x_4 + 0.7x_5 = 0.3 ,$$

$$0.6x_1 + 0.8x_2 + 0.5x_3 + 0.2x_4 + 0.3x_5 = 0.7 ,$$

$$x_1 + x_2 + x_3 + x_4 + x_5 = 1 .$$

Note that the three constraint equations are not independent, since the linear combination of any two of these equations may produce the third one.

It may be reasoned intuitively that we should take as much as possible from pit No. 3 for which the cost is the lowest, and having a higher percentage of sand than needed, the mix can be balanced by that obtained from pit No. 2 for which the cost is the next lowest. (Pit No. 4 is ruled out because it will produce even a higher percentage of sand.) It is not surprising, therefore, that the set of variables leading to minimum cost is $x_1 = x_4 = x_5 = 0$, $x_2 = \frac{2}{3}$ and $x_3 = \frac{1}{3}$. Of course, this is, based on the assumption that the supplies at various pits are unlimited; hence some pits may not be used at all so long as the others are available and produce deposits at lower cost. Thus min $z = \frac{5}{3}$.

Example 3–3. A construction job requires a mixture of aggregates composed of 50% coarse gravel, 30% fine gravel, and 20% sand. There are four sources for the material near the location of job. The percentage of each grade in the material from each pit and the unit costs of the material delivered to the job are given in Table 3–2. If excess material is delivered, the unit cost for its removal is $ 2 for all grades of aggregates. Determine the amount to be purchased from each pit in order to obtain a unit of the required aggregate at the minimum cost.

Table 3–2

MIXTURE OF SAND AND AGGREGATES

	Pit number				Desired aggregate
	1	2	3	4	
Percent coarse	0	50	40	40	50
Percent fine	60	40	30	20	30
Percent sand	40	10	30	40	20
Unit cost	$8	$10	$12	$11	Min z

Let us first assume that no excess material is delivered to the job, since it appears to be logical to ship no more than necessary. Let x_j denote the amount of material taken from Pit j ($j = 1, 2, 3, 4$). Then for all $x_j \geq 0$,

$$0.5x_2 + 0.4x_3 + 0.4x_4 = 0.5\,,$$

$$0.6x_1 + 0.4x_2 + 0.3x_3 + 0.2x_4 = 0.3\,,$$

$$0.4x_1 + 0.1x_2 + 0.3x_3 + 0.4x_4 = 0.2\,,$$

and

$$8x_1 + 10x_2 + 12x_3 + 11x_4 = \text{Min } z\,.$$

If we try to find a solution for this problem, we will soon discover that a feasible solution does not exist. This means that there is no possible way to combine the nonnegative outputs of these pits to get the exact proportions that we specify; it does not mean that the physical problem has no solution if we allow the excess material of any grade to be removed from the job site.

To pose this problem properly, we introduce a set of nonnegative variables x_5, x_6, and x_7, representing the excesses of coarse gravel, fine gravel, and sand, respectively. Physically, at least one, but no more than two, of the three grades may be excessive. Thus for all $x_j \geq 0$,

$$\begin{aligned} 0.5x_2 + 0.4x_3 + 0.4x_4 - x_5 \qquad\qquad &= 0.5\,, \\ 0.6x_1 + 0.4x_2 + 0.3x_3 + 0.2x_4 \qquad - x_6 \qquad &= 0.3\,, \\ 0.4x_1 + 0.1x_2 + 0.3x_3 + 0.4x_4 \qquad\qquad - x_7 &= 0.2\,, \end{aligned}$$

and

$$8x_1 + 10x_2 + 12x_3 + 11x_4 + 2x_5 + 2x_6 + 2x_7 = \text{Min}\, z\,.$$

In this case, an optimal solution has been found to be $x_1 = x_3 = x_5 = x_7 = 0$, and

$$x_2 = \tfrac{3}{4}\,, \qquad x_4 = \tfrac{5}{16}\,, \qquad x_6 = \tfrac{1}{16}\,, \qquad \min z = 11\tfrac{1}{16}\,.$$

In other words, the most economical solution is to purchase $\frac{3}{4}$ unit from Pit 2 and $\frac{5}{16}$ unit from Pit 4. There will be $\frac{1}{16}$ unit of excess fine gravel. The total cost, including the cost of removing the excess, is $11\frac{1}{16}$.

3-2 STANDARD FORM

All linear programming problems can be expressed in a general form of m linear equations involving n variables $x_1, x_2, \ldots, x_n$, which are nonnegative. The objective of the program is to minimize the figure of merit z, which is also a linear function of the variables. Thus the problem can be stated as follows:

$$\begin{aligned} a_{11}x_1 + a_{12}x_2 + \cdots + a_{1n}x_n &= b_1\,, \\ a_{21}x_1 + a_{22}x_2 + \cdots + a_{2n}x_n &= b_2\,, \\ &\vdots \\ a_{m1}x_1 + a_{m2}x_2 + \cdots + a_{mn}x_n &= b_m\,; \end{aligned} \tag{3-1}$$

and

$$x_j \geq 0\,, \qquad j = 1, 2, \ldots, n\,, \tag{3-2}$$

$$c_1x_1 + c_2x_2 + \cdots + c_nx_n = \text{Min}\, z\,, \tag{3-3}$$

in which the constants c_j ($j = 1, 2, \ldots, n$) specifying criteria values are called *cost coefficients*; the constants b_i ($i = 1, 2, \ldots, m$) defining the constraint requirements are called *stipulations*, and the constants a_{ij} ($i = 1, 2, \ldots, m$ and $j = 1, 2, \ldots, n$) are called *structual coefficients*. This form of a linear programming problem is called the *standard form*.

The optimization of a linear program may either be maximization or minimization. If the objective is maximization such that

$$\text{Max } A = c_1x_1 + c_2x_2 + \cdots + c_nx_n ,$$

a minimization problem may be formed by multiplying the function by -1 as follows:

$$\text{Min } (-A) = -c_1x_1 - c_2x_2 - \cdots - c_nx_n ,$$

because the maximization of a quantity can be replaced by the minimization of the negative of that quantity. By letting $z = -A$, we have

$$\text{Min } z = -c_1x_1 - c_2x_2 - \cdots - c_nx_n .$$

Although only equality signs are used for (3-1), the actual constraint conditions in general may include inequalities of both greater-than and less-than signs or a mixture of inequalities and equations. In order to replace inequality constraint conditions by equations, we introduce the concept of slack variables. If the ith condition involving q variables has a less-than sign in the following form

$$a_{i1}x_1 + a_{i2}x_2 + \cdots + a_{iq}x_q \leq b_i ,$$

it can be replaced by

$$a_{i1}x_1 + a_{i2}x_2 + \cdots + a_{iq}x_q + x_{q+i} = b_i ,$$

where x_{q+i} is a nonnegative slack variable. On the other hand, if it has a greater-than sign, the inequality is replaced by

$$a_{i1}x_1 + a_{i2}x_2 + \cdots + a_{iq}x_q - x_{q+i} = b_i ,$$

where x_{q+i} again is a nonnegative slack variable.

If some of the variables in (3-1) are not physically constrained to be nonnegative, they can always be treated as the differences of two nonnegative variables. For instance, consider, the variable x_j, which is not restricted in sign. We may express it as

$$x_j = x'_j - x''_j ,$$

where

$$x'_j \geq 0 \quad \text{and} \quad x''_j \geq 0 .$$

Since x'_j may be greater or smaller than x''_j, the sign of x_j may be positive or negative. Thus the problem can again be expressed in the standard form by breaking up a nonrestricted variable into two nonnegative vaiables.

Example 3-4. Reduce the following linear programming problem to the standard form:

$$\text{Max} \quad A = 2x_1 + x_2,$$

subject to

$$\begin{aligned} -2x_1 + x_2 &\leq 4, \\ x_1 + 3x_2 &\geq 3, \\ x_1 + x_2 &\leq 7, \end{aligned}$$

and

$$x_1 \geq 0, \qquad x_2 \geq 0.$$

The problem can be expressed in the standard form as follows:

$$\begin{aligned} -2x_1 + x_2 + x_3 &= 4, \\ x_1 + 3x_2 - x_4 &= 3, \\ x_1 + x_2 + x_5 &= 7, \end{aligned}$$

$$x_j \geq 0, \quad j = 1, 2; \quad x_{2+i} \geq 0, \quad i = 1, 2, 3;$$

$$-2x_1 - x_2 = \text{Min } z.$$

Example 3-5. Reduce the previous example to the standard form given that only x_1 is nonnegative but x_2 is nonrestricted.

Let

$$x_2 = x_2' - x_2'',$$

where both x_2' and x_2'' are nonnegative. Thus the standard form of the problem becomes:

$$\begin{aligned} -2x_1 + x_2' - x_2'' + x_3 &= 4, \\ x_1 + 3x_2' - 3x_2'' - x_4 &= 3, \\ x_1 + x_2' + x_2'' + x_5 &= 7, \end{aligned}$$

$$x_1 \geq 0, \quad x_2' \geq 0, \quad x_2'' \geq 0, \quad x_3 \geq 0, \quad x_4 \geq 0, \quad x_5 \geq 0,$$

$$-2x_1 - x_2' + x_2'' = \text{Min } z.$$

3-3 INEQUALITY FORMS

Linear programming problems can also be expressed in inequality form with all constraints having the same inequality sign. Consider the minimization problem:

$$x_j \geq 0\,, \qquad j = 1, 2, \ldots, q\,, \tag{3–4}$$

$$\begin{aligned} a_{11}x_1 + a_{12}x_2 + \cdots + a_{1q}x_q &\geq b_1\,, \\ a_{21}x_1 + a_{22}x_2 + \cdots + a_{2q}x_q &\geq b_2\,, \\ &\vdots \\ a_{p1}x_1 + a_{p2}x_2 + \cdots + a_{pq}x_q &\geq b_p\,, \end{aligned} \tag{3–5}$$

$$c_1x_1 + c_2x_2 + \cdots + c_qx_q = \text{Min}\, z\,. \tag{3–6}$$

Note that the greater-than sign is used for all constraints. This orientation of inequality sign is consistent with the minimization problem, since the objective function itself can be interpreted as having a greater-than relation if z becomes minimum.

If a system of given constraint conditions consists of some equations, then each of them, say the ith equation,

$$a_{i1}x_1 + a_{i2}x_2 + \cdots + a_{iq}x_q = b_i$$

can be replaced by two inequalities,

$$a_{i1}x_1 + a_{i2}x_2 + \cdots + a_{iq}x_q \geq b_i\,,$$

and

$$a_{i1}x_1 + a_{i2}x_2 + \cdots + a_{iq}x_q \leq b_i\,.$$

Furthermore, any greater-than sign can be changed to less-than sign, and conversely, by multiplying through the inequality by -1. Thus for

$$a_{i1}x_1 + a_{i2}x_2 + \cdots + a_{iq}x_q \leq b_i\,,$$

we get

$$-a_{i1}x_1 - a_{i2}x_2 - \cdots - a_{iq}x_q \geq -b_i\,,$$

which is consistent with the inequalities in (3–5).

On the other hand, a maximization problem can be formed instead of a minimization problem if (3–5) and (3–6) are multiplied through by -1. Thus

$$\begin{aligned} -a_{11}x_1 - a_{12}x_2 - \cdots - a_{1q}x_q &\leq -b_1\,, \\ -a_{21}x_1 - a_{22}x_2 - \cdots - a_{2q}x_q &\leq -b_2\,, \\ &\vdots \\ -a_{p1}x_1 - a_{p2}x_2 - \cdots - a_{pq}x_q &\leq -b_q\,, \\ -c_1x_1 - c_2x_2 - \cdots - c_qx_q &= \text{Max}\,(-z)\,. \end{aligned}$$

Replacing $-a_{ij}$ by a'_{ij}, $-b_i$ by b'_i, $-c_j$ by c'_j and $-z$ by A, the maximization problem is seen to be as follows:

$$\begin{aligned}
a'_{11}x_1 + a'_{12}x_2 + \cdots + a'_{1q}x_q &\leq b'_1\,, \\
a'_{21}x_1 + a'_{22}x_2 + \cdots + a'_{2q}x_q &\leq b'_2\,, \\
&\vdots \\
a'_{p1}x_1 + a'_{p2}x_2 + \cdots + a'_{pq}x_q &\leq b'_p\,, \\
c'_1x_1 + c'_2x_2 + \cdots + c'_qx_q &= \text{Max } A\,.
\end{aligned}$$

Note also that the less-than sign is used for all constraints in connection with the maximization problem.

The inequality form for the minimization problem can easily be converted to the standard form. Let $x_{q+1}, x_{q+2}, \ldots, x_{q+p}$ be nonnegative slack variables. Then

$$\begin{aligned}
a_{11}x_1 + a_{12}x_2 + \cdots + a_{1q}x_q - x_{q+1} \qquad\qquad &= b_1\,, \\
a_{21}x_1 + a_{22}x_2 + \cdots + a_{2q}x_q \qquad - x_{q+2} \qquad &= b_2\,, \\
&\vdots \\
a_{p1}x_1 + a_{p2}x_2 + \cdots + a_{pq}x_q \qquad\qquad + x_{q+p} &= b_p\,,
\end{aligned} \tag{3-7}$$

$$\begin{aligned}
x_j &\geq 0\,, \qquad j = 1, 2, \ldots, q\,; \\
x_{q+i} &\geq 0\,, \qquad i = 1, 2, \ldots, p\,;
\end{aligned} \tag{3-8}$$

$$c_1x_1 + c_2x_2 + \cdots + c_qx_q = \text{Min } z\,, \tag{3-9}$$

Similarly, the inequality form for the maximization problem can be converted to the standard form by adding instead of subtracting a nonnegative slack variable to each of the inequalities. In either case, a system of m equations in n variables is obtained from p inequalities in q variables. (In this particular case, $m = p$ and $n = p + q$.) We can therefore regard the constraints of any linear general programming problem as either p inequalities in q variables or as m equations in n variables without losing generality. In the inequality form, either p or q may be the larger of the two numbers; in the standard form, however, m is usually smaller than n because of the addition of slack variables to the original system of inequalities.

Example 3-6. Express the following linear programming problem in the inequality form for minimization:

$$\begin{aligned}
x_1 \geq 0\,, \qquad & x_2 \geq 0\,, \\
2x_1 + 4x_2 &\leq 3\,, \\
x_1 + x_2 &\geq 1\,, \\
2x_1 \qquad &= 1\,, \\
-2x_1 + 3x_2 &= \text{Max } A\,.
\end{aligned}$$

The result is given in the following inequality form which is self-explanatory.

$$\begin{aligned} x_1 \geq 0\,, \qquad & x_2 \geq 0\,, \\ -2x_1 - 4x_2 &\geq -3\,, \\ x_1 + x_2 &\geq 1\,, \\ 2x_1 \qquad &\geq 1\,, \\ -2x_1 \qquad &\geq -1, \\ 2x_1 - 3x_2 &= \text{Min } z\,. \end{aligned}$$

3–4 GENERAL APPROACH IN SEEKING AN OPTIMAL SOLUTION

Let us return to the standard form represented by (3–1), (3–2), and (3–3). In general, the equations in (3–1) may include slack variables as well as the decision variables of the original problem, since all inequality constraints in the original problem have been replaced by equations in the standard form. Both the decision variables and the slack variables are called *admissible variables* and are treated indiscriminately in the solution of the system of linear equations in (3–1). If the number of variables n equals the number of equations m, that is, $n = m$ for m linearly independent equations, the unique solution of this set of equations is the optimal solution; if $n < m$, there can be no solution unless $(m - n)$ equations are linearly dependant or redundent. Hence we are primarily interested in the optimal solution of the problem for $n > m$.

If we select arbitrarily $(n - m)$ variables for the equations in (3–1) and set them equal to any arbitrary constant values, we can solve for the remaining m variables from the resulting set of m independent equations in m unknowns. If these $(n - m)$ variables is set equal to zero, the set of remaining m variables is said to form a *basis*; the m variables on the basis are called *basic variables*, and the $(n - m)$ variables set equal to zero are called *nonbasic variables*. The set of values for the basic variables is called a *basic solution*. Since any set of m variables can be chosen, there are $n(n-1)\cdots(n-m+1)/m!$ possible basic solutions. The region of feasible solutions in the problem is bounded by the equations in (3–1) and the nonnegative constraints in (3–2); and the basic solution represents the "vertices" or "extreme points" of a piecewise linear boundary where the optimal solution is expected to be found. Thus a basic solution obtained from (3–1) which also satisfies (3–2) is called a *basic feasible solution*, and the basic feasible solutions represent a subset of feasible solutions from which the optimal solution is to be chosen according to the criterion in (3–3).

The method of search for an optimal solution can therefore be reduced from total enumeration among all feasible solutions to the enumeration of a subset of feasible solutions, i.e., basic feasible solutions. However, since the number of basic feasible solutions can be very large for most problems, rules must be provided for a systematic search for the optimal solution. One obvious method of approach is first to find an initial basic feasible solution, then to test whether the solution is optimal according to criterion of optimality, and if necessary, to improve the solution by changing the basis until an optimal solution is reached. We shall delay the discussion of an iterative procedure by which the solution can be improved successively until an optimal solution is obtained after a finite number of steps. At present, we shall examine the determination of an optimal solution by enumerating all basic feasible solutions for a simple example.

Example 3–7. Find the optimal solution for the following linear programming problem:

$$5x_1 + 10x_2 \geq 8 ,$$
$$x_1 + \quad x_2 \leq 1 ,$$
$$x_1, x_2 \geq 0 ,$$
$$2x_1 + \quad x_2 = \text{Min } z .$$

By introducing slack variables x_3 and x_4, the standard form is given by

$$5x_1 + 10x_2 - x_3 \qquad = 8 , \tag{a}$$
$$x_1 + \quad x_2 \qquad + x_4 = 1 , \tag{b}$$
$$x_j \geq 0 , \qquad j = 1, 2, 3, 4 ,$$
$$2x_1 + \quad x_2 \qquad\qquad = \text{Min } z . \tag{c}$$

Since there are two constraint equations and four admissible variables, two of the four variables are chosen as basic variables to form a basis, and the remaining two become nonbasic variables. There are altogether six possible combinations of basic variables. Without any guideline for such a selection, let us exhaust all possibilities and express each combination of basic variables in terms of nonbasic variables and constants. Also, express the objective function in terms of the basic variables in each case. When the nonbasic variables are set equal to zero, we then have the basic solution and the corresponding value of z. For example, if we let x_1 and x_2 be the basic variables, we can divide Eq. (a) through by 5 and eliminate x_1 from Eq. (b) to get

$$x_1 + 2x_2 - 0.2x_3 \qquad = 1.6 , \tag{d}$$
$$x_2 - 0.2x_3 - x_4 = 0.6 . \tag{e}$$

Eliminating x_2 from Eq. (d), we have

$$x_1 \quad + 0.2x_3 + 2x_4 = 0.4\,, \tag{f}$$

$$x_2 - 0.2x_3 - \quad x_4 = 0.6\,. \tag{g}$$

Using Eqs. (f) and (g), we can eliminate x_1 and x_2 from Eq. (c):

$$-0.2x_3 - 3x_4 = z - 1.4\,. \tag{h}$$

When $x_3 = x_4 = 0$, we have $x_1 = 0.4$, $x_2 = 0.6$, and $z = 1.4$. Hence all six possible cases of different bases can be listed below:

Case (A). x_3 and x_4 $(x_1 = x_2 = 0)$:

$$\begin{aligned} x_3 \quad - 5x_1 - 10x_2 &= -8\,, \\ x_4 + x_1 + \quad x_2 &= 1\,, \\ 2x_1 + \quad x_2 &= z\,. \end{aligned}$$

Case (B). x_2 and x_4 $(x_1 = x_3 = 0)$:

$$\begin{aligned} x_2 \quad + 0.5x_1 - 0.1x_3 &= 0.8\,, \\ x_4 + 0.5x_1 + 0.1x_3 &= 0.2\,, \\ 1.5x_1 + 0.1x_3 &= z - 0.8\,. \end{aligned}$$

Case (C). x_2 and x_3 $(x_1 = x_4 = 0)$:

$$\begin{aligned} x_2 \quad + \ x_1 + \quad x_4 &= 1\,, \\ x_3 + 5x_1 + 10x_4 &= 2\,, \\ x_1 - \quad x_4 &= z - 1\,. \end{aligned}$$

Case (D). x_1 and x_4 $(x_2 = x_3 = 0)$:

$$\begin{aligned} x_1 \quad + 2x_2 - 0.2x_3 &= 1.6\,, \\ x_4 - \quad x_2 + 0.2x_3 &= -0.6\,, \\ -3x_2 + 0.4x_3 &= z - 3.2\,. \end{aligned}$$

Table 3-3

BASIC SOLUTIONS

Variable	A	B	C	D	E	F
x_1	0	0	0	1.6	1	0.4
x_2	0	0.8	1	0	0	0.6
x_3	−8	0	2	0	−3	0
x_4	1	0.2	0	−0.6	0	0
z	0	0.8	1	3.2	2	1.4

Case (E). x_1 and x_3 ($x_2 = x_4 = 0$):

$$\begin{aligned} x_1 \quad + x_2 + x_4 &= 1\,, \\ x_3 - 5x_2 + 5x_4 &= -3\,, \\ -x_2 - 2x_4 &= z - 2\,. \end{aligned}$$

Case (F). x_1 and x_2 ($x_3 = x_4 = 0$):

$$\begin{aligned} x_1 \quad + 0.2x_3 + 2x_4 &= 0.4\,, \\ x_2 - 0.2x_3 - x_4 &= 0.6\,, \\ -0.2x_3 - 3x_4 &= z - 1.4\,. \end{aligned}$$

The results of these basic solutions are summarized in Table 3–3.

Since all x_j ($j = 1, 2, 3, 4$) are constrained to be nonnegative, any solution which has a negative value for any of its variables is not a feasible solution. From Table 3–3, we see that cases A, D, and E are infeasible solutions, while cases B, C, and F are feasible solutions. Since these solutions are also basic, cases B, C, and F are basic feasible solutions. Among these latter three solutions, case B has a minimum value of z, and is therefore the optimal solution.

3–5 GEOMETRICAL INTERPRETATION

The geometrical interpretation of a linear programming problem can best be illustrated by simple problems involving only two decision variables although the generalized concept is applicable to problems involving any number of variables. One such example has been given by Example 2–14 in Chapter 2, which is also solved algebraically in Example 3–7. Let us examine the graphic solution in detail and compare it with the algebraic solution. The constraint conditions,

$$\begin{aligned} 5x_1 + 10x_2 &\geq 8\,, \\ x_1 + x_2 &\leq 1\,, \end{aligned}$$

are shown in parts (a) and (b) of Fig. 3–1 while the nonnegative constraints $x_1 \geq 0$ and $x_2 \geq 0$ are given in parts (c) and (d) of the same figure. Note that each equality divides the x_1x_2-plane into two half-planes, and that each inequality refers to a half-plane corresponding to the inequality sign. When the graphs of these constraints are superimposed on the same plane, as shown in Fig. 3–2, the triangle *BCF* represents the region of feasible solutions, since any point in the triangle or on its boundary satisfies the constraints. (The nonnegative constraint $x_2 \geq 0$ is automatically satisfied.) This region of feasible solutions is a *convex set* which, by definition, is a set of points in which a line segment joining any two

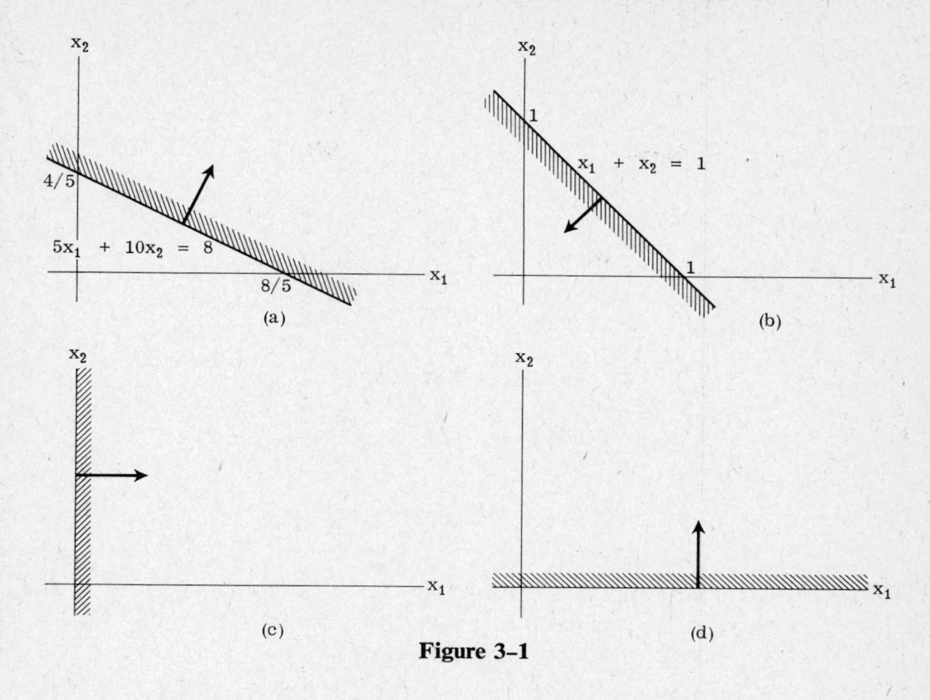

Figure 3–1

points in the set must lie entirely within the set. The vertices B, C, and F of the triangle are extreme points of the set. If we regard z as a constant in the objective function

$$2x_1 + x_2 = z ,$$

we see that the slope of this line is $dx_2/dx_1 = -2$. In other words, if we plot this line on the x_1x_2-plane by assigning an arbitrary value for z, then any line representing this equation and having a different value of z must be parallel to this line. Two such lines for $z = 0.5$ and $z = 0.8$ are shown in Fig. 3–2. If we move the line $2x_1 + x_2 = z$ by assigning different

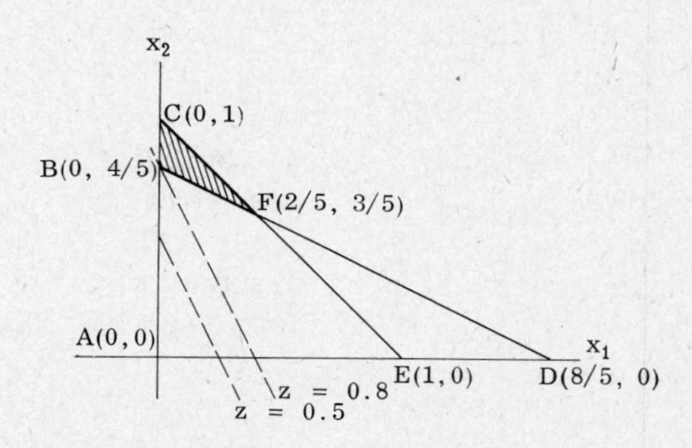

Figure 3–2

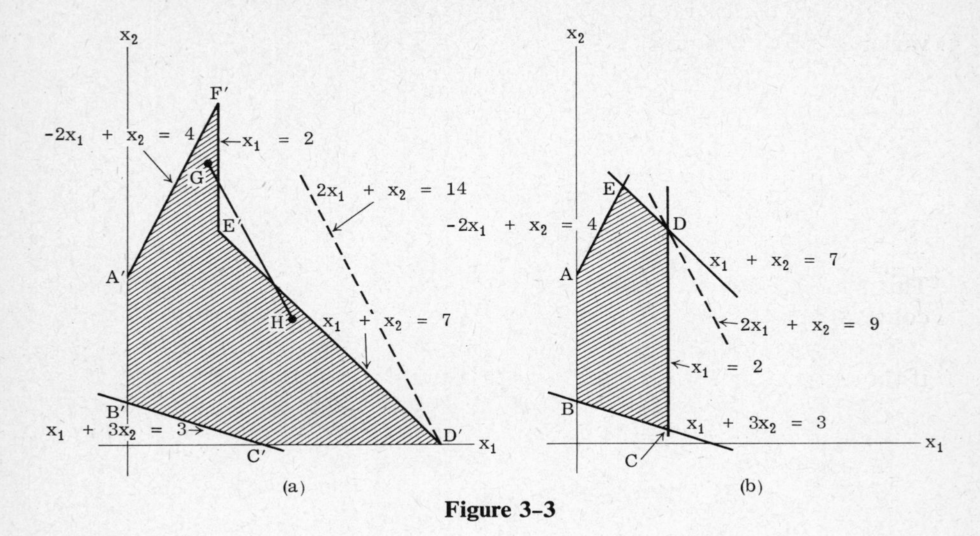

Figure 3-3

values for z, we can expect to reach eventually a position (point B in this case) in the region of feasibility where z is the required minimum.

The set of constraints (including nonnegative constraints) in a linear programming problem always constitutes a convex set. This can be seen from another example,

$$x_1 \geq 0\,, \qquad x_2 \geq 0\,,$$

$$-2x_1 + x_2 \leq 4\,,$$

$$x_1 + 3x_2 \geq 3\,,$$

$$x_1 + x_2 \leq 7\,,$$

$$x_1 \leq 2\,,$$

$$\text{Max } A = 2x_1 + x_2\,.$$

If these constraints were plotted, as shown in Fig. 3-3(a), the resulting polygon, $A'B'C'D'E'$, would not be a convex set, since part of the line segment joining points G and H lies outside of the polygon. However, part (a) does not depict a correct representation of the constraints, as the region of feasibility is actually given by the polygon $ABCDE$ as shown in part (b). Thus we can conclude that a set of feasible solutions which are defined as the constraints in a linear programming problem constitutes a convex set whose extreme points correspond to basic feasible solutions.

A linear programming problem may not even have a feasible solution if the constraint conditions are inconsistent, or if the points satisfying the constraint conditions do not satisfy the nonnegative restrictions on the variables. For example,

$$5x_1 + 10x_2 \geq 8 ,$$
$$2x_1 + \ \ 2x_2 \leq 1 ,$$
$$x_1, x_2 \geq 0 ,$$
$$2x_1 + \quad x_2 = \text{Min } z .$$

This problem has no feasible solution since the constraint conditions are contradicting, as indicated by Fig. 3–4.

The solution of a linear programming problem is said to be *unbounded* if the optimal value is infinite. Consider the example

$$x_1 \geq 0 , \qquad x_2 \geq 0 ,$$
$$4x_1 + 2x_2 \geq 3 ,$$
$$x_1 + \quad x_2 \geq 1 ,$$
$$\text{Max } A = 12x_1 + 8x_2 .$$

The solution of this problem is unbounded, since A can be made infinitely large by increasing x_1 and x_2 indefinitely. However, it is not necessary that all variables be made infinite in order to obtain an unbounded solution, or that arbitrarily large values of the variables must lead to an infinite optimal value of the objective function. Consider the example

$$y_1 \geq 0 , \qquad y_2 \geq 0 ,$$
$$2y_1 - y_2 \geq 2 ,$$
$$y_2 \leq 3 ,$$
$$\text{Max } A = y_1 - y_2 .$$

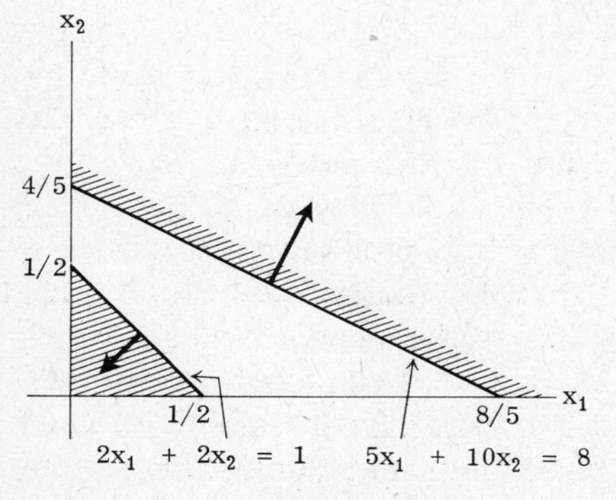

Figure 3–4

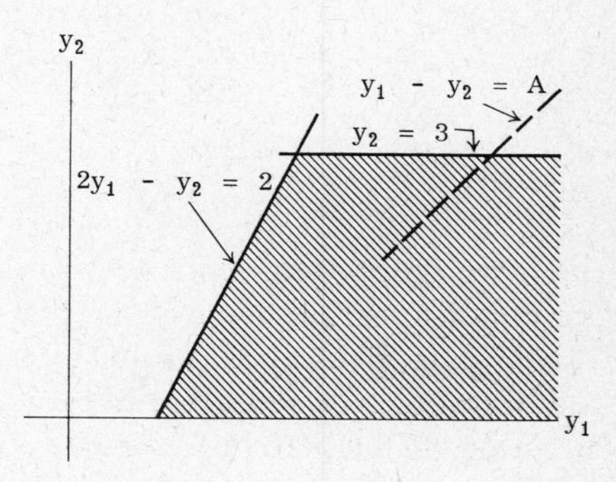

Figure 3–5

Then A is unbounded even if y_2 is finite, as shown in Fig. 3–5. On the other hand, consider the example

$$y_1 \geq 0 , \qquad y_2 \geq 0 ,$$

$$2y_1 - y_2 \geq 2 ,$$

$$y_1 - 2y_2 \geq -4 ,$$

$$\text{Max } A = -2y_1 + 4y_2 .$$

Since the line representing the objective function for Max A coincides with one of the constraint conditions, there are infinitely many sets of y_1 and y_2, including sets of arbitrarily large values which may lead to the optimal value of A, as shown in Fig. 3–6.

We can restate the geometrical interpretation of a linear programming problem in more general terms. The set of points corresponding to the feasible solutions constitutes a convex set. A basic feasible solution corresponds to an extreme point in the convex set. If the objective function has a finite optimum, and if the optimum (maximum or minimum) is unique, then the optimal solution is a basic feasible solution. A reasonable search procedure is to move from one extreme point to another along the boundary of the convex set until an optimal solution is reached, since the optimum will not occur at interior points. Furthermore, once an optimum is reached at the boundary, it is also a global maximum or minimum.

The detailed procedure for finding a graphical solution of simple problems with two decision variables is illustrated by examples.

Example 3–8. Determine graphically the optimal solution for Example 3–4.

$$x_1 \geq 0 , \qquad x_2 \geq 0 ,$$

$$-2x_1 + x_2 \leq 4 , \tag{a}$$

$$x_1 + 3x_2 \geq 3 , \tag{b}$$

$$x_1 + x_2 \leq 7 , \tag{c}$$

$$\text{Max } A = 2x_1 + x_2 . \tag{d}$$

First, we ignore temporarily the inequalities and plot the equations for the conditions, as shown in Fig. 3–7. Next, we shall decide which side of each line gives a valid solution according to the inequality sign. For condition (a), for example, the line $-2x_1 + x_2 = 4$ can move only in the direction of the arrowhead if the constant on the right-hand side of the condition is restricted to 4. The same reasoning can be applied to other constraint conditions as indicated. Since the slope of the line $A = 2x_1 + x_2$

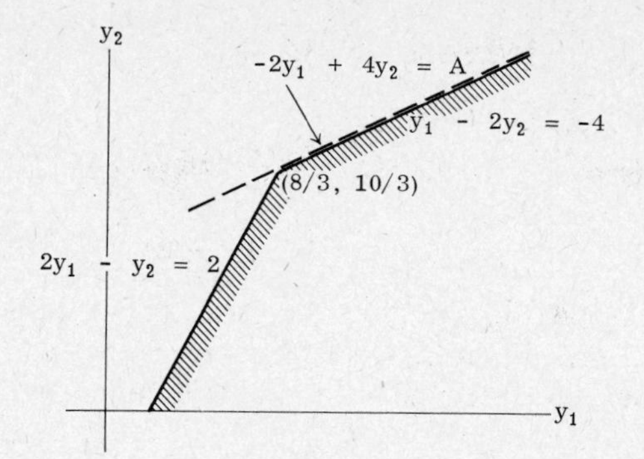

Figure 3–6

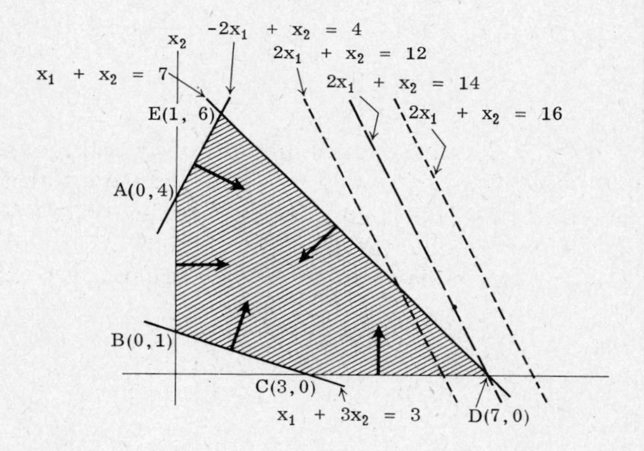

Figure 3–7

representing the objective function is independent of the value of A, the line can be tentatively plotted for an arbitrary value of A. In order to obtain a unique solution, this line is moved back and forth in a parallel position until it hits a single point in the shaded polygon. This situation is possible only if the line passes through one of the vertices of the polygon, such as vertex D in Fig. 3–7. Otherwise, the line will either intersect with too many points in the polygon or have no intersection at all. Hence, for $x_1 = 7$ and $x_2 = 0$, a maximum $A = 14$ is obtained.

3–6 BASIC SOLUTIONS OF CONSTRAINT EQUATIONS

Since the system of linear equations in (3–1) generally has more unknowns than equations ($n > m$), a basic solution can be obtained by setting $n - m$ variables equal to zero and solving for the remaining m variables, provided that the resulting m equations are *linearly independent*. Let $x_1, x_2, \ldots, x_m$ be the basic variables, and $x_{m+1} = x_{m+2} = \cdots = x_n = 0$ be the nonbasic variables. Then the system of resulting equations becomes

$$\begin{aligned}
a_{11}x_1 + a_{12}x_2 + \cdots + a_{1m}x_m &= b_1 , \\
a_{21}x_1 + a_{22}x_2 + \cdots + a_{2m}x_m &= b_2 , \\
&\vdots \\
a_{m1}x_1 + a_{m2}x_2 + \cdots + a_{mm}x_m &= b_m .
\end{aligned} \tag{3-10}$$

By successive elimination, i.e., by eliminating x_1 from all except the first equation, x_2 from all except the first two equations, x_3 from all except the first three equations, etc., we finally obtain a triangular form as follows:

$$\begin{aligned}
x_1 + a'_{12}x_2 + a'_{13}x_3 + \cdots + a'_{1m}x_m &= b'_1 , \\
x_2 + a'_{23}x_3 + \cdots + a'_{2m}x_m &= b'_2 , \\
x_3 + \cdots + a'_{3m}x_m &= b'_3 , \\
&\vdots \\
x_m &= b'_m ,
\end{aligned} \tag{3-11}$$

in which the coefficients a_{ij} are linear combinations of a'_{ij} and the coefficients b'_i are linear combinations of a_{ij} and b_j. Thus the values of all m variables can be determined by back substitution, starting from the last of Eqs. (3-11). If all except one variable are eliminated in each of Eqs. (3-10) such that only x_i remains in the ith equation, we obtain a canonical form as follows:

$$\begin{aligned}
x_1 &= \alpha_{11}b_1 + \alpha_{12}b_2 + \cdots + \alpha_{1m}b_m = \bar{b}_1 , \\
x_2 &= \alpha_{21}b_1 + \alpha_{22}b_2 + \cdots + \alpha_{2m}b_m = \bar{b}_2 , \\
&\vdots \\
x_m &= \alpha_{m1}b_1 + \alpha_{m2}b_2 + \cdots + \alpha_{mm}b_m = \bar{b}_m ,
\end{aligned} \tag{3-12}$$

in which α_{ij} ($i = 1, 2, \ldots, m$; $j = 1, 2, \ldots, m$) are linear combinations of a_{ij} during the process of elimination. The coefficients $\bar{b}_i$ are linear combinations of α_{ij} and b_i. Thus the values of m variables are uniquely determined, provided that the determinant of the coefficients a_{ij} ($i = 1, 2, \ldots, m$; $j = 1, 2, \ldots, m$) does not vanish.

In general, the nonbasic variables can be retained temporarily in the process of elimination. Let us select m basic variables, say $x_1, x_2, \ldots, x_m$ ($m < n$). Then by successive elimination we can obtain a *triangular form* as follows:

$$\begin{aligned}
x_1 + a'_{12}x_2 + a'_{13}x_3 + \cdots + a'_{1m}x_m + \cdots + a'_{1n}x_n &= b'_1 , \\
x_2 + a'_{23}x_3 + \cdots + a'_{2m}x_m + \cdots + a'_{2n}x_n &= b'_2 , \\
x_3 + \cdots + a'_{3m}x_m + \cdots + a'_{3n}x_n &= b'_3 , \\
&\vdots \\
x_m + \cdots + a'_{mn}x_n &= b'_m .
\end{aligned} \tag{3-13}$$

If a set of values is assigned to $x_{m+1}, x_{m+2}, \ldots, x_n$, the values $x_m, x_{m-1}, \ldots, x_2, x_1$ may be obtained successively by back substitution. However, by further elimination, we can reduce the system to the *canonical form* as follows:

$$\begin{aligned}
x_1 \qquad\qquad\quad + \bar{a}_{1,m+1}x_{m+1} + \cdots + \bar{a}_{1n}x_n &= \bar{b}_1 , \\
x_2 \qquad\quad + \bar{a}_{2,m+1}x_{m+1} + \cdots + \bar{a}_{2n}x_n &= \bar{b}_2 , \\
x_3 \quad + \bar{a}_{3,m+1}x_{m+1} + \cdots + \bar{a}_{3n}x_n &= \bar{b}_3 , \qquad (3\text{–}14) \\
&\ \ \vdots \\
x_m + \bar{a}_{m,m+1}x_{m+1} + \cdots + \bar{a}_{mn}x_n &= \bar{b}_m .
\end{aligned}$$

Thus the values of $x_1, x_2, \ldots, x_m$ can be obtained directly from the canonical form once a set of values is assigned to $x_{m+1}, x_{m+2}, \ldots, x_n$. The process of elimination is straightforward in principle, but is quite tedious in the sense that many arithmetical operations in addition, subtraction, multiplication and division are required.

Analogous to the elimination of the basic variables in the constraint equations in (3–1), the objective function in (3–3), namely

$$c_1x_1 + c_2x_2 + \cdots + c_mx_m + c_{m+1}x_{m+1} + \cdots + c_nx_n = \text{Min } z ,$$

can also be reduced to a form without the basic variables as follows:

$$\bar{c}_{m+1}x_{m+1} + \bar{c}_{m+2}x_{m+2} + \cdots + \bar{c}_nx_n = \text{Min } (z - \bar{z}) , \qquad (3\text{–}15)$$

in which $\bar{c}_j$ ($j = m + 1, m + 2, \ldots, n$) and $\bar{z}$ are to be defined later. This can be accomplished by multiplying the first equation of (3–14) by c_1, the second equation by c_2, ... and the mth equation by c_m and then subtracting the products from (3–3). Hence in (3–15),

$$\begin{aligned}
\bar{c}_{m+1} &= c_{m+1} - \sum_{i=1}^{m} c_i\bar{a}_{i,m+1} , \\
\bar{c}_{m+2} &= c_{m+2} - \sum_{i=1}^{m} c_i\bar{a}_{i,m+2} , \\
&\ \ \vdots \\
\bar{c}_n &= c_n - \sum_{i=1}^{m} c_i\bar{a}_{i,n} ,
\end{aligned}$$

and

$$\bar{z} = \sum_{i=1}^{m} c_i\bar{b}_i .$$

Thus when the nonbasic variables are set equal to zero, we immediately obtain $z = \bar{z}$.

The numerical computation has previously been illustrated in Example 3–7 in which the constraint equations (a) and (b) have been reduced to a triangular form in Eq. (d) and (e), and to a canonical form in Eqs. (f) and (g). The objective function (c) has been reduced to the desired form in Eq. (h).

3-7 BASIC CONCEPTS OF THE SIMPLEX METHOD

We shall now consider an iterative procedure for finding an optimal solution to a linear programming problem. This method, as well as a number of modifications and variations, was first developed by Dantzig.

Let us suppose that the linear programming problem defined by (3-1), (3-2) and (3-3) has been reduced to canonical form by selecting arbitrarily $x_1, x_2, \ldots, x_m$ as basic variables. The canonical form as shown in (3-14) and the corresponding objective function in (3-15) are written in this form:

$$\begin{aligned}
x_1 \quad + \bar{a}_{1,m+1}x_{m+1} + \cdots + \bar{a}_{1s}x_s + \cdots + \bar{a}_{1n}x_n &= \bar{b}_1 , \\
x_2 \quad + \bar{a}_{2,m+1}x_{m+1} + \cdots + \bar{a}_{2s}x_s + \cdots + \bar{a}_{2n}x_n &= \bar{b}_2 , \\
&\vdots \\
x_r \quad + \bar{a}_{r,m+1}x_{m+1} + \cdots + \bar{a}_{rs}x_s + \cdots + \bar{a}_{rn}x_n &= \bar{b}_r , \\
&\vdots \\
x_m + \bar{a}_{m,m+1}x_{m+1} + \cdots + \bar{a}_{ms}x_s + \cdots + \bar{a}_{mn}x_n &= \bar{b}_m ,
\end{aligned} \tag{3-16}$$

$$\bar{c}_{m+1}x_{m+1} + \cdots + \bar{c}_s x_s + \cdots + \bar{c}_n x_n = \text{Min}\,(z - \bar{z}) . \tag{3-17}$$

Thus the basic solution corresponding to this canonical form is seen as

$$\begin{aligned}
&x_1 = \bar{b}_1 , \quad x_2 = \bar{b}_2 , \quad \ldots, \quad x_m = \bar{b}_m , \\
&x_{m+1} = x_{m+2} = \cdots = x_n = 0 , \\
&z = \bar{z} .
\end{aligned}$$

If $\bar{b}_1, \bar{b}_2, \ldots, \bar{b}_m$ are nonnegative, then the basic solution is also feasible. However, if one or more $\bar{b}_i$ $(i = 1, 2, \ldots, m)$ are zero, the basic solution is said to be *degenerate* even though $\bar{b}_i$ are still nonnegative. We shall assume, at this point, that we have a basic feasible solution which is not degenerate. The question is how to decide that $z = \bar{z}$ thus obtained is or is not a minimum.

We can answer this question by examining the variation of z in (3-17) when we move away from the original basis. If the coefficients $\bar{c}_j$ $(j = m + 1, m + 2, \ldots, n)$ are all nonnegative, then we cannot make the value of z smaller than $\bar{z}$, since all x_j $(j = m + 1, m + 2, \ldots, n)$ are nonnegative. On the other hand, if at least one of the coefficients, say $\bar{c}_s$, is negative, then z can be made smaller than $\bar{z}$ if x_s takes on a positive value instead of zero. In order to maintain a basic solution, however, we must select a variable, say x_r, from the original set of basic variables, $x_1, x_2, \ldots, x_r, \ldots, x_m$, and set it equal to zero because a basic solution can have only m variables which are not set equal to zero, m being the number of constraint equations. In essence, if $\bar{c}_s$ is negative, the original basis does not lead to an optimal solution, and a new basis can be formed by introducing a new variable x_s and deleting an existing variable x_r. Our next problem is to determine which, of all existing basic variables, should be deleted from the original basis.

Again, the solution can be found by examining the variation of z in (3–17) when x_s is increased from zero to a positive quantity. Since z is to be minimized, we wish to increase x_s as much as possible if $\bar{c}_s$ is negative, provided that the new basic solution thus obtained remains feasible. Thus

$$x_i \geq 0 \qquad \text{for} \quad i = 1, 2, \ldots, m\,,$$

except for the one variable which will be set equal to zero, and

$$x_i = 0 \qquad \text{for} \quad i = m + 1\,, \quad m + 2, \ldots, n$$

except for variable x_s which is positive. Transposing the x_s terms in (3–16) to the right-hand side, we note that, for all positive a_{is},

$$\begin{aligned}
x_1 &= \bar{b}_1 - \bar{a}_{1s}x_s\,, & \bar{b}_1/\bar{a}_{1s} - x_s &\geq 0\,,\\
x_2 &= \bar{b}_2 - \bar{a}_{2s}x_s\,, & \bar{b}_2/\bar{a}_{2s} - x_s &\geq 0\,,\\
&\vdots \\
x_r &= \bar{b}_r - \bar{a}_{rs}x_s\,, & \bar{b}_r/\bar{a}_{rs} - x_s &\geq 0\,,\\
&\vdots \\
x_m &= \bar{b}_m - \bar{a}_{ms}x_s\,, & \bar{b}_m/\bar{a}_{ms} - x_s &\geq 0\,.
\end{aligned} \tag{3–18}$$

As the value of x_s is increased, it will be limited by one of the inequalities in (3–18), which will cause x_i to become zero first, or

$$\bar{b}_i/\bar{a}_{is} - x_s = 0\,, \qquad x_s = \bar{b}_i/\bar{a}_{is}\,.$$

Suppose that the inequality for $i = r$ represents such a case. Then $\bar{b}_r/\bar{a}_{is}$ must be the smallest for all positive $\bar{a}_{is}$. On the other hand, if all $\bar{a}_{is}$ are negative, no upper limit exists for x_s, as indicated by the equations in (3–18). If some $\bar{a}_{is}$ are positive and some are negative, we need be concerned only with the positive $\bar{a}_{is}$ and select $\bar{a}_{rs}$ among them such that

$$x_s = \bar{b}_r/\bar{a}_{rs} = \min_{\bar{a}_{is}>0} \bar{b}_i/\bar{a}_{is} > 0\,, \qquad i = 1, 2, \ldots, m\,. \tag{3–19}$$

Hence the selection of $\bar{a}_{rs}$ on the basis of (3–19) will determine the departing basic variable x_r from the old basis as well as the value of the entering variable x_s for the new basis. The coefficient $\bar{a}_{rs}$ thus selected is called the *pivotal element* in changing the basis.

After x_s and x_r are selected as entering and departing variables, respectively, a canonical form corresponding to the new basis may be obtained by first dividing the equation for $i = r$ in (3–16) by $\bar{a}_{rs}$ and then eliminating x_s from all other equations ($i \neq r$) in (3–16) and from the objective function in (3-17). The resulting equations become:

$$
\begin{aligned}
x_1 \quad + \bar{a}'_1 x_r \qquad & + \bar{a}'_{1,m+1} x_{m+1} + \cdots + 0 + \cdots + \bar{a}'_{1n} x_n = \bar{b}'_1 , \\
x_2 + \bar{a}'_2 x_r \qquad & + \bar{a}'_{2,m+1} x_{m+1} + \cdots + 0 + \cdots + \bar{a}'_{2n} x_n = \bar{b}'_2 , \\
& \vdots \\
\bar{a}'_r x_r \qquad & + \bar{a}'_{r,m+1} x_{m+1} + \cdots + x_s + \cdots + \bar{a}'_{rn} x_n = \bar{b}'_r , \\
& \vdots \\
\bar{a}'_m x_r + x_m & + \bar{a}'_{m,m+1} x_{m+1} + \cdots + 0 + \cdots + \bar{a}'_{mn} x_n = \bar{b}'_m ,
\end{aligned}
\tag{3-20}
$$

$$
\bar{c}'_r x_r \qquad + \bar{c}'_{m+1} x_{m+1} + \cdots + 0 + \cdots + \bar{c}'_n x_n = \text{Min}\,(z - \bar{z}) . \tag{3-21}
$$

Thus, the entire cycle of operations may be repeated until an optimal solution is obtained.

This method of solution, called the *simplex method*, is an iterative procedure of improving one basic feasible solution to obtain another which has a lower value of z. The new basis consisting of $x_1, x_2, \ldots, x_i, \ldots, x_m$ $(i \neq r)$, plus x_s, differs from the old basis consisting of $x_1, x_2, \ldots, x_r, \ldots,$ x_m only by one variable as x_s enters and x_r departs. Geometrically, this method is equivalent to moving from one extreme point in the convex set to an adjacent extreme point having a lower value of z. The procedure describing the simplex method can be recapitulated as follows:

1. Examine $\bar{c}_j$ in the canonical form of a basic feasible solution as given by (3-16) and (3-17). If all $\bar{c}_j$ are nonnegative, the existing solution is optimal. If one or more $\bar{c}_j$ are negative, select the $\bar{c}_s$ which is the smallest of $\bar{c}_j$, that is, the most negative one.

2. Examine coefficients $\bar{a}_{is}$ in column s. If all $\bar{a}_{is}$ are nonpositive, then the optimal solution is unbounded. If one or more $\bar{a}_{is}$ are positive, set up the ratios $\bar{b}_i/\bar{a}_{is}$ for $\bar{a}_{is} > 0$, and determine the smallest value as stated in (3-19). In the case of a tie, the choice is arbitrary.

3. Using $\bar{a}_{rs}$ as a pivotal element, obtain a new basic feasible solution as shown in (3-20) and (3-21) by having x_s as an entering variable and x_r as a departing variable.

4. The procedure is repeated for as many cycles as needed until all $\bar{c}_j$ in a new solution become positive, and such a solution is the optimal solution.

Example 3-9. Consider the linear programming problem

$$
\begin{aligned}
5x_1 + 10x_2 - x_3 \qquad &= 8 , \\
x_1 + \quad x_2 \qquad + x_4 &= 1 , \\
x_j \geq 0 , \qquad j = 1, 2, 3, 4 &, \\
2x_1 + \quad x_2 \qquad &= \text{Min}\, z ,
\end{aligned}
$$

which is known to have a basic feasible solution

$$x_1 \quad + 0.2x_3 + 2x_4 = 0.4\,, \tag{a}$$

$$x_2 - 0.2x_3 - x_4 = 0.6\,, \tag{b}$$

$$-0.2x_3 - 3x_4 = z - 1.4\,. \tag{c}$$

Find the optimal solution by the simplex method.

Note that $\bar{c}_3 = -0.2$ and $\bar{c}_4 = -3$ in the objective function. Thus, the existing basic feasible solution is not optimal. Since $\bar{c}_4 = \text{Min}\,\bar{c}_j$, we examine the coefficients of x_4 and note that $\bar{a}_{14} = 2\ (> 0)$. Hence for $\bar{a}_{rs} > 0$,

$$\text{Min}\,\frac{\bar{b}_r}{\bar{a}_{rs}} = \frac{0.4}{2} = 0.2\,.$$

Using $\bar{a}_{14}$ as the pivotal element, we shall bring x_4 into the basis and eliminate x_1 from the basis. Dividing Eq. (a) by $\bar{a}_{14} = 2$ and then applying the result to eliminate x_4 in Eqs. (b) and (c), we get

$$0.5x_1 \quad + 0.1x_3 + x_4 = 0.2\,, \tag{d}$$

$$0.5x_1 + x_2 - 0.1x_3 \quad = 0.8\,, \tag{e}$$

$$1.5x_1 \quad + 0.1x_3 \quad = z - 0.8\,. \tag{f}$$

Note that Eq. (d) is the result of dividing Eq. (a) by 2, Eq. (e) is obtained by adding Eq. (b) to Eq. (d), and Eq. (f) is the sum of Eq. (c) and three times Eq. (d).

Since all $\bar{c}_j$ are positive, the new basic feasible solution is optimal. No repetition of the procedure is necessary. Hence

$$x_1 = x_3 = 0\,, \qquad x_4 = 0.2\,, \qquad x_2 = 0.8\,, \qquad \text{and} \qquad \text{Min}\,z = 0.8\,.$$

3-8 DETERMINATION OF AN INITIAL BASIC FEASIBLE SOLUTION

In the discussion of the simplex method in the previous section, we have avoided the question of how to find an initial basic feasible solution in order to start the problem. We shall now consider a procedure which not only will lead to such a solution if it exists, but also will eliminate the redundancy of given constraints, or will terminate the iteration whenever a feasible solution does not exist.

We shall begin with a linear programming problem in the standard form as represented by (3–1), (3–2), and (3–3), in which $x_j\ (j = 1, 2, \ldots, n)$ are *admissible* variables. We shall specify that all $b_i\ (i = 1, 2, \ldots, m)$ must be nonnegative. If some of them are negative, the equations corresponding to the negative b_i are multiplied by -1 before the procedure is applied. Thus the assumption of nonnegativity does not cause the loss of generality in the procedure.

We now introduce a set of m nonnegative *artificial* variables, x_{n+1}, $x_{n+2}, \ldots, x_{n+m}$, one for each of the m constraint equations as follows:

$$\begin{aligned}
a_{11}x_1 + a_{12}x_2 + \cdots + a_{1n}x_n + x_{n+1} \qquad\qquad &= b_1 , \\
a_{21}x_1 + a_{22}x_2 + \cdots + a_{2n}x_n \qquad + x_{n+2} \qquad &= b_2 , \\
\vdots \qquad\qquad & \\
a_{m1}x_1 + a_{m2}x_2 + \cdots + a_{mn}x_n \qquad\qquad + x_{n+m} &= b_m ,
\end{aligned} \tag{3-22}$$

If the sum of the artificial variables is denoted by

$$x_{n+1} + x_{n+2} + \cdots + x_{n+m} = w , \tag{3-23}$$

then w must equal zero for a linear programming problem which has a feasible solution, since each of the artificial variables must vanish. The equation in (3-23) may therefore be regarded as an objective function in which w is to be minimized; it is called the *infeasibility form*. If Min $w = 0$, a feasible solution exists; otherwise no feasible solution exists.

If we subtract each equation in (3-22) from (3-23), we obtain a modified objective function for minimization as follows:

$$d_1x_1 + d_2x_2 + \cdots + d_nx_n = w - w_0 , \tag{3-24}$$

in which

$$\begin{aligned}
d_1 &= -\sum_{i=1}^{m} a_{i1} , \\
d_2 &= -\sum_{i=1}^{m} a_{i2} , \\
&\vdots \\
d_n &= -\sum_{i=1}^{m} a_{in} , \\
w_0 &= \sum_{i=1}^{m} b_i .
\end{aligned}$$

In comparing (3-22) and (3-24) with (3-20) and (3-21), it becomes obvious that (3-22) and (3-24) represent an enlarged linear programming problem consisting of $n + m$ variables which has to be expressed in a cannonical form directly solvable by the simplex method itself. However, the artificial variables must eventually be driven out of the basis if the condition $w = 0$ is to be satisfied. Thus the procedure can be used to test the consistency and redundancy of the constraint equations as well as to find an initial basic feasible solution.

Since the constraint equations often contain slack variables in the admissible variables, we can take advantage of those slack variables with a coefficient of positive one, and we can thus reduce the number of artificial variables needed to form an initial basic feasible solution. However, we must also observe that, for any basic variable x_j, the corresponding d_j must be zero. For example, consider the constraint equations in Example 3–9, which upon the introduction of artificial variables x_5 and x_6 become

$$5x_1 + 10x_2 - x_3 \qquad + x_5 \qquad = 8\,, \tag{a}$$

$$x_1 + \quad x_2 \qquad + x_4 \qquad + x_6 = 1\,. \tag{b}$$

Subtracting these equations from the infeasibility form $x_5 + x_6 = w$, we get

$$-6x_1 - 11x_2 + x_3 - x_4 = \text{Min}\,(w - 9)\,. \tag{c}$$

Since the slack variable x_4 has a coefficient of positive one, it can be used as a basic variable in Eq. (b) instead of the artificial variable x_6. On the other hand, x_3 in Eq. (a) cannot be used as a basic variable in the canonical form because if the sign of the coeffiicient of x_3 were changed from -1 to $+1$, the value of b_i in Eq. (a) would become -8, and an infeasible solution would result. Furthermore, the coefficients d_j of the objective function for w should be determined by subtracting from the infeasibility form only those equations which contain artificial variables in this problem, that is, subtracting only Eq. (a) from $x_5 = w$. Thus the problem can be posed as follows:

$$5x_1 + 10x_2 - x_3 \qquad + x_5 = 8\,, \tag{d}$$

$$x_1 + \quad x_2 \qquad + x_4 \qquad = 1\,, \tag{e}$$

$$-5x_1 - 10x_2 + x_3 \qquad = w - 8\,. \tag{f}$$

In general, if only k $(0 < k < m)$ artificial variables are introduced in the m constraint equations, we have the infeasibility form

$$x_{m+1} + x_{m+2} + \cdots + x_{m+k} = w\,. \tag{3–25}$$

Subtracting each of the k equations containing an artificial variable from (3–25), we obtain only

$$d_1x_1 + d_2x_2 + \cdots + d_{n-k}x_{n-k} = w - w_0\,, \tag{3–26}$$

in which

$$\begin{aligned}
d_1 &= -\sum_{i=1}^{k} a_{i1}\,,\\
d_2 &= -\sum_{i=1}^{k} a_{i2}\,,\\
&\vdots\\
d_{n-k} &= -\sum_{i=1}^{k} a_{i,n-k}\,,\\
w_0 &= \sum_{i=1}^{k} b_i\,.
\end{aligned}$$

Note that the k equations containing artificial variables need not be in consecutive order in the m constraint equations since (3-25) and (3-26) are applicable to any k equations, each of which contains an artificial variable.

Example 3-10. Consider the linear programming problem

$$
\begin{aligned}
5x_1 + 10x_2 &\geq 8\,, \\
2x_1 + 2x_2 &\leq 1\,, \\
x_1 \geq 0\,, \quad & x_2 \geq 0\,, \\
2x_1 + x_2 &= \text{Min}\ z\,.
\end{aligned}
$$

The problem can be expressed in the standard form by the introduction of slack variables x_3 and x_4. Thus

$$
\begin{aligned}
5x_1 + 10x_2 - x_3 &= 8\,, \\
2x_1 + 2x_2 + x_4 &= 1\,, \\
2x_1 + x_2 &= \text{Min}\ z\,.
\end{aligned}
$$

Introduce an artificial variable x_5 and the infeasibility form $x_5 = w$. Then we have from the consideration of feasibility a new problem, namely,

$$
\begin{aligned}
5x_1 + 10x_2 - x_3 + x_5 &= 8\,, &\text{(g)} \\
2x_1 + 2x_2 + x_4 &= 1\,, &\text{(h)} \\
-5x_1 - 10x_2 + x_3 &= w - 8\,. &\text{(i)}
\end{aligned}
$$

We observe that $\min d_j = d_2 = -10$ in Eq. (i); and $a_{12} = 10$ and $a_{22} = 2$ are both positive. By comparing $b_1/a_{12} = \frac{8}{10} = 0.8$ and $b_2/a_{12} = \frac{1}{2} = 0.5$, we select a_{22} as the pivotal element. Hence x_2 is the entering variable, and x_4, which is the basic variable in Eq. (h), is the departing variable. The fact that x_1 is not a basic variable in Eq. (h) is obvious, and it is not necessary to rearrange Eqs. (g), (h), and (i) in exactly the same order of (3-23) and (3-24) if the correct meaning in the latter is observed. The new basis obtained by the simplex method will be as follows:

$$
\begin{aligned}
-5x_1 - x_3 - 5x_4 + x_5 &= 3\,, \\
x_1 + x_2 + 0.5x_4 &= 0.5\,, \\
5x_1 + x_3 + 5x_4 &= w - 3\,.
\end{aligned}
$$

All d_j are now positive, and the solution for w ends after we assign zero value to the nonbasic bariables which appear in the infeasible form, that is, $x_1 = x_3 = x_4 = 0$. Since $\min w = 3$, the original problem is infeasible and there is no feasible solution for this problem.

If the original problem were feasible, it would be necessary to express the original objective function in terms of the nonbasic variables x_1, x_3, and x_4. This could have been done along with the elimination of basic variables in the infeasible form. Thus in this cycle we have

$$x_1 - 0.5x_4 = z - 0.5 \; .$$

The reason why this problem has no feasible solution can easily be detected if the original inequality constraints are represented graphically as in Fig. 3–4. These constraints are shown to be inconsistent; hence no solution is possible in order to satisfy both conditions.

3–9 TABLEAU FOR TWO PHASES OF SIMPLEX METHOD

We may regard the simplex method as a two-phase operation. In phase I, artificial variables and the infeasibility form are introduced for obtaining an initial basic feasible solution. The original objective does not enter into the solution, but may be carried along for the purpose of eliminating nonbasic variables in it at each cycle. *If* min $w \neq 0$, *the problem has no feasible solution, and the procedure ends. If* min $w = 0$, *an initial basic feasible solution is obtained for the next phase.* In Phase II, the same procedure is applied for as many cycles as needed to obtain the optimal solution. The computation can be carried out efficiently in tabulated form by listing the eoefficients separately from the variables. A schematic simplex tableau showing the starting cycles of both Phases I and II is given in Table 3–4. Note that the original objectve function for z as well as the infeasible form for w is included in Phase I, and that the values of z and $w - w_0$ are listed in the table in the same column as b_i. When Phase I reaches its last cycle k, Phase II can start from the same tableau if min $w = 0$ and the infeasibility form for w can thus be eliminated.

It can be seen from the tableau for cycle 0 of Phase I that d_q is assumed to be the most negative of all d_j, and b_p/a_{pq} is the smallest ratio for all positive a_{iq}. Hence x_q is the entering variable, and x_{n+p} is the departing variable, as indicated by the upward and downward arrow heads, respectively, in the last row of the tableau. At the kth cycle of Phase I, when all $\bar{d}_j$ in the infeasibility form become positive, the artificial variables can be dropped after they have departed from the basis. These variables are marked by a check ($\surd$) in their respective columns in the table. Hence $w - \bar{w}_0 = 0$; or min $w = \bar{w}_0$. If $w_0 \neq 0$, the problem terminates at this point; otherwise Phase II can begin. Again, $\bar{c}_s$ is assumed to be the smallest of all $\bar{c}_j$, and $\bar{b}_r/\bar{a}_{rs}$ is the smallest ratio for all positive $\bar{a}_{is}$. Hence x_s is the entering variable, and x_r is the departing variable, as indicated by the entering and departing arrow heads, respectively. Several numerical examples are given to illustrate the use of simplex tableau.

Table 3–4

SIMPLEX TABLEAU FOR START CYCLES

Phase and cycle	Basic variables	Admissible variables	Artificial variables	Values
		$x_1 \quad x_2 \; \cdots \; x_q \; \cdots \; x_n$	$x_{n+1} \quad x_{n+2} \cdots x_{n+p} \cdots x_{n+m}$	
I (0)	x_{n+1}	$a_{11} \quad a_{12} \cdots a_{1q} \cdots x_{1n}$	1	b_1
	x_{n+2}	$a_{21} \quad a_{22} \cdots a_{2q} \cdots x_{2n}$	1	b_2
	$\vdots$	$\vdots$	$\ddots$	$\vdots$
	x_{n+p}	$a_{p1} \quad a_{p2} \cdots \boxed{a_{pq}} \cdots a_{pn}$	$\boxed{1}$	b_p
	$\vdots$	$\vdots$	$\ddots$	$\vdots$
	x_{n+m}	$a_{m1} \quad a_{m2} \cdots a_{mq} \cdots a_{mn}$	1	b_m
	z-form	$c_1 \quad c_2 \cdots c_q \cdots c_n$		z
	w-form	$d_1 \quad d_2 \cdots d_q \cdots d_n$		$w - w_0$
	Change	$\uparrow$ (under x_q)	$\downarrow$ (under x_{n+p})	
		$x_1 \cdots x_r \cdots x_m \cdots x_s \cdots x_n$	$x_{n+1} \quad x_{n+2} \cdots x_{n+p} \cdots x_{n+m}$	
I and II (k)	x_1	$1 \quad \bar{a}_{1s} \quad \bar{a}_{1n}$	$\bar{a}_{1,n+1} \quad \bar{a}_{1,n+2} \quad \bar{a}_{1,n+p} \quad \bar{a}_{1,n+m}$	$\bar{b}_1$
	$\vdots$	$\vdots$	$\vdots$	$\vdots$
	x_r	$\boxed{1} \quad \boxed{\bar{a}_{rs}} \quad \bar{a}_{rn}$	$\bar{a}_{r,n+1} \quad \bar{a}_{r,n+2} \quad \bar{a}_{r,n+p} \quad \bar{a}_{r,n+m}$	b_r
	$\vdots$	$\vdots$	$\vdots$	$\vdots$
	x_m	$1 \quad \bar{a}_{ms} \quad \bar{a}_{mn}$	$\bar{a}_{m,n+1} \quad \bar{a}_{m,n+2} \quad \bar{a}_{m,n+p} \quad \bar{a}_{m,n+m}$	b_m
	z-form	$\bar{c}_s \quad \bar{c}_n$		$z - \bar{z}$
	w-form		$\bar{d}_{n+1} \quad \bar{d}_{n+2} \quad \bar{d}_{n+p} \quad \bar{d}_{n+m}$	$w - w_0$
	Change	$\downarrow$ (under x_r) $\quad \uparrow$ (under x_s)	$\checkmark \quad \checkmark \quad \checkmark \quad \checkmark$	

Table 3–5

SOLUTION FOR EXAMPLE 3–11

Basis	x_1	x_2	x_3	x_4	x_5	Values
x_5	5	$\boxed{10}$	−1		1	8
x_4	1	1		1		1
	2	1				z
	−5	−10	1			$w - 8$
0/I		↑			↓	
x_2	0.5	1	−0.1			0.8
x_4	0.5		0.1	1		0.2
	1.5		0.1			$z - 0.8$
						w
1/(I − II)					✓	

Example 3–11. Repeat the solution of the linear programming problem in Example 3–9 by the use of two phases of the simplex method in the tableau.

The complete solution is given in Table 3–5 which is self-explanatory if the processes of elimination in Example 3–9 are followed. When we get to be familiar with the procedure, we need not identify the basic variables, admissible variables, and artificial variables by their headings, and the cycle number and phase number are given at the lower left corner at the end of each cycle.

Since x_5 is the only artificial variable introduced in Phase I, the infeasibility form is $x_5 = w$. At the end of cycle 2, when all $\bar{d}_j$ become positive, Min $w = 0$, since both x_4 and x_5 are nonbasic variables and can be assigned zero value. Hence x_5 is dropped before entering into Phase II.

All $\bar{c}_j$ in the objective function for z happen to be positive; hence Phase II also terminates at this cycle, and the optimal solution gives $z^* = 0.8$.

Example 3–12. Repeat the solution of the linear programming problem in Example 3–10 using simplex tableau.

The simplex procedure is carried out in Table 3–6, which is self-explanatory.

Example 3–13. Solve the problem in Example 3–2 by the simplex method.

The redundant equation in the example is deliberately included, and the solution of the problem is given in Table 3–7. It is sufficient to point out that as soon as an artificial variable moves out of the basis, it can be dropped immediately, since it will be set equal to zero in the initial basic feasible solution later anyway. Thus in Table 3–7, x_7 is dropped at the end

Table 3-6

SOLUTION FOR EXAMPLE 3-12

Basis	x_4	x_2	x_3	x_4	x_5	Values
x_5	5	10	-1		1	8
x_4	2	$\boxed{2}$		1		1
	2	1				z
	-5	-10	$+1$			$w - 8$
0/I		↑		↓		
x_5	-5		-1	-5	1	3
x_2	1	-1		0.5		0.5
	1			-0.5		$z - 0.5$
	5		1	5		$w - 3$
1/I						

of the cycle 1 and no further computation need be carried in the column for x_7; similarly, x_6 is dropped at the end of cycle 2, which is also the end of Phase I. However, x_6 does not drop out of the basis at the end of Phase I; instead, it takes on a value of $x_8 = 0$, which in essence also eliminates itself from further consideration. Hence in cycle 3 we have two instead of three constraint equations, the redundant equation being eliminated at the end of Phase I by virtue of $x_8 = 0$.

It may be of interest to note that the x_4 enters the basis on account of $\bar{c}_4$ being the smallest at the end of cycle 2 but departs at the end of cycle 3. Cycle 3 contributes nothing to the optimal solution of the problem, and could have been eliminated if x_3 had been chosen instead of x_4 as the entering variable at the end of cycle 2. This illustrates that while the rules of simplex method for improving an initial basis will eventually lead to an optimal solution if such a solution exists, the procedure does not necessarily give the quickest way to arrive at the solution.

Example 3-14. Solve the problem in Example 3-1 by the simplex method.

The solution of this problem is given in Table 3-8 which is self-explanatory.

Example 3-15. Solve the following linear programming problem by the simplex method:

$$
\begin{aligned}
5x_2 + 4x_3 + 4x_4 &= 5\,, \\
6x_1 + 4x_2 + 3x_3 + 2x_4 &= 3\,, \\
4x_1 + x_2 + 3x_3 + 4x_4 &= 2\,, \\
8x_1 + 10x_2 + 12x_3 + 11x_4 &= \text{Min } z\,.
\end{aligned}
$$

Table 3–7

SOLUTION FOR EXAMPLE 3–13

Basis	x_1	x_2	x_3	x_4	x_5	x_6	x_7	x_8	Values
x_6	1	1	1	1	1	1			1
x_7	$\boxed{0.4}$	0.2	0.5	0.8	0.7		1		0.3
x_8	0.6	0.8	0.5	0.2	0.3			1	0.7
	3	2	1	1.5	2.5				z
	−2	−2	−2	−2	−2				$w - 2$
0/I	↑						↓		
x_6		$\boxed{\frac{1}{2}}$	$-\frac{1}{4}$	−1	$-\frac{3}{4}$	1			$\frac{1}{4}$
x_1	1	$\frac{1}{2}$	$\frac{5}{4}$	2	$\frac{7}{4}$				$\frac{3}{4}$
x_8		$\frac{1}{2}$	$-\frac{1}{4}$	−1	$-\frac{3}{4}$			1	$\frac{1}{4}$
		$\frac{1}{2}$	$-\frac{11}{4}$	$-\frac{9}{2}$	$-\frac{11}{4}$				$z - \frac{9}{4}$
		−1	$\frac{1}{2}$	2	$\frac{3}{2}$				$w - \frac{1}{2}$
1/I		↑				↓	✓		
x_2		1	$-\frac{1}{2}$	−2	$-\frac{3}{2}$				$\frac{1}{2}$
x_1	1		$\frac{3}{2}$	$\boxed{3}$	$\frac{5}{2}$				$\frac{1}{2}$
x_8								1	0
			$-\frac{5}{2}$	$-\frac{7}{2}$	−2				$z - \frac{5}{2}$
									w
2/(I − II)	↓			↑		✓	✓		
x_2	$\frac{2}{3}$	1	$\frac{1}{2}$		$\frac{1}{6}$				$\frac{5}{6}$
x_4	$\frac{1}{3}$		$\boxed{\frac{1}{2}}$	1	$\frac{5}{6}$				$\frac{1}{6}$
	$\frac{7}{6}$		$-\frac{3}{4}$		$\frac{11}{12}$				$z - \frac{23}{12}$
3/II			↑	↓		✓	✓	✓	
x_2	$\frac{1}{3}$	1		−1	$-\frac{2}{3}$				$\frac{2}{3}$
x_3	$\frac{2}{3}$		1	2	$\frac{5}{3}$				$\frac{1}{3}$
	$\frac{5}{3}$			$\frac{3}{2}$	$\frac{13}{6}$				$z - \frac{5}{3}$
4/II									

By introducing artificial variables x_5, x_6, and x_7, we have the infeasibility form

$$x_5 + x_6 + x_7 = w\,.$$

The solution can be carried out as far as shown in Table 3–9. Obviously, the solution is infeasible and ends at Phase I when all cost coefficients beome nonnegative. The table is self-explanatory except that z and w are transposed to the left-hand side of their respective equations, and $-z$

Table 3-8

SOLUTION FOR EXAMPLE 3-14

Basis	x_1	x_2	x_3	x_4	x_5	x_6	x_7	Values
x_6	1	1	-1			1		1
x_7	2	$\boxed{4}$		-1			1	3
x_5	3	7			1			6
	4	5						z
	-3	-5	1	1				$w - 4$
0/I		↑					↓	
x_6	$\boxed{\frac{1}{2}}$		-1	$\frac{1}{4}$		1		$\frac{1}{4}$
x_2	$\frac{1}{2}$	1		$-\frac{1}{4}$				$\frac{3}{4}$
x_5	$-\frac{1}{2}$			$\frac{7}{4}$	1			$\frac{3}{4}$
	$\frac{3}{2}$			$\frac{5}{4}$				$z - \frac{15}{4}$
	$-\frac{1}{2}$		1	$-\frac{1}{4}$				$w - \frac{1}{4}$
1/I	↑					↓	✓	
x_1	1		-2	$\frac{1}{2}$				$\frac{1}{2}$
x_2		1	1	$-\frac{1}{2}$				$\frac{1}{2}$
x_5			-1	2	1			1
			3	$\frac{1}{2}$				$z - \frac{9}{2}$
								w
2/(I − II)						✓	✓	

and $-w$ are treated as basic variables in computational manipulation. By comparing this problem with the first part of Example 3-3, it confirms that the latter example has no feasible solution if we do not allow the excess material of any grade to be removed from the job site.

3-10 DEGENERACY

If some basic variables in a basic feasible solution of a linear programming problem turn out to be zero, the basis is said to be *degenerate* and such variables are called *degenerate variables*. This situation occurs quite frequently at some stage of many problems, since it is caused by a tie in the minimum of the ratios $\bar{b}_i/\bar{a}_{is}$ (for $\bar{a}_{is} > 0$). As shown in the solution of Example 3-15 (Table 3-9), the ratios $\frac{3}{6}$ and $\frac{2}{4}$ are tied for x_6 and x_7 as x_1 enters at the end of cycle 0. When x_7 is chosen as the departing variable, x_6 remains in the basis for the next cycle; however, the value of x_6 in cycle 1 is seen to be zero as must be the value of all basic variables which have the same minimum $\bar{b}_i/\bar{a}_{is}$ ratio as the departing variable at the end of the previous cycle. It is apparent that because $x_6 = 0$ in cycle 1, the

Table 3–9

SOLUTION FOR EXAMPLE 3–15

Basis	x_1	x_2	x_3	x_4	x_5	x_6	x_7	$-z$	$-w$	Value
x_5	0	5	4	4	1					5
x_6	6	4	3	2		1				3
x_7	$\boxed{4}$	1	3	4			1			2
$-z$	8	10	12	11				1		0
$-w$	-10	-10	-10	-10					1	-10
0/I	↑						↓			
x_5		5	4	4	1					5
x_6		$\boxed{\frac{5}{2}}$	$-\frac{3}{2}$	-4		1				0
x_1	1	$\frac{1}{4}$	$\frac{3}{4}$	1						$\frac{1}{2}$
$-z$		8	6	3				1		-4
$-w$		$-\frac{15}{2}$	$-\frac{5}{2}$	0					1	-5
1/I		↑				↓	✓			
x_5			7	12	1					5
x_2		1	$-\frac{3}{5}$	$-\frac{8}{5}$						0
x_1	1		$\frac{9}{10}$	$\boxed{\frac{7}{5}}$						$\frac{1}{2}$
$-z$			$\frac{54}{5}$	$\frac{79}{5}$				1		-4
$-w$			-7	-12					1	-5
2/I	↓			↑		✓	✓			
x_5	$-\frac{60}{7}$		$-\frac{5}{7}$		1					$\frac{5}{7}$
x_2	$\frac{8}{7}$	1	$\frac{3}{7}$							$\frac{4}{7}$
x_4	$\frac{5}{7}$		$\frac{9}{14}$	1						$\frac{5}{14}$
$-z$	$-\frac{79}{7}$		$\frac{9}{14}$					1		$-\frac{135}{14}$
$-w$	$\frac{60}{7}$		$\frac{5}{7}$						1	$-\frac{5}{7}$
3/I						✓	✓			

value of w cannot be improved in the next cycle. Hence it is pertinent to ask what effect, if any, is the degeneracy on the iterative procedure.

Our major concern in degeneracy is the possibility of the circling of entering and departing variables without their improving the solution. If circling should occur periodically in a solution, the iterative procedure will break down unless some devices are introduced to avoid degeneracy, which is the cause of circling. It has been found that circling is a very rare phenomenon in practice. Hence it is usually not necessary to provide devices to avoid degeneracy even though such devices have been developed and are available.

Degeneracy may also occur initially in a given problem, as indicated in the example that follows.

Table 3–10

SOLUTION FOR EXAMPLE 1–16

Basis	x_1	x_2	x_3	x_4	x_5	Values
x_5	1	$\boxed{1}$	-1		1	20
x_4	$\frac{1}{2}$	-1		1		0
	$\frac{1}{2}$	1				z
	-1	-1	1			$w - 20$
0/I		↑			↓	
x_2	1	1	-1			20
x_4	$\boxed{\frac{3}{2}}$		-1	1		20
	$-\frac{1}{2}$		1			$z - 20$
						w
1/(I − II)	↑			↓	✓	
x_2		1	$-\frac{1}{3}$			$\frac{20}{3}$
x_1	1		$-\frac{2}{3}$			$\frac{40}{3}$
			$\frac{1}{3}$			$z - \frac{40}{3}$
2/II						

Example 3–16. Solve the linear programming problem stated in Problem P1–4 by the simplex method.

The solution of this problem is given in Table 3–10, which is self-explanatory.

3–11 LOWER AND UPPER BOUNDS

In the formulation of linear programming problems, we may encounter cases in which the physical requirements impose positive lower and/or upper bounds on the variables. Let $L_j \geq 0$ and $U_j \geq 0$ be the lower bounds and upper bounds, respectively, for x_j ($j = 1, 2, \ldots, n$). Then the permissible range of x_j may be governed by one of the following cases:

$$\text{(a)} \quad x_j \geq L_j\,, \tag{3–27}$$

$$\text{(b)} \quad x_j \leq U_j\,, \tag{3–28}$$

$$\text{(c)} \quad L_j \leq x_j \leq U_j\,. \tag{3–29}$$

A linear programming problem with positive lower bounds in (3–27) in place of nonnegative constraints in (3–2) can be reduced to the standard form in a set of new variables:

$$y_j = x_j - L_j \geq 0\,. \tag{3–30}$$

Since the conditions $y_j \geq 0$ automatically satisfy $x_j \geq 0$, the substitution of

$$x_j = L_j + y_j \qquad \text{for} \qquad y_j \geq 0\,, \tag{3-31}$$

into the original problem leads to a new problem in y_j, which has nonnegative constraints instead of positive lower bounds. After a set of y_j leading to an optimal solution is obtained, we can determine the corresponding set of x_j from (3-31).

A problem with positive upper bounds in (3-28), in addition to nonnegative constraints, is not as simple. If we let

$$y_j = U_j - x_j \geq 0\,, \tag{3-32}$$

we must still maintain $x_j \geq 0$, or

$$U_j - y_j \geq 0\,. \tag{3-33}$$

In other words, we have added constraints (the number corresponding to the number of upper bounds) to the original problem, whether or not we change the variables from x_j to y_j. Hence the solution of the problem is inherently more complex, since up to n more constraint conditions may be involved. A short-cut procedure for treating upper bounds has been developed, but it will not be treated in this elementary text.

When a problem has both positive lower and upper bounds, we can introduce

$$y_j = \frac{U_j - x_j}{U_j - L_j} \tag{3-34}$$

and

$$1 \geq y_j \geq 0\,. \tag{3-35}$$

This substitution will reduce the lower bounds to nonnegative constraints, but the upper bounds will remain as additional constraint conditions to the problem.

Example 3-17. Given that the nonnegative constraints in Example 3-1 are replaced by $x_1 \geq 0.2$ and $x_2 \geq 0.3$, restate the problem using nonnegative constraints instead of positive lower bounds.

Let $x_1 = 0.2 + y_1$ and $x_2 = 0.3 + y_2$. Then upon substitution into the original inequality constraints, we have

$$\begin{aligned} y_1 + y_2 &\geq 0.5\,, \\ 2y_1 + 4y_2 &\geq 1.4\,, \\ 3y_1 + 7y_2 &\leq 3.3\,, \end{aligned}$$

$$y_1 \geq 0\,, \qquad y_2 \geq 0\,,$$

$$\text{Min } z = 4y_1 + 5y_2 + 2.3\,.$$

Example 3-18. Given that an upper bound $x_2 \leq 0.6$ and the lower bounds $x_1 \geq 0.2$ and $x_2 \geq 0.3$ are introduced for Example 3-1, restate the problem using nonnegative constraints instead of positive lower bounds.

Let $y_1 = x_1 - 0.2$ as in the previous example, and for $1 \geq y_2 \geq 0$ let

$$y_2 = \frac{0.6 - x_2}{0.6 - 0.3} \qquad \text{or} \qquad x_2 = 0.3(2 - y_2)\,.$$

Upon substitution of x_1 and x_2 into the original problem, we have

$$\begin{aligned} y_1 - 0.3y_2 &\geq 0.2\,, \\ 2y_1 - 1.2y_2 &\geq 0.2\,, \\ 3y_1 - 2.1y_2 &\leq 1.2\,, \\ y_2 &\leq 1\,, \\ y_1 \geq 0\,, \qquad y_2 &\geq 0\,, \\ \text{Min } z = 4y_1 - 1.5y_2 &+ 3.8\,. \end{aligned}$$

3-12 ADDITIONAL EXAMPLES OF APPLICATION

In this chapter, the basic concepts of linear programming have been introduced. In general, a linear programming problem can be expressed in the standard form, which is condensed in the summation notation as follows:

$$\sum_{j=1}^{n} a_{ij}x_j = b_i\,, \qquad i = 1, 2, \ldots, m\,; \tag{3-36}$$

$$x_j \geq 0\,, \qquad j = 1, 2, \ldots, n\,; \tag{3-37}$$

$$\sum_{j=1}^{n} c_j x_j = \text{Min } z\,. \tag{3-38}$$

The optimal solution is given by the set

$$z^*, x_1^*, x_2^*, \ldots, x_m^*\,,$$
$$x_{m+1}^* = x_{m+2}^* = \cdots = x_n^* = 0\,,$$

in which z^* is the minimum value of z^*; $x_1^*, x_2^*, \ldots, x_m^*$ are the values of m basic variables in the optimal basis; and $x_{m+1}^*, x_{m+2}^*, \ldots, x_n^*$ are the nonbasic variables in the optimal basis. So far, very simple examples have been used for illustration. The broad range of applicability of linear programming can be seen from additional examples of practical interset. However, even these examples are necessarily simplified and adapted from the real situations in order to describe the problems concisely, and a discussion on the solutions for such problems is beyond the scope of this introductory text. Detailed descriptions of similar problems may be found in the references cited at the end of this chapter.†

Example 3–19. A contractor has entered into an agreement with the state highway department on the grading operation for two miles of interstate highway. The contract involves (1) clearing and grubbing, and (2) excavation and fill. The right-of-way must be cleared and grubbed in preparation for excavation and fill. A total of 16 acres must be cleared and grubbed, and 28,250 yd³ of earth must be excavated and filled. There will be no waste or borrow on this job, i.e., the volume of cuts equals the volume of fills; and the distances for earthmoving between cuts and fills are within the limits such that excavation, transportation, and compaction from cuts to fills will be regarded as one operation. This job will be done in three phases. Phase I consists of clearing and grubbing stations 0 through 50; phase 2 consists of clearing and grubbing the remaining stations, and excavation and fill from stations 0 through 50; and phase 3 consists of excavation and fill for the remainder of the job.

Table 3–11

QUANTITIES OF WORK

Phase	Clearing and grubbing		Excavation and fill	
i	Location (station nos.)	Quantity b_i (acres)	Location (station nos.)	Quantity e_i (yd³)
1	0 – 50.0	7	—	—
2	50.0–105.6	9	0 – 50.0	8,000
3	—	—	50.0–105.6	20,250
Total		16		28,250

Let i denote the ith phase of operation for a total of q phases ($q = 3$ in this case). Thus b_i is the number of acres for clearing and grubbing in phase i, and e_i is the number of cubic yards for excavation and fill in phase i. The quantities of work for both b_i and e_i ($i = 1, 2, 3$) are given in Table 3–11. Note that clearing and grubbing must precede excavation and fill by at least one phase.

The types of equipment available for the project are shown in Table 3–12. They may be used either singly or jointly as production units, with a total of 23 possible combinations. For example, clearing and grubbing can be done with a unit of A or B or a combination of both. Excavation,

† See References 3–4 and 3–5 in connection with Examples 3–19 and 3–20, respectively.

Table 3–12

TYPES OF AVAILABLE EQUIPMENT

Classification of equipment	Type and description	Number available
Tractors	A (with blade)	1
	B (with blade)	1
	C (without blade)	4
	D (without blade)	1
	E (without blade)	2
Scrapers	F (pulled by C)	3
	G (pulled by D)	1
Tractor-scrapers (fixed combination)	H (tractor)	5
	I (scraper)	5
Tandem rollers	J	4

hauling, dumping, and spreading can be done with equipment sets $C + F$, $D + G$, $H + I + C$, etc. These combinations are well known to the contractor although they are not listed here. Let k denote the kth equipment type among the s types available ($s = 10$ for types A to J inclusive in this case). Then N_k is the number of units of the kth type equipment, and P_k is the expected available time for each unit of the kth type. Furthermore, let j denote the jth combination for a total of n possible combinations of equipment units ($n = 23$ in this case). Hence a_j is the expected rate of production for the jth combination of equipment units, and m_{kj} is the number of equipment units of the kth type in the jth combination.

The amount of capital layout required for the jth equipment combination per unit time is t_j, and the direct cost of using the jth equipment combination for unit time is c_j. The maximum amount of capital assigned to the project is r.

Assuming that all the above quantities are known, let x_j be the length of time that the jth equipment combination is used, and z be the total direct cost for the project. Subject to the given constraints, what combinations of equipment should be used and for what lengths of time and in what sequence and location should they be used such that the total direct cast may be minimized?

The job requirement for clearing and grubbing for the first $(q - 1)$ phases can be stated by the equations:

$$\sum_{j=1}^{n} a_j x_j = b_i \,, \qquad i = 1, 2, \ldots, (q - 1) \,. \tag{3–39}$$

The job requirement for excavation, hauling, fill, and compaction in the subsequent phases can be given by the equations:

$$\sum_{j=1}^{n} a_j x_j = e_i\,, \qquad i = 2, 3, \ldots, q\,. \tag{3–40}$$

The equipment costraints can be expressed as

$$\sum_{j=1}^{n} m_{kj} x_j \leq N_k P_k\,, \qquad k = 1, 2, \ldots, s\,. \tag{3–41}$$

The capital constraint is given by

$$\sum_{j=1}^{n} t_j x_j \leq r\,. \tag{3–42}$$

In addition, we have the nonnegative constraints

$$x_j \geq 0\,. \tag{3–43}$$

The objective function is given in the form

$$\text{Min } z = \sum_{j=1}^{n} c_j x_j\,. \tag{3–44}$$

Example 3–20. In the rigid frame shown in Fig. 3–8(a), the ultimate moment capacity of each member is governed by the plastic moment M_0. The three possible collapse mechanisms of the frame are represented by (b), (c) and (d) of Fig. 3–8, in which plastic hinges are indicated at locations of maximum positive and/or negative moments. Determine the value of load F which will cause the plastic collapse of the rigid frame.

In the plastic analysis of rigid frames, the relationships between external forces and internal moments can be obtained from the consideration of the balance of external work and internal work in each of the deformed shapes, as indicated by the collapse mechanisms. Thus from (b), (c), and (d) of Fig. 3–8 we have, respectively,

$$\begin{aligned} -M_2\theta + 2M_3\theta - M_4\theta &= 2FL\theta\,, \\ -M_1\theta + M_2\theta \qquad\qquad - M_4\theta &= 6FL\theta\,, \\ -M_1\theta \qquad\qquad + 2M_3\theta - 2M_4\theta &= 8FL\theta\,, \end{aligned}$$

in which M_1, M_2, M_3, and M_4 are moments at joints 1, 2, 3, and 4, respectively, and θ is the angle of deformation which can be eliminated from this set of equations. Each of these moments M_i ($i = 1, 2, 3, 4$) can take only a value between $-M_0$ and M_0. Hence the constraints for moments M_i may be written as

$$-M_0 \leq M_i \leq M_0\,, \qquad i = 1, 2, 3, 4\,.$$

The problem is to maximize load F such that the loads causing the plastic collapse of the rigid frame may be determind.

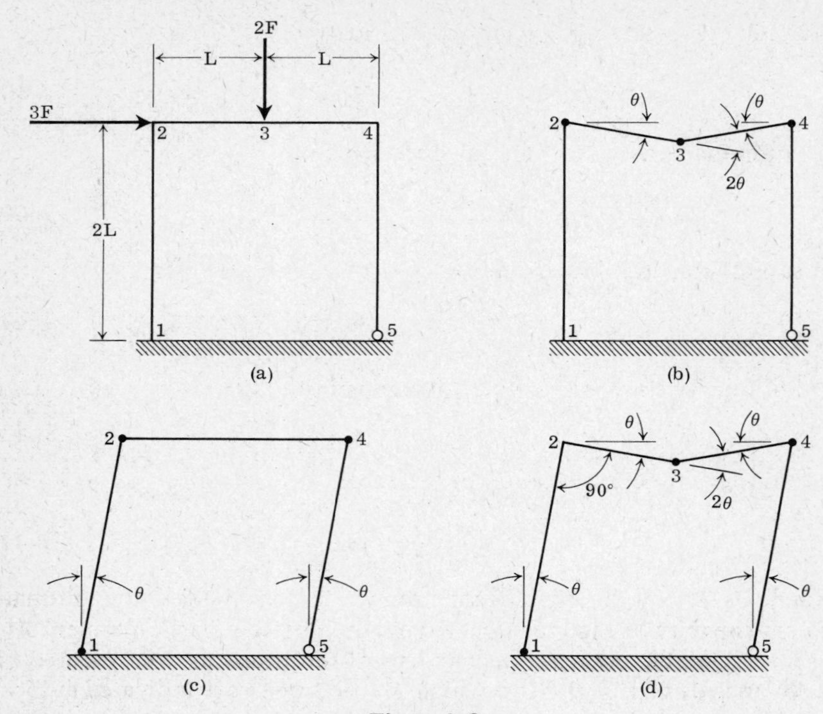

Figure 3-8

We can reduce the equations to dimensionless form by introducing the variables

$$M_i' = M_i/M_0 \qquad \text{and} \qquad f = FL/M_0 \,.$$

Hence the set of equations representing the collapse mechanisms becomes

$$\begin{aligned} 2f \qquad\quad + M_2' - 2M_3' + \ M_4' &= 0 \,, \\ 6f + M_1' - M_2' \qquad\qquad + \ M_4' &= 0 \,, \\ 8f + M_1' \qquad\quad - 2M_3' + 2M_4' &= 0 \,. \end{aligned}$$

The moment constraints are $-1 \leq M_i' \leq 1$, $i = 1, 2, 3, 4$. The objective function becomes

$$\text{Max } A = f \,,$$

where f is related to M_i by the set of equations representing the collapse mechanisms of the rigid frame.

In order to reduce the problem to the form having nonnegative constraints, we change the variables by introducing

$$x_i = M_i' + 1 \,, \qquad i = 1, 2, 3, 4 \,.$$

Then the set of equations is reduced to

$$\begin{aligned} 2f \qquad + x_2 - 2x_3 + \ \ x_4 &= 0\,, \\ 6f + x_1 - x_2 \qquad\quad + \ \ x_4 &= 1\,, \\ 8f + x_1 \qquad\ - 2x_3 + 2x_4 &= 1\,, \end{aligned}$$

and

$$0 \leq x_i \leq 2\,, \qquad i = 1, 2, 3, 4\,.$$

Therefore

$$\text{Max}\, A = f\,.$$

Note that the variables in this problem have upper bounds as well as nonnegative constraints, and that f can also be treated as a nonnegative variable, since load F is not allowed to reverse its direction on the rigid frame.

REFERENCES

3–1. DANTZIG, G. B., *Linear Programming and Extensions*, Princeton University Press, Princeton, N. J., 1963.

3–2. CHARNES, A., and W. W. COOPER, *Management Models and Industrial Applications of Linear Programming*, Vols. I and II, Wiley, New York, 1961.

3–3. GARVIN, W. W., *Introduction to Linear Programming*, McGraw-Hill, New York, 1960.

3–4. SHAFFER, L. R., "Analytical Methods in Transportation: Planning a Grading Operation for Least Total Cost," *Journal of Engineering Mechanics Division, ASCE*, **89,** No. EM6 (1963), 47–66.

3–5. HODGE, PHILIP G., JR., *Plastic Analysis of Structures*, McGraw-Hill, New York, 1959, 42–46.

3–6. RUBINSTEIN, M. F., and J. KARAGOZIAN, "Building Design Using Linear Programming," *Journal of Structural Division, ASCE*, **92,** No. ST–6 (1966), 223–245.

PROBLEMS

P3–1. Solve the following linear programming problem by the graphical method:

$$y_1 \geq 0\,, \qquad y_2 \geq 0\,,$$

$$\begin{aligned} -4y_1 + 2y_2 &\leq 1\,, \\ 5y_1 - 4y_2 &\leq 3\,, \end{aligned}$$

$$8y_1 - 2y_2 = \text{Min}\, z\,.$$

P3–2. Solve the following linear programming problem by the graphical method:

$$x_1 \geq 0\,, \qquad x_2 \geq 0\,,$$

$$4x_1 - 5x_2 \leq 8\,,$$
$$2x_1 - 4x_2 \geq 2\,,$$

$$x_1 + 3x_2 = \text{Min } z\,.$$

P3–3. Solve the following linear programming problem by the graphical method:

$$x_1 \geq 0\,, \qquad x_2 \geq 0\,,$$

$$x_1 - 2x_2 \leq 2\,,$$
$$x_1 \qquad \leq 9\,,$$
$$-x_1 + x_2 \leq 5\,,$$

$$x_1 + 2x_2 = \text{Max } A\,.$$

P3–4. Solve the following linear programming problem by the graphical method:

$$x_1 \geq 0\,, \qquad x_2 \geq 0\,,$$

$$8x_1 + x_2 \geq 8\,,$$
$$2x_1 + x_2 \geq 6\,,$$
$$x_1 + 2x_2 \geq 6\,,$$
$$x_1 + 8x_2 \geq 8\,,$$

$$\text{Min } z = 3x_1 + 3x_2\,,$$

P3–5. Solve the following linear programming problem by the graphical method:

$$x_1 \geq 0\,, \qquad x_2 \geq 0\,,$$

$$3x_1 + 2x_2 \geq 3\,,$$
$$x_1 + 4x_2 \geq 4\,,$$
$$x_1 + x_2 \leq 5\,,$$

$$5x_1 + 8x_2 = \text{Min } z\,.$$

P3–6. Solve Problem P1–5 in Chapter 1 by the graphical method.

P3–7 through P3–12. Solve each of the linear programming problems corresponding to Problems P3–1 through P3–6, respectively, by the simplex method.

P3–13. Solve the following linear programming problem:

$$y_1 \geq 0\,, \qquad y_2 \geq 0\,, \qquad y_3 \geq 0\,,$$

$$3y_1 + y_2 - y_3 \leq 5\,,$$
$$2y_1 + 4y_2 - y_3 \leq 8\,,$$

$$-3y_1 - 4y_2 + 5y_3 = \text{Min } z\,.$$

P3–14. Solve the following linear programming problem:

$$y_1 \geq 0\,, \quad y_2 \geq 0\,, \quad y_3 \geq 0\,,$$

$$\begin{aligned} y_1 + y_2 - y_3 &\geq 1\,, \\ -2y_1 \qquad + y_3 &\geq 2\,, \end{aligned}$$

$$2y_1 + 9y_2 + 5y_3 = \text{Min } z\,.$$

P3–15. Solve the following problem if a feasible solution exists.

$$x_1 \geq 0\,, \quad x_2 \geq 0\,,$$

$$\begin{aligned} x_1^3 x_2^2 &\geq e^3\,, \\ x_1 x_2^4 &\geq e^4\,, \\ x_1^2 x_2^3 &\leq e\,, \end{aligned}$$

$$x_1 x_2 = \text{Min } z\,,$$

where e is the base of natural logarithms.

P3–16. The planting area in a public park is to be seeded with three different species (A, B, and C) of summer plants blended together randomly. Two types of prepackaged mixed seeds are commercially available. The type I package contains 3 lb of seed A, 2 lb of seed B and 1 lb of seed C; the type II package contains 1 lb of seed A and 3 lb each of seed B and seed C. For each acre of land, it is desirable to have at least 4 lb of seed A, 9 lb of seed B, and 6 lb of seed C. The cost of the type I package is \$ 20 each, and that of the type II package is \$ 10 each. What is the combination of types I and II packages which will produce the minimum cost? Given that the total planting area in the park is 7 acres, determine the minimum cost. Solve this problem by the simplex method.

P3–17. A manufacturing company requires the daily consumption of 4 million gallons of water of a quality such that the concentration of a certain mineral pollutant must be kept below 100 mg/l. The water can be supplied from two sources: (1) purchase from a local company at the cost of \$ 100 per million gallons, and (2) pumping from a nearby stream at the cost of \$ 50 per million gallons. The concentration of the pollutant from the first source is 50 mg/l, and that from the second source is 200 mg/l. The water from the two sources is completely mixed before it is used. Now the company plans a new expansion which will at least double the daily consumption of water of the same quality. The local water company can further supply up to 10 million gallons daily, but the additional quantity to be pumped from the nearby stream cannot exceed 2 million gallons daily. Determine the additional quantities of water to be obtained from these two sources to meet the need of the new expansion such that the total cost will be minimum. What is the daily cost? Solve this problem by the graphical method.

P3–18. Construct a computer flowchart for Phase I of the simplex method.

P3–19. Construct a computer flowchart for both Phases I and II of the simplex method.

CHAPTER 4

SIMPLEX MULTIPLIERS AND DUALITY

4-1 ELEMENTARY EXAMPLES

In the development of an algorithm for the solution of the linear programming problems in Chapter 3, we deliberately avoided the discussion of some subtle implications which might have confused the basic procedure. Having mastered the use of the simplex method, we are now in a position to examine certain aspects of computation which can be simplified or clarified. In particular, we shall introduce the *concept of simplex multipliers* in the solution of linear programming problems. We shall also examine the *concept of duality* and its application to the solution of linear programming problems. If a solution is to be useful for a realistic problem, we must also know the sensitivity of the solution to the variation of various coefficients in the problem, which often are not rigidly specified in the real world. A few elementary examples will illustrate the type of problem to be treated in this chapter.

Example 4-1. A small steel fabricator supplying two types of steel plate products (I and II) to a contractor uses two machines (A and B) in the fabricating process. On the average, it takes 1 hr of operation on machine A and 1 hr on machine B in fabricating one ton of type I product. On the other hand, it requires 4 hr of operation on machine A and 2 hr on machine B for one ton of type II product. Machine A is run no more than 12 hr/day and machine B, no more than 8 hr/day. Given that the fabricator makes a profit of \$1/ton of type I product and \$3/ton of type II product, and that he can sell all he produces, determine the quantities of products in types I and II that he should produce in order to maximize the profit per day.

Let y_1 and y_2 be the quantities of products in tons to be fabricated in types I and II, respectively, and let A be the total profit in dollars per day. Then

$$\text{Max } A = y_1 + 3y_2 \,,$$

subject to the constraints imposed by machines A and B:

$$y_1 + 4y_2 \leq 12\,,$$
$$y_1 + 2y_2 \leq 8\,,$$
$$y_1 \geq 0\,, \qquad y_2 \geq 0\,.$$

Example 4–2. If a contractor offers to rent the machines from the steel fabricator in the previous example, what rental rate (in dollars per hour) should the contractor pay for renting each machine in order to minimize his total cost? It is understood, of course, that the fabricator will not agree on anything leading to less than his current rate of profit (in dollars per ton for each type of product), and that the machines will be made available for the same number of hours per day as before.

Let x_1 and x_2 be the rental rates (in dollars per hour) for renting machines A and B, respectively, and let z be the total cost in dollars per day. Then

$$\text{Min } z = 12x_1 + 8x_2\,,$$

subject to the conditions that the fabricator will agree to rent only if the current profit rate for each product is matched or improved, i.e.,

$$x_1 + x_2 \geq 1\,,$$
$$4x_1 + 2x_2 \geq 3\,,$$
$$x_1 \geq 0\,, \qquad x_2 \geq 0\,.$$

Examples 4–1 and 4–2 are examples of what would be defined later as *primal and dual problems*, respectively. The coefficients of one problem are the transpose of those of the other and can be tabulated as in Table 4–1. The graphical solutions of these problems are shown in Figs. 4–1 and 4–2, respectively. It can be seen that $A^* = z^* = \$10$ is the optimal solution for both the primal and dual problems.

Table 4–1

COEFFICIENTS FOR EXAMPLES 4–1 AND 4–2

a_{ij} (hr/ton)		x_j (\$/hr)		b_i (\$/ton)
		x_1	x_2	
y_i (tons)	y_1	1	1	1
	y_2	4	2	3
c_j	(hr)	12	8	z or A (\$)

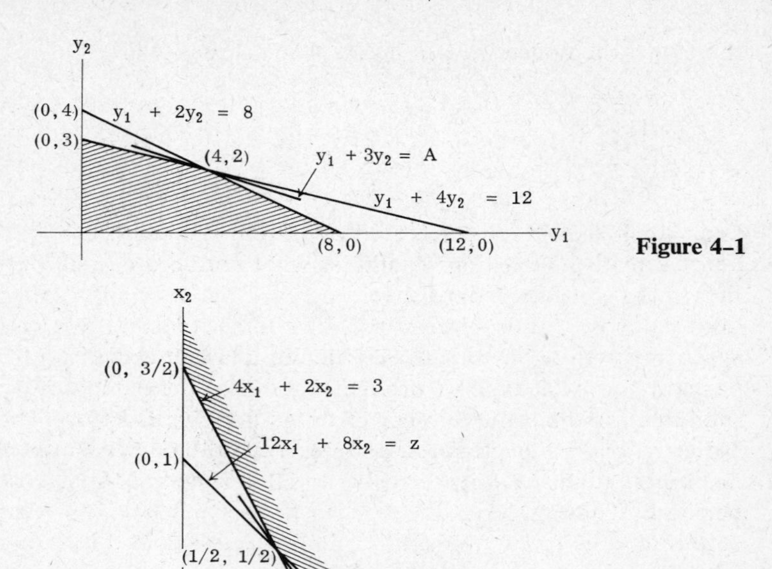

Figure 4-1

Figure 4-2

Example 4-3. Consider the solution of the following linear programming problem:

$$
\begin{aligned}
-x_1 - x_2 + x_3 \qquad\qquad &= -1\,, \\
-2x_1 - 4x_2 \qquad + x_4 \qquad &= -3\,, \\
3x_1 + 7x_2 \qquad\qquad + x_5 &= 6\,, \\
4x_1 + 5x_2 = z \quad (\text{Min}), & \\
x_j \geq 0\,, \qquad \text{for all } j\,. &
\end{aligned}
$$

In this problem, the admissible variables x_3 and x_4 cannot be used as a part of an initial basis for the simplex procedure because the stipulations in the first two constraint equations are negative. Although the problem can still be solved by using the simplex method if we first multiply these two constraint equations by -1 and then introduce artificial variables x_6 and x_7, one for each of these equations, such an approach is indirect and lengthy. We may note that while some of the stipulations b_i are negative, all the cost coefficients c_j are positive. Thus a procedure may be developed by reversing the roles of stipulations and cost coefficients in the solution.

Example 4-4. Suppose that in Example 3-1 of Chapter 3 the requirement per ton of raw material is changed to no less than 1 lb of manganese, nor less than 2.5 lb of silicon, nor more than 5 lb of carbon, while the ore supplies remain unchanged. How would the optimal solution be affected by this change?

Using the same notation in Example 3-1, the product-mix problem becomes:

$$x_1 \geq 0\,, \qquad x_2 \geq 0\,,$$
$$x_1 + \ \ x_2 \geq 1\,,$$
$$2x_1 + 4x_2 \geq 2.5\,,$$
$$3x_1 + 7x_2 \leq 5\,,$$
$$4x_1 + 5x_2 = \text{Min } z\,.$$

Note that the only change in this example, in comparison with Example 3-1, is made in the stipulations of the inequality constraints. Of course, this problem can be solved by the simplex method with the optimal solution $x_1^* = \frac{3}{4}$, $x_2^* = \frac{1}{4}$, and $z^* = 4\frac{1}{4}$ as opposed to the optimal solution $x_1^* = 0.5$, $x_2^* = 0.5$, and $z^* = 4.5$ in Example 3-1. However, if a number of slightly different mixes can produce the same desired properties, it will be unwieldy to determine which one requires the lowest cost by repeating the solution for every slight change in the mix. It is interesting to note that, although the stipulations b_i are changed with the change of the mix, the cost coefficients c_j remain unchanged. Hence the duality relationship also appears to be useful in the study of the sensitivity of the solution of a linear programming problem.

4-2 MATRIX REPRESENTATION OF LINEAR EQUATIONS

The systems of linear equations encountered in linear programming problems can be expressed conveniently in matrix form. Although matrix notation is not used in this text, it is desirable to introduce the elementary definitions of matrix algebra in order to provide better understanding of the basic concept in the *revised simplex method*. Since only a bare minimum of matrix operations are summarized here for this special purpose, the student is referred to textbooks on matrix algebra for further discussion or review.

A set of numbers or elements arranged in a rectangular array of m rows and n columns in the form

$$\mathbf{A} = \begin{bmatrix} a_{11} & a_{12} & \cdots & a_{1n} \\ a_{21} & a_{22} & \cdots & a_{2n} \\ \vdots & & & \\ a_{m1} & a_{m2} & \cdots & a_{mn} \end{bmatrix}$$

is called an $m \times n$ *matrix*. The numbers m and n are called the dimensions of the matrix. It is usually convenient to denote the matrix $\mathbf{A}$ in the abbreviated form.

$$\mathbf{A} = [a_{ij}], \qquad i = 1, 2, \ldots, m, \quad j = 1, 2, \ldots, n,$$

in which a_{ij} denotes the element in the ith row and the jth column of $\mathbf{A}$. Matrices composed of a single row are called *row vectors*, and matrices composed of a single column are called *column vectors*. A row vector of $1 \times n$ is denoted by $[a_1\, a_2 \cdots a_n]$; a column vector of $m \times 1$ is denoted by $\{a_1\, a_2 \cdots a_m\}$. If the dimensions of a matrix are $m = n$, the matrix is called a *square matrix* of order m. The elements $a_{11}, a_{22}, \ldots, a_{mm}$ are called the *elements of the principal diagonal*. A square matrix whose elements of the principal diagonal are unity and whose other elements all equal zero is called a *unit matrix* and is denoted by $\mathbf{I}$. A matrix of any dimensions whose elements all equal zero is called a *zero matrix*, and is designated by $\mathbf{0}$.

Two matrices $\mathbf{X}$ and $\mathbf{Y}$ are equal only if they have the same dimensions and if their corresponding elements are equal. A matrix $\mathbf{Z}$ whose element are the sums of the corresponding elements of $\mathbf{X}$ and $\mathbf{Y}$, having the same dimensions, is called the sum of $\mathbf{X}$ and $\mathbf{Y}$. Thus for $\mathbf{Z} = \mathbf{X} + \mathbf{Y}$,

$$\begin{bmatrix} z_{11} & z_{12} & \cdots & z_{1n} \\ z_{21} & z_{22} & \cdots & z_{2n} \\ \vdots & & & \\ z_{m1} & z_{m2} & \cdots & z_{mn} \end{bmatrix} = \begin{bmatrix} x_{11} + y_{11} & x_{12} + y_{12} & \cdots & x_{1n} + y_{1n} \\ x_{21} + y_{21} & x_{22} + y_{22} & \cdots & x_{2n} + y_{2n} \\ \vdots & & & \\ x_{m1} + y_{m1} & x_{m2} + y_{m2} & \cdots & x_{mn} + y_{mn} \end{bmatrix}.$$

A matrix $\mathbf{W}$ whose elements are obtained by multiplying each of the elements of matrix $\mathbf{X}$ by a scalar k is called the product of the scalar k and the matrix $\mathbf{X}$. Thus for $\mathbf{W} = k\mathbf{X}$,

$$\begin{bmatrix} w_{11} & w_{12} & \cdots & w_{1n} \\ w_{21} & w_{22} & \cdots & w_{2n} \\ \vdots & & & \\ w_{m1} & w_{m2} & \cdots & w_{mn} \end{bmatrix} = \begin{bmatrix} kx_{11} & kx_{12} & \cdots & kx_{1n} \\ kx_{21} & kx_{22} & \cdots & kx_{2n} \\ \vdots & & & \\ kx_{m1} & kx_{m2} & \cdots & kx_{mn} \end{bmatrix}.$$

Multiplication of two matrices $\mathbf{X}$ and $\mathbf{Y}$ is defined only if the number of columns of the former matrix equals the number of rows of the latter. Let $\mathbf{X}$ be an $m \times n$ matrix and $\mathbf{Y}$ be an $n \times p$ matrix. Then $\mathbf{X}$ can be *post-multiplied* by $\mathbf{Y}$ and the product $\mathbf{Z} = \mathbf{XY}$ is an $m \times p$ matrix. The elements of $\mathbf{Z}$ are defined by

$$\begin{aligned} z_{ij} &= [x_{i1}x_{i2} \cdots x_{in}]\{y_{1j}y_{2j} \cdots y_{nj}\} \\ &= x_{i1}y_{1j} + x_{i2}y_{2j} + \cdots + x_{1n}y_{nj} = \sum_{k=1}^{n} x_{ik}y_{kj}, \\ & \qquad i = 1, 2, \ldots, m; \quad j = 1, 2, \ldots, p. \end{aligned}$$

In general, the commutative law for multiplication of scalars does not hold for matrix multiplication; and in the above example, $\mathbf{Y}$ cannot be postmultiplied by $\mathbf{X}$. However, for square matrices of the same order, there are exceptions to this general rule. For example, if $\mathbf{X}$ is an $m \times m$ matrix and if $\mathbf{I}$ is a unit matrix of the same order, it can be shown that

$$\mathbf{XI} = \mathbf{IX} = \mathbf{X}\,.$$

If a square matrix A is post-multiplied by a square matrix $\mathbf{B}$ of the same order, such that $\mathbf{AB} = \mathbf{I}$, then $\mathbf{B}$ is said to be the *inverse* of $\mathbf{A}$, and is denoted by $\mathbf{B} = \mathbf{A}^{-1}$. Furthermore,

$$\mathbf{AA}^{-1} = \mathbf{A}^{-1}\mathbf{A} = \mathbf{I}\,.$$

In order to construct the inverse of $\mathbf{A}$, we shall first define several pertinent terms. The interchange of rows and columns in a matrix is called the transposition of the matrix. The transpose of an $m \times n$ matrix $\mathbf{X} = [x_{ij}]$ is given by an $n \times m$ matrix, $\mathbf{X}^T = [x_{ji}]$, or explicitly:

$$\mathbf{X} = \begin{bmatrix} x_{11} & x_{12} & \cdots & x_{1n} \\ x_{21} & x_{22} & \cdots & x_{2n} \\ \vdots & & & \\ x_{m1} & x_{m2} & \cdots & x_{mn} \end{bmatrix}; \qquad \mathbf{X}^T = \begin{bmatrix} x_{11} & x_{21} & \cdots & x_{m1} \\ x_{12} & x_{22} & \cdots & x_{m2} \\ \vdots & & & \\ x_{1n} & x_{2n} & \cdots & x_{mn} \end{bmatrix}.$$

For a square matrix of order m, the transpose is also a square matrix of the same order. The determinant whose elements are the elements of a square matrix $\mathbf{A}$ without disarrangement is said to be the *determinant of the matrix*, and is denoted by $|\mathbf{A}|$, that is,

$$|\mathbf{A}| = \begin{vmatrix} a_{11} & a_{12} & \cdots & a_{1m} \\ a_{21} & a_{22} & \cdots & a_{2m} \\ \vdots & & & \\ a_{m1} & a_{m2} & \cdots & a_{mm} \end{vmatrix}.$$

The *adjoint* of $\mathbf{A}$ is defined as the transpose of the matrix formed by the cofactors (minors with signs) in place of the original element. Let $\mathbf{C}$ be the adjoint of $\mathbf{A}$. Then

$$\mathbf{C} = \begin{bmatrix} c_{11} & c_{21} & \cdots & c_{m1} \\ c_{12} & c_{22} & \cdots & c_{m2} \\ \vdots & & & \\ c_{1m} & c_{2m} & \cdots & c_{mm} \end{bmatrix},$$

where

$$c_{ij} = (-1)^{i+j} \times \text{minor of } a_{ij}\,.$$

From a theorem on the expansion of determinants, it can be shown that for $|\mathbf{A}| \neq 0$.

$$\mathbf{A}^{-1} = \frac{\mathbf{C}}{|\mathbf{A}|}.$$

It should also be noted that the transpose of an inverse is identical to the inverse of the transpose of a matrix.

For a system of m linearly independent equations in m variables given in the form

$$\begin{aligned} a_{11}x_1 + a_{12}x_2 + \cdots + a_{1m}x_m &= b_1, \\ a_{21}x_1 + a_{22}x_2 + \cdots + a_{2m}x_m &= b_2, \\ &\vdots \\ a_{m1}x_1 + a_{m2}x_2 + \cdots + a_{mm}x_m &= b_m, \end{aligned}$$

the coefficients can be detached from the variables in the matrix representation as follows:

$$\begin{bmatrix} a_{11} & a_{12} & \cdots & a_{1m} \\ a_{21} & a_{22} & \cdots & a_{2m} \\ \vdots & & & \\ a_{m1} & a_{m2} & \cdots & a_{mm} \end{bmatrix} \begin{bmatrix} x_1 \\ x_2 \\ \vdots \\ x_m \end{bmatrix} = \begin{bmatrix} b_1 \\ b_2 \\ \vdots \\ b_m \end{bmatrix}.$$

Let $[\alpha_{ij}]$ be the inverse of the detached coefficients $[a_{ij}]$ such that

$$\begin{bmatrix} a_{11} & a_{12} & \cdots & a_{1m} \\ a_{21} & a_{22} & \cdots & a_{2m} \\ \vdots & & & \\ a_{m1} & a_{m2} & \cdots & a_{mm} \end{bmatrix} \begin{bmatrix} \alpha_{11} & \alpha_{12} & \cdots & \alpha_{1m} \\ \alpha_{21} & \alpha_{22} & \cdots & \alpha_{2m} \\ \vdots & & & \\ \alpha_{m1} & \alpha_{m2} & \cdots & \alpha_{mm} \end{bmatrix} = \begin{bmatrix} 1 & 0 \cdots 0 \\ 0 & 1 \cdots 0 \\ \vdots & \\ 0 & 0 \cdots 1 \end{bmatrix}.$$

Then the variables in the system of equations are given correspondingly by

$$\begin{bmatrix} x_1 \\ x_2 \\ \vdots \\ x_m \end{bmatrix} = \begin{bmatrix} 1 & 0 \cdots 0 \\ 0 & 1 \cdots 0 \\ \vdots & \\ 0 & 0 \cdots 1 \end{bmatrix} \begin{bmatrix} x_1 \\ x_2 \\ \vdots \\ x_m \end{bmatrix}$$

$$= \begin{bmatrix} \alpha_{11} & \alpha_{12} & \cdots & \alpha_{1m} \\ \alpha_{21} & \alpha_{22} & \cdots & \alpha_{2m} \\ \vdots & & & \\ \alpha_{m1} & \alpha_{m2} & \cdots & \alpha_{mm} \end{bmatrix} \begin{bmatrix} b_1 \\ b_2 \\ \vdots \\ b_m \end{bmatrix} = \begin{bmatrix} \bar{b}_1 \\ \bar{b}_2 \\ \vdots \\ \bar{b}_m \end{bmatrix},$$

in which

$$x_i = \bar{b}_i = \sum_{k=1}^{m} \alpha_{ik} b_k, \qquad i = 1, 2, \ldots, m.$$

4–3 SIMPLEX MULTIPLIERS

Let us return to the linear programming problem stated in the standard form ($n > m$):

$$\begin{aligned} a_{11}x_1 + a_{12}x_2 + \cdots + a_{1n}x_n &= b_1\,, \\ a_{21}x_1 + a_{22}x_2 + \cdots + a_{2n}x_n &= b_2\,, \\ &\vdots \\ a_{m1}x_1 + a_{m2}x_2 + \cdots + a_{mn}x_n &= b_m\,, \end{aligned} \tag{4–1}$$

and

$$x_j \geq 0\,, \qquad j = 1, 2, \ldots, n\,, \tag{4–2}$$

$$c_1x_1 + c_2x_2 + \cdots + c_nx_n = \text{Min}\, z\,. \tag{4–3}$$

In the simplex method, we seek a basic feasible solution for a specific basis by reducing the constraint equations to canonical form and then test whether such a solution is optimal. For example, if $x_1, x_2, \ldots, x_m$ constitute the current basis, they are the basic variables in the canonical form. Furthermore, the modified objective function will contain only the nonbasic variables as follows:

$$\bar{c}_{m+1}x_{m+1} + \bar{c}_{m+2}x_{m+2} + \cdots + \bar{c}_nx_n = \text{Min}\,(z - \bar{z})\,. \tag{4–4}$$

Thus we may examine the values of $\bar{c}_j$ ($j = m + 1, m + 2, \ldots, n$) in the test of optimality for the current basic feasible solution.

However, we may approach the solution of the problem in a slightly different manner. Let us select as the initial basis the variables $x_1, x_2, \ldots, x_m$. Let the detached coefficients corresponding to the basic variables be denoted by $A = [a_{ij}]$ and let the inverse of A be denoted by $A^{-1} = [\alpha_{ij}]$ for $i = 1, 2, \ldots, m$ and $j = 1, 2, \ldots, m$. We speak of A as the basis and A^{-1} as the *inverse of the basis*, referring to the following matrices:

$$A = \begin{bmatrix} a_{11} & a_{12} & \cdots & a_{1m} \\ a_{21} & a_{22} & \cdots & a_{2m} \\ \vdots & & & \\ a_{m1} & a_{m2} & \cdots & a_{mm} \end{bmatrix}, \qquad A^{-1} = \begin{bmatrix} \alpha_{11} & \alpha_{12} & \cdots & \alpha_{1m} \\ \alpha_{21} & \alpha_{22} & \cdots & \alpha_{2m} \\ \vdots & & & \\ \alpha_{m1} & \alpha_{m2} & \cdots & \alpha_{mm} \end{bmatrix}.$$

Then the values of these basic variables can be determined by matrix inversion. Thus

$$\begin{bmatrix} x_1 \\ x_2 \\ \vdots \\ x_m \end{bmatrix} = \begin{bmatrix} \alpha_{11} & \alpha_{12} & \cdots & \alpha_{1m} \\ \alpha_{21} & \alpha_{22} & \cdots & \alpha_{2m} \\ \vdots & & & \\ \alpha_{m1} & \alpha_{m2} & \cdots & \alpha_{mm} \end{bmatrix} \begin{bmatrix} b_1 \\ b_2 \\ \vdots \\ b_m \end{bmatrix} = \begin{bmatrix} \bar{b}_1 \\ \bar{b}_2 \\ \vdots \\ \bar{b}_m \end{bmatrix}. \tag{4–5}$$

To determine the values of $\bar{c}_j$ $(j = m + 1, m + 2, \ldots, n)$, we shall introduce the concept of *simplex multipliers.* Suppose that we multiply the first equation of (4-1) by a number π_1, the second equation by a number π_2, and so forth down to the mth equation by a number π_m. If we subtract the sum of these products from the objective function in (4-3), we obtain

$$\left(c_1 - \sum_{i=1}^{m} a_{i1}\pi_i\right) x_1 + \left(c_2 - \sum_{i=1}^{m} a_{i2}\pi_i\right) x_2 + \cdots$$
$$+ \left(c_m - \sum_{i=1}^{m} a_{im}\pi_m\right) x_m + \cdots$$
$$+ \left(c_n - \sum_{i=1}^{m} a_{in}\pi_i\right) x_n = \text{Min}\left(z - \sum_{i=1}^{m} b_i\pi_i\right).$$

The numbers π_i $(i = 1, 2, \ldots, m)$ thus defined are called the *simplex multipliers*. By introducing the notation

$$\bar{c}_j = c_j - \sum_{i=1}^{m} a_{ij}\pi_i\,, \qquad j = 1, 2, \ldots, n\,, \tag{4-6}$$

and

$$\bar{z} = \sum_{i=1}^{m} b_i\pi_i\,, \tag{4-7}$$

we have

$$\bar{c}_1 x_1 + \bar{c}_2 x_2 + \cdots + \bar{c}_m x_m + \cdots + \bar{c}_n x_n = \text{Min}\,(z - \bar{z})\,. \tag{4-8}$$

In order that the modified objective function (4-8) be identical to (4-4), which has been obtained by direct elimination in the simplex method, it is necessary that

$$\bar{c}_j = c_j - \sum_{i=1}^{m} a_{ij}\pi_i = 0\,, \qquad j = 1, 2, \ldots, m\,. \tag{4-9}$$

Since there are m unknowns π_i $(i = 1, 2, \ldots, m)$ in m equations $(j = 1, 2, \ldots, m)$, we can solve for π_i, which satisfy the conditions in (4-9). The multipliers π_i can now be used to determine $\bar{c}_j$ for $j = m + 1, m + 2, \ldots, n$ from (4-6), and $\bar{z}$ from (4-7).

By expanding the equations in (4-9), we have the following system of linear equations:

$$\begin{aligned}
a_{11}\pi_1 + a_{21}\pi_2 + \cdots + a_{m1}\pi_m &= c_1\,,\\
a_{12}\pi_1 + a_{22}\pi_2 + \cdots + a_{m2}\pi_m &= c_2\,,\\
&\vdots\\
a_{1m}\pi_1 + a_{2m}\pi_2 + \cdots + a_{mm}\pi_m &= c_m\,.
\end{aligned}$$

Note that the matrix representing the detached coefficients of this set of equations is the transpose of that of the basis, i.e.,

$$A^T = [a_{ij}]^T = \begin{bmatrix} a_{11} & a_{21} & \cdots & a_{m1} \\ a_{12} & a_{22} & \cdots & a_{m2} \\ \vdots & & & \\ a_{1m} & a_{2m} & \cdots & a_{mm} \end{bmatrix}.$$

Since the transpose of an inverse is identical to the inverse of the transpose, it follows that

$$[A^T]^{-1} = [A^{-1}]^T = [\alpha_{ij}]^T = \begin{bmatrix} \alpha_{11} & \alpha_{21} & \cdots & \alpha_{m1} \\ \alpha_{12} & \alpha_{22} & \cdots & \alpha_{m2} \\ \vdots & & & \\ \alpha_{1m} & \alpha_{2m} & \cdots & \alpha_{mm} \end{bmatrix}.$$

Thus the simplex multipliers can be determined from the following relation:

$$\begin{bmatrix} \pi_1 \\ \pi_2 \\ \vdots \\ \pi_m \end{bmatrix} = \begin{bmatrix} \alpha_{11} & \alpha_{21} & \cdots & \alpha_{m1} \\ \alpha_{12} & \alpha_{22} & \cdots & \alpha_{m2} \\ \vdots & & & \\ \alpha_{1m} & \alpha_{2m} & \cdots & \alpha_{mm} \end{bmatrix} \begin{bmatrix} c_1 \\ c_2 \\ \vdots \\ c_m \end{bmatrix}, \tag{4-10}$$

or

$$\pi_i = \sum_{k=1}^{m} \alpha_{ki} c_k, \qquad i = 1, 2, \ldots, m. \tag{4-11}$$

We have just described a procedure for obtaining a basic feasible solution and the corresponding modified objective function from the original data. Once a basis is selected, the values of the basic variables and the coefficients for the modified objective function can be obtained from the inverse of the basis. In fact, the procedure may further be simplified if the objective function in (4-3) is treated as an additional equation by transposing z to the left-hand side as follows:

$$-z + c_1x_1 + c_2x_2 + \cdots + c_nx_n = 0. \tag{4-12}$$

Then for a given basis of $x_1, x_2, \ldots, x_m$, together with $x_0 = -z$, the values of the basic variables and the objective z can be determined from the following:

$$\begin{bmatrix} a_{11} & a_{12} & \cdots & a_{1m} & 0 \\ a_{21} & a_{22} & \cdots & a_{2m} & 0 \\ \vdots & & & & \\ a_{m1} & a_{m2} & \cdots & a_{mm} & 0 \\ c_1 & c_2 & \cdots & c_m & 1 \end{bmatrix} \begin{bmatrix} x_1 \\ x_2 \\ \vdots \\ x_m \\ x_0 \end{bmatrix} = \begin{bmatrix} b_1 \\ b_2 \\ \vdots \\ b_m \\ 0 \end{bmatrix}, \tag{4-13}$$

in which the square matrix represents the detached coefficients of the constraint equations and the objective function. Then by matrix inversion,

$$\begin{bmatrix} x_1 \\ x_2 \\ \vdots \\ x_m \\ x_0 \end{bmatrix} = \begin{bmatrix} \alpha_{11} & \alpha_{12} & \cdots & \alpha_{1m} & 0 \\ \alpha_{21} & \alpha_{22} & \cdots & \alpha_{2m} & 0 \\ \vdots & & & & \\ \alpha_{m1} & \alpha_{m2} & \cdots & \alpha_{mm} & 0 \\ \alpha_{01} & \alpha_{02} & \cdots & \alpha_{0m} & 1 \end{bmatrix} \begin{bmatrix} b_1 \\ b_2 \\ \vdots \\ b_m \\ 0 \end{bmatrix} = \begin{bmatrix} \bar{b}_1 \\ \bar{b}_2 \\ \vdots \\ \bar{b}_m \\ \bar{b}_0 \end{bmatrix}, \tag{4-14}$$

in which the square matrix is the inverse of that in (4-13). Let the expanded matrix of the basis in (4-13) and its inverse in (4-14) be denoted by A_0 and A_0^{-1}, respectively. From the relation $A_0 A_0^{-1} = I$, we see that when the last row of A_0 is multiplied by the ith column of A_0^{-1}, we have

$$\sum_{k=1}^{m} c_k \alpha_{ki} + \alpha_{0i} = 0\,, \qquad i = 1, 2, \ldots, m\,.$$

Comparing this result with (4-11), we note that

$$\pi_i = -\alpha_{0i}\,, \qquad i = 1, 2, \ldots, m\,. \tag{4-15}$$

In other words, we can obtain the simplex multipliers π_i directly from (4-14) without separately carrying out the matrix multiplication in (4-11). From (4-14), we note also that

$$\bar{b}_0 = \sum_{i=1}^{m} \alpha_{0i} b_i\,.$$

Comparing the above with (4-7) and noting (4-15), we have

$$\bar{b}_0 = -\sum_{i=1}^{m} \pi_i b_i = -\bar{z}\,. \tag{4-16}$$

This procedure provides a direct method for obtaining and testing the optimality of the solution for a given basis. If the selected basis is not optimal, we need further guidelines for determining the new basis. Although the coefficients for the modified objective function obtained either by (4-7) or from (4-14) may be used to determine the departing variable x_s, since

$$\bar{c}_s = \text{Min}\, \bar{c}_j\,,$$

we still must select the entering variable x_r from the pivotal element $\bar{a}_{rs}$ such that

$$x_s = \frac{\bar{b}_r}{\bar{a}_{rs}} = \underset{a_{is}>0}{\text{Min}} \frac{\bar{b}_i}{\bar{a}_{is}} > 0\,, \qquad i = 1, 2, \ldots, m\,.$$

In order to compute $\bar{a}_{is}$, we can imagine that the terms a_{is} $(i = 1, 2, \ldots, m)$ of (4-1) and c_s of (4-12) have been transposed to the right-hand side of the equations. Since x_s is a nonbasic variable, the transposition does not affect the matrix inversion of the basis. Then, instead of (4-14), the solution of these equations becomes

$$\begin{bmatrix} x_1 \\ x_2 \\ \vdots \\ x_m \\ -z \end{bmatrix} = \begin{bmatrix} \alpha_{11} & \alpha_{12} & \cdots & \alpha_{1m} & 0 \\ \alpha_{21} & \alpha_{22} & \cdots & \alpha_{2m} & 0 \\ \vdots & & & & \\ \alpha_{m1} & \alpha_{m2} & \cdots & \alpha_{mm} & 0 \\ -\pi_1 & -\pi_2 & \cdots & -\pi_m & 1 \end{bmatrix} \begin{bmatrix} b_1 & -a_{1s}x_s \\ b_2 & -a_{2s}x_s \\ \vdots & \\ b_m & -a_{ms}x_s \\ 0 & -c_s x_s \end{bmatrix}$$

$$= \begin{bmatrix} \bar{b}_1 & -\bar{a}_{1s}x_s \\ \bar{b}_2 & -\bar{a}_{2s}x_s \\ \vdots & \\ \bar{b}_m & -\bar{a}_{ms}x_s \\ -\bar{z} & -\bar{c}_s x_s \end{bmatrix}. \tag{4-17}$$

In view of (4-14), we conclude that

$$\begin{bmatrix} \bar{a}_{1s} \\ \bar{a}_{2s} \\ \vdots \\ \bar{a}_{ms} \\ \bar{c}_s \end{bmatrix} = \begin{bmatrix} \alpha_{11} & \alpha_{12} & \cdots & \alpha_{1m} & 0 \\ \alpha_{21} & \alpha_{22} & \cdots & \alpha_{2m} & 0 \\ \vdots & & & & \\ \alpha_{m1} & \alpha_{m2} & \cdots & \alpha_{mm} & 0 \\ -\pi_1 & -\pi_2 & \cdots & -\pi_m & 1 \end{bmatrix} \begin{bmatrix} a_{1s} \\ a_{2s} \\ \vdots \\ a_{ms} \\ c_s \end{bmatrix}, \tag{4-18}$$

or

$$\bar{a}_{is} = \sum_{k=1}^{m} \alpha_{ik} a_{ks}, \qquad i = 1, 2, \ldots, m, \tag{4-19}$$

and

$$\bar{c}_s = -\sum_{k=1}^{m} \pi_k a_{ks} + c_s. \tag{4-20}$$

The last equation is a restatement of (4-6) for $j = s$.

Thus we can compute $\bar{a}_{is}$ from (4-19) and determine $\bar{a}_{rs}$ in the same manner as in the simplex method. Since both the entering and departing variables from the current basis have been determined, we can select a new basis and apply the same procedure repeatedly until an optimal solution is obtained.

Example 4-5. Find a basic feasible solution for the basis x_2 and x_4 by matrix inversion, and test the optimality of the solution by using simplex multipliers for the following linear programming problem:

$$5x_1 + 10x_2 - x_3 \qquad = 8, \tag{a}$$

$$x_1 + x_2 \qquad + x_4 = 1, \tag{b}$$

$$x_j \geq 0, \qquad j = 1, 2, 3, 4,$$

$$2x_1 + x_2 = \text{Min } z. \tag{c}$$

Table 4-2

COEFFICIENTS FOR EXAMPLES 4-5

Equation	x_1	x_2	x_3	x_4	$-z$	Values
(a)	5	10	-1	0	0	8
(b)	1	1	0	1	0	1
(c)	2	1	0	0	1	0

Let the detached coefficients for the constraint equations and the objective function be arranged as shown in Table 4-2.

First, let us obtain the basic solution by using (4-5) and (4-11). Then for the basis of x_2 and x_4,

$$A = \begin{bmatrix} a_{12} & a_{14} \\ a_{22} & a_{24} \end{bmatrix} = \begin{bmatrix} 10 & 0 \\ 1 & 1 \end{bmatrix},$$

$$A^{-1} = \frac{1}{10}\begin{bmatrix} 1 & 0 \\ -1 & 10 \end{bmatrix} = \begin{bmatrix} 0.1 & 0 \\ -0.1 & 1 \end{bmatrix},$$

$$[A^{-1}]^T = \begin{bmatrix} 0.1 & -0.1 \\ 0 & 1 \end{bmatrix}.$$

Hence from (4-5),

$$\begin{bmatrix} x_2 \\ x_4 \end{bmatrix} = \begin{bmatrix} 0.1 & 0 \\ -0.1 & 1 \end{bmatrix} \begin{bmatrix} 8 \\ 1 \end{bmatrix} = \begin{bmatrix} 0.8 \\ 0.2 \end{bmatrix},$$

and from (4-11),

$$\begin{bmatrix} \pi_1 \\ \pi_2 \end{bmatrix} = \begin{bmatrix} 0.1 & -0.1 \\ 0 & 1 \end{bmatrix} \begin{bmatrix} 1 \\ 0 \end{bmatrix} = \begin{bmatrix} 0.1 \\ 0 \end{bmatrix}.$$

The alternative procedure of using (4-13) and (4-14) leads to the same results, since the matrix A_0 and its inverse A_0^{-1} are given by

$$A_0 = \begin{bmatrix} a_{12} & a_{14} & 0 \\ a_{22} & a_{24} & 0 \\ c_2 & c_4 & 1 \end{bmatrix} = \begin{bmatrix} 10 & 0 & 0 \\ 1 & 1 & 0 \\ 1 & 0 & 1 \end{bmatrix},$$

and

$$A_0^{-1} = \frac{1}{10}\begin{bmatrix} 1 & 0 & 0 \\ -1 & 10 & 0 \\ -1 & 0 & 10 \end{bmatrix} = \begin{bmatrix} 0.1 & 0 & 0 \\ -0.1 & 1 & 0 \\ -0.1 & 0 & 1 \end{bmatrix}.$$

Hence from (4–14),

$$\begin{bmatrix} x_2 \\ x_4 \\ -z \end{bmatrix} = \begin{bmatrix} 0.1 & 0 & 0 \\ -0.1 & 1 & 0 \\ -0.1 & 0 & 1 \end{bmatrix} \begin{bmatrix} 8 \\ 1 \\ 0 \end{bmatrix} = \begin{bmatrix} 0.8 \\ 0.2 \\ -0.8 \end{bmatrix}.$$

Furthermore, from (4–15) and (4–16),

$$\pi_1 = 0.1\,, \qquad \pi_2 = 0\,, \qquad z = 0.8\,.$$

The coefficients $\bar{c}_j$ for the nonbasic variables ($j = m+1, m+2, \ldots, n$) are determined from (4–6) as follows:

$$\bar{c}_1 = c_1 - (a_{11}\pi_1 + a_{21}\pi_2) = 2 - [(5)(0.1) + (1)(0)] = 1.5\,,$$
$$\bar{c}_3 = c_3 - (a_{13}\pi_1 + a_{23}\pi_2) = 0 - [(-1)(0.1) + (0)(0)] = 0.1\,.$$

Also, $\bar{z}$ may be found from (4–7) as follows:

$$\bar{z} = b_1\pi_1 + b_2\pi_2 = (8)(0.1) + (1)(0) = 0.8\,.$$

Thus the modified objective function is

$$1.5x_1 + 0.1x_3 = z - 0.8 \quad \text{(Min)}\,.$$

Since all $\bar{c}_j > 0$, the basic solution is optimal, i.e.,

$$x_2^* = 0.8\,, \qquad x_4^* = 0.2\,, \qquad x_1^* = x_3^* = 0\,, \qquad z^* = 0.8\,.$$

Example 4–6. Find the optimal solution of the above problem by first selecting x_1 and x_2 as the basis, and then changing the basis later when it becomes necessary.

Using the relations in (4–13) and (4–14), we have the expanded matrix for the basis and its inverse as follows:

$$A_0 = \begin{bmatrix} a_{11} & a_{12} & 0 \\ a_{21} & a_{22} & 0 \\ c_1 & c_2 & 1 \end{bmatrix} = \begin{bmatrix} 5 & 10 & 0 \\ 1 & 1 & 0 \\ 2 & 1 & 1 \end{bmatrix}$$

and

$$A_0^{-1} = -\frac{1}{5}\begin{bmatrix} 1 & -10 & 0 \\ -1 & 5 & 0 \\ -1 & 15 & -5 \end{bmatrix} = \begin{bmatrix} -0.2 & 2 & 0 \\ 0.2 & -1 & 0 \\ 0.2 & -3 & 1 \end{bmatrix}.$$

Hence

$$\begin{bmatrix} x_1 \\ x_2 \\ -z \end{bmatrix} = \begin{bmatrix} -0.2 & 2 & 0 \\ 0.2 & -1 & 0 \\ 0.2 & -3 & 1 \end{bmatrix} \begin{bmatrix} 8 \\ 1 \\ 0 \end{bmatrix} = \begin{bmatrix} 0.4 \\ 0.6 \\ -1.4 \end{bmatrix}.$$

Furthermore, from (4-15) and (4-16) we have

$$\pi_1 = -0.2\,, \qquad \pi_2 = 3\,, \qquad \bar{z} = 1.4\,.$$

The coefficients $\bar{c}_3$ and $\bar{c}_4$ are determined from (4-6) as follows:

$$\bar{c}_3 = c_3 - (a_{13}\pi_1 + a_{23}\pi_2) = 0 - [(-1)(-0.2) + (0)(3)] = -0.2\,,$$
$$\bar{c}_4 = c_4 - (a_{14}\pi_1 + a_{24}\pi_2) = 0 - [(0)(-0.2) + (1)(3)] = -3\,.$$

The modified objective function is therefore

$$-0.2x_3 - 3x_4 = z - 1.4 \quad \text{(Min)}\,,$$

which indicates that the basic solution is not optimal. Since $\bar{c}_4 = \text{Min}\,\bar{c}_j$, we attempt to compute $\bar{a}_{is}$ from (4-18) as follows:

$$\begin{bmatrix} \bar{a}_{14} \\ \bar{a}_{24} \\ \bar{c}_4 \end{bmatrix} = \begin{bmatrix} -0.2 & 2 & 0 \\ 0.2 & -1 & 0 \\ 0.2 & -3 & 1 \end{bmatrix} \begin{bmatrix} 0 \\ 1 \\ 0 \end{bmatrix} = \begin{bmatrix} 2 \\ -1 \\ -3 \end{bmatrix}.$$

Hence $\bar{a}_{14} = 2$ and $\bar{a}_{24} = -1$. The coefficient $\bar{c}_4$ serves as a check. Since $\bar{a}_{14}$ is the only $\bar{a}_{is} > 0$, the departing variable is x_1, while the entering variable is x_4.

We can now repeat the procedure from the original data by using x_2 and x_4 as the basis. The results will be the same as those obtained in Example 4-5. We can terminate the procedure when the optimal solution is found for the new basis. Thus

$$x_2^* = 0.8\,, \qquad x_4^* = 0.2\,, \qquad x_1^* = x_3^* = 0\,, \qquad z^* = 0.8\,.$$

4-4 INVERSE OF THE NEW BASIS

The procedure of using the inverse of the basis and simplex multipliers described in the previous section has the characteristics of computing the necessary quantities at each cycle from the original data (at cycle 0). To the extent that $\bar{a}_{is}$ at the end of any cycle can be computed directly from the original data, the procedure is very attractive because we can avoid computing other $\bar{a}_{ij}$ for the nonbasic variables not associated with the change of basis in all previous cycles as required in the simplex method. On the other hand, if the inverse of the basis at each cycle must be obtained directly from the original data, the amount of computation will become prohibitively large. Fortunately, this need not be the case if we observe that the inverse of the new basis can be obtained with relatively few computational steps from the inverse in the previous cycle, and that the simplex multipliers for the new basis can also be obtained accordingly. In this section, we shall consider the procedure of obtaining the inverse—and hence the simplex multipliers—of the new basis from that of the current basis.

Suppose that at the end of a cycle, (say Cycle t), x_s is found to be the entering variable and x_r is the departing variable from the current basis in order to form a new basis. Let us denote each of the coefficients in the new basis with a prime ($'$), and each of those in the current basis without a prime. If the values of x_i for the current basis have been obtained from (4–14), together with the value of z expressed in terms of relations in (4–15) and (4–16), then the new values of x_i and z at the start of Cycle $t+1$ can be obtained directly from the original data in (4–13) for Cycle t as follows:

$$\begin{bmatrix} x_1 \\ x_2 \\ \vdots \\ x_m \\ -z \end{bmatrix} = \begin{bmatrix} \alpha'_{11} & \alpha'_{12} & \cdots & \alpha'_{1m} & 0 \\ \alpha'_{21} & \alpha'_{22} & \cdots & \alpha'_{2m} & 0 \\ \vdots & & & & \\ \alpha'_{m1} & \alpha'_{m2} & \cdots & \alpha'_{mm} & 0 \\ -\pi'_1 & -\pi'_2 & \cdots & -\pi'_m & 1 \end{bmatrix} \begin{bmatrix} b_1 \\ b_2 \\ \vdots \\ b_m \\ 0 \end{bmatrix} = \begin{bmatrix} \bar{b}'_1 \\ \bar{b}'_2 \\ \vdots \\ \bar{b}'_m \\ -\bar{z}' \end{bmatrix}, \tag{4–21}$$

or

$$\bar{b}'_i = \sum_{k=1}^{m} \alpha'_{ik} b_k \tag{4–22}$$

and

$$\bar{z}' = \sum_{i=1}^{m} \pi'_i b_i \,. \tag{4–23}$$

On the other hand, if we also compute the inverse of the new basis from that of the current basis, we can establish the relation between the two inverses by comparing the derived results with (4–21). At the end of Cycle t, we denote the entering variable x_s, which is obtained from

$$x_s = \frac{\bar{b}_r}{\bar{a}_{rs}} = \bar{b}'_r \,. \tag{4–24}$$

Then the new values of other variables remaining on the basis may be obtained from (4–17) for $i \neq r$:

$$x_i = \bar{b}_i - \bar{a}_{is}\bar{b}'_r = \bar{b}'_i \,, \tag{4–25}$$

and $x_r = 0$, since it is the departing variable. The new value of z can also be obtained from (4–17):

$$-z = -\bar{z} - \bar{c}_s\bar{b}'_r = -\bar{z}' \,. \tag{4–26}$$

From (4–14), we note that for $i = 1, 2, \ldots, m$,

$$\bar{b}_i = \sum_{k=1}^{m} \alpha_{ik} b_k \,. \tag{4–27}$$

For $i = r$, we can substitute the value of $\bar{b}_r$ in (4-27) into (4-24) and obtain

$$\bar{b}'_r = \frac{\bar{b}_r}{\bar{a}_{rs}} = \frac{1}{\bar{a}_{rs}} \sum_{k=1}^{m} \alpha_{rk} b_k = \sum_{k=1}^{m} \frac{\alpha_{rk}}{\bar{a}_{rs}} b_k \,. \tag{4-28}$$

Then for $i \neq r$, the substitution of the values of $\bar{b}_i$ in (4-27) and $\bar{b}'_r$ in (4-28) into (4-25) leads to

$$\bar{b}'_i = \sum_{k=1}^{m} \alpha_{ik} b_k - \frac{\bar{a}_{is}}{\bar{a}_{rs}} \sum_{k=1}^{m} \alpha_{rk} b_k = \sum_{k=1}^{m} \left[\alpha_{ik} - \bar{a}_{is} \frac{\alpha_{rk}}{\bar{a}_{rs}} \right] b_k \,. \tag{4-29}$$

Furthermore, by substituting (4-16) and (4-28) into (4-26), we get

$$-\bar{z}' = -\sum_{i=1}^{m} \pi_i b_i - \frac{\bar{c}_s}{\bar{a}_{rs}} \sum_{k=1}^{m} \alpha_{rk} b_k = -\sum_{i=1}^{m} \left[\pi_i + \bar{c}_s \frac{\alpha_{ri}}{\bar{a}_{rs}} \right] b_i \,. \tag{4-30}$$

Comparing (4-28) with (4-22) for $i = r$, where the entering variable x_s takes the value of $\bar{b}'_r$, we note that

$$\alpha'_{rk} = \frac{\alpha_{rk}}{\bar{a}_{rs}} \,. \tag{4-31}$$

Similarly, comparing (4-29) with (4-22) for $i \neq r$, and noting the relation in (4-31), we have

$$\alpha'_{ik} = \alpha_{ik} - \bar{a}_{is} \alpha'_{rk} \,. \tag{4-32}$$

Also, comparing (4-30) with (4-22) and noting (4-31), we get

$$-\pi'_i = -\pi_i - \bar{c}_s \alpha'_{ri} \,. \tag{4-33}$$

Example 4-7. Find the basic solution for the basic variables x_2 and x_4 in Example 4-6 from the current basis x_1 and x_2, instead of the original data.

In Example 4-6, we found at the end of the first cycle that x_1 is the departing variable and x_4 is the entering variable. Thus from (4-24) we have for $r = 1$

$$x_4 = \bar{b}'_1 = \frac{\bar{b}_1}{\bar{a}_{14}} = \frac{0.4}{2} = 0.2 \,.$$

Also, from (4-31) we have for $r = 1$ and $k = 1, 2$,

$$\alpha'_{11} = \frac{\alpha_{11}}{\bar{a}_{14}} = \frac{-0.2}{2} = -0.1 \,,$$

$$\alpha'_{12} = \frac{\alpha_{12}}{\bar{a}_{14}} = \frac{2}{2} = 1 \,.$$

From (4-32), we have for $i = 2$ and $k = 1, 2$,

$$\alpha'_{21} = \alpha_{21} - \bar{a}_{24}\alpha'_{11} = 0.2 - (-1)(-0.1) = 0.1\,,$$
$$\alpha'_{22} = \alpha_{22} - \bar{a}_{24}\alpha'_{12} = -1 - (-1)(1) = 0\,.$$

From (4-33), we get for $i = 1, 2$,

$$-\pi'_1 = -\pi_1 - \bar{c}_4\alpha'_{11} = -(-0.2) - (-3)(-0.1) = -0.1\,,$$
$$-\pi'_2 = -\pi_2 - \bar{c}_4\alpha'_{12} = -(3) - (-3)(1) = 0\,.$$

Thus the new values of the basic variables, including the new value of z, are given by

$$\begin{bmatrix} x_4 \\ x_2 \\ -z \end{bmatrix} = \begin{bmatrix} \alpha'_{11} & \alpha'_{12} & 0 \\ \alpha'_{21} & \alpha'_{22} & 0 \\ -\pi'_1 & -\pi'_2 & 1 \end{bmatrix} \begin{bmatrix} b_1 \\ b_2 \\ 1 \end{bmatrix} = \begin{bmatrix} -0.1 & 1 & 0 \\ 0.1 & 0 & 0 \\ -0.1 & 0 & 1 \end{bmatrix} \begin{bmatrix} 8 \\ 1 \\ 0 \end{bmatrix} = \begin{bmatrix} 0.2 \\ 0.8 \\ -0.8 \end{bmatrix}.$$

The results thus obtained are equivalent to those obtained directly from the original data in Example 4-5.

4-5 REVISED SIMPLEX METHOD

We shall now consider the systematic application of the procedure of using the inverse of the basis and the simplex multipliers in the solution of linear programming problems described in the last two sections. The basic concept underlying the procedure is similar to that of the simplex method. Hence the procedure is generally known as the *revised simplex method*.

Since the revised simplex method involves the selection of a basis at the beginning, it is again logical to assume a set of artificial variables to start the solution, as in the simplex method. For a set of m constraint equations with n admissible variables, let the artificial variables be $x_{n+1}, x_{n+2}, \ldots, x_{n+m}$. We then have the infeasibility form

$$x_{n+1} + x_{n+2} + \cdots + x_{n+m} = w\,.$$

Hence the set of constraint equations containing the artificial variables and the modified objective function for w, as obtained in Chapter 3, are as follows:

$$\begin{aligned} a_{11}x_1 + a_{12}x_2 + \cdots + a_{1n}x_n + x_{n+1} \qquad\qquad &= b_1 \\ a_{21}x_1 + a_{22}x_2 + \cdots + a_{2n}x_n \qquad + x_{n+2} \qquad &= b_2 \\ \vdots \qquad\qquad\qquad\qquad & \\ a_{m1}x_1 + a_{m2}x_2 + \cdots + a_{mn}x_n \qquad\qquad + x_{n+m} &= b_m \end{aligned} \tag{4-34}$$

$$d_1x_1 + d_2x_2 + \cdots + d_nx_n = w - w_0\,. \tag{4-35}$$

By transposing w to the left-hand side of the equation, we have

$$-w + d_1x_1 + d_2x_2 + \cdots + d_nx_n = -w_0 . \tag{4-36}$$

This is analogous to (4-12) for z except that the constant on the right-hand side generally is not zero, whereas in (4-12)

$$-z + c_1x_1 + c_2x_2 + \cdots + c_nx_n = 0 .$$

By introducing σ_i $(i = 1, 2, \ldots, m)$ as the simplex multipliers for equations (4-34), and by following the derivation of (4-8) for the modified objective function for z, we have

$$\bar{d}_1x_1 + \bar{d}_2x_2 + \cdots + \bar{d}_nx_n + \cdots + \bar{d}_{n+m}x_{n+m} = \text{Min}\,(w - \bar{w}) , \tag{4-37}$$

in which

$$\bar{d}_j = d_j - \sum_{i=1}^{m} a_{ij}\sigma_i \tag{4-38}$$

and

$$\bar{w} = w_0 + \sum_{i=1}^{m} b_i\sigma_i . \tag{4-39}$$

Furthermore, if the variables x_i $(i = 1, 2, \ldots, m)$ are selected as the basis, then σ_i may be obtained from a set of equations analogous to (4-11) for π_i. Thus

$$\sigma_i = \sum_{k=1}^{m} \alpha_{ki}d_k . \tag{4-40}$$

Then $\bar{d}_j$ $(j = m + 1, m + 2, \ldots, m + n)$ for the nonbasic variables can be determined from (4-38).

In general, the basic solution of the problem involving w can be expressed in the same way as (4-14) for z. As a matter of fact, if we include both (4-12) for z and (4-36) for w in the solution of (4-34), the inverse of the expanded matrix becomes

$$\begin{bmatrix} x_1 \\ x_2 \\ \vdots \\ x_m \\ -z \\ -w \end{bmatrix} = \begin{bmatrix} \alpha_{11} & \alpha_{12} & \cdots & \alpha_{1m} & 0 & 0 \\ \alpha_{21} & \alpha_{22} & \cdots & \alpha_{2m} & 0 & 0 \\ \vdots & & & & & \\ \alpha_{m1} & \alpha_{m2} & \cdots & \alpha_{mm} & 0 & 0 \\ -\pi_1 & -\pi_2 & \cdots & -\pi_m & 1 & 0 \\ -\sigma_1 & -\sigma_2 & \cdots & -\sigma_m & 0 & 1 \end{bmatrix} \begin{bmatrix} b_1 \\ b_2 \\ \vdots \\ b_m \\ 0 \\ -w_0 \end{bmatrix} = \begin{bmatrix} \bar{b}_1 \\ \bar{b}_2 \\ \vdots \\ \bar{b}_m \\ -\bar{z} \\ -\bar{w} \end{bmatrix} . \tag{4-41}$$

However, if we select the set of artificial variables $x_{n+1}, x_{n+2}, \ldots, x_{n+m}$ as the initial basis, the initial solution will be greatly simplified. First, the detached coefficients of the basic variables constitute a unit matrix of order m, and the inverse of the basis is also a unit matrix of order m. Second, both d_j and c_j for $j = n+1, n+2, \ldots, n+m$, corresponding to the coefficients of the basic variables, are zero; hence from (4–40) and (4–11), we conclude that $\sigma_i = 0$ and $\pi_i = 0$. Thus at the beginning cycle of the solution we have

$$\begin{bmatrix} x_{n+1} \\ x_{n+2} \\ \vdots \\ x_{n+m} \\ -z \\ -w \end{bmatrix} = \begin{bmatrix} 1 & 0 \cdots 0 & 0 & 0 \\ 0 & 1 \cdots 0 & 0 & 0 \\ \vdots & & & \\ 0 & 0 \cdots 1 & 0 & 0 \\ 0 & 0 \cdots 0 & 1 & 0 \\ 0 & 0 \cdots 0 & 0 & 1 \end{bmatrix} \begin{bmatrix} b_1 \\ b_2 \\ \vdots \\ b_m \\ 0 \\ -w_0 \end{bmatrix} = \begin{bmatrix} b_1 \\ b_2 \\ \vdots \\ b_m \\ 0 \\ -w_0 \end{bmatrix}. \tag{4–42}$$

Then from (4–38) we have $\bar{d}_j = d_j$ for $\sigma_i = 0$. In other words, if the initial basis is in the canonical form, the inverse of the unit matrix at the beginning cycle has the effect of reproducing the original data, as indicated by (4–42). This same effect holds even if only some of the basic variables are artificial so long as the basis is in the canonical form and the modified objective for w in (4–35) is obtained for the partially artificial basis.

The computation of $\bar{a}_{is}$ for optimizing w is identical to that in (4–19) for optimizing z. In order to check d_s and c_s when we are computing $\bar{a}_{is}$, we can expand the matrix multiplication in (4–18) to the following:

$$\begin{bmatrix} \bar{a}_{1s} \\ \bar{a}_{2s} \\ \vdots \\ \bar{a}_{ms} \\ \bar{c}_s \\ \bar{d}_s \end{bmatrix} = \begin{bmatrix} \alpha_{11} & \alpha_{12} & \cdots & \alpha_{1m} & 0 & 0 \\ \alpha_{21} & \alpha_{22} & \cdots & \alpha_{2m} & 0 & 0 \\ \vdots & & & & & \\ \alpha_{m1} & \alpha_{m2} & \cdots & \alpha_{mm} & 0 & 0 \\ -\pi_1 & -\pi_2 & \cdots & -\pi_m & 1 & 0 \\ -\sigma_1 & -\sigma_2 & \cdots & -\sigma_m & 0 & 1 \end{bmatrix} \begin{bmatrix} a_{1s} \\ a_{2s} \\ \vdots \\ a_{ms} \\ c_s \\ d_s \end{bmatrix}. \tag{4–43}$$

At the beginning cycle, in which the inverse of the basis is a unit matrix, it is seen that $\bar{a}_{is} = a_{is}$.

The derivation of the inverse of the new basis from that of the current basis for optimizing w is the same as that in (4–31) and (4–32) for optimizing z. The computation of new simplex multipliers σ'_i is similar to that in (4–33) for computing π'_i. Thus, if x_s is the entering variable and x_r the departing variable, we get

$$-\sigma'_i = -\sigma_i - \bar{d}_s \alpha'_{ri}\,. \tag{4–44}$$

Furthermore, analogous to z in (4–26), we have

$$-w = -\bar{w} - \bar{d}_s \bar{b}'_r = -\bar{w}'\,. \tag{4–45}$$

The revised simplex method can therefore be treated as a two-phase operation as in the simplex method. In Phase I, artificial variables are introduced to start the problem. *If* min $w \neq 0$, *no feasible solution exists for the original problem, and the iterative procedure terminates.* If min $w = 0$, we can drop the data associated with the artificial variables at the end of Phase I and proceed to find the optimal solution in Phase II.

Example 4-8. Repeat Example 4-5 by the use of artificial variables in starting the iterative procedure.

Let us introduce artificial variable x_5 in the first constraint equation. Together with the infeasibility form $w = x_5$, we obtain the following:

$$\begin{aligned}
5x_1 + 10x_2 - x_3 \qquad\quad + x_5 &= 8 \\
x_1 + \quad x_2 \qquad + x_4 \qquad &= 1 \\
2x_1 + \quad x_2 \qquad\qquad\qquad &= z \quad \text{(Min)} \\
-5x_1 - 10x_2 + x_3 \qquad\qquad &= w - 8 \quad \text{(Min)}
\end{aligned}$$

At Cycle 0, we simply write

$$\begin{bmatrix} x_5 \\ x_4 \\ -z \\ -w \end{bmatrix} = \begin{bmatrix} 1 & 0 & 0 & 0 \\ 0 & 1 & 0 & 0 \\ 0 & 0 & 1 & 0 \\ 0 & 0 & 0 & 1 \end{bmatrix} \begin{bmatrix} 8 \\ 1 \\ 0 \\ -8 \end{bmatrix} = \begin{bmatrix} 8 \\ 1 \\ 0 \\ -8 \end{bmatrix}.$$

Since $\sigma_i = 0$ for $i = 1, 2$, we have $\bar{d}_j = d_j$. In examining d_j in the original objective function for w, we note that $\text{Min}\, d_j = d_2 = -10$. Hence x_2 is the entering variables whose coefficients in the constraint equations are

$$\bar{a}_{12} = a_{12} = 10\,, \qquad \bar{a}_{22} = a_{22} = 1\,.$$

Note also that because

$$\min_{a_{is}>0} \frac{\bar{b}_i}{\bar{a}_{is}} = \frac{\bar{b}_1}{\bar{a}_{12}} = \frac{8}{10}\,,$$

the departing variable is x_5 in the first equation.

In Cycle 1, we compute the elements for the inverse of the basis according to (4-31), (4-32), (4-33), and (4-44). From (4-31), we have for $r = 1$ and $k = 1, 2$,

$$\alpha'_{11} = \frac{\alpha_{11}}{\bar{a}_{12}} = \frac{1}{10} = 0.1\,,$$

$$\alpha'_{12} = \frac{\alpha_{12}}{\bar{a}_{12}} = \frac{0}{10} = 0\,.$$

From (4-32), we have for $i = 2$ and $k = 1, 2$,

$$\alpha'_{21} = \alpha_{21} - \bar{a}_{22}\alpha'_{11} = 0 - (1)(0.1) = -0.1\,,$$
$$\alpha'_{22} = \alpha_{22} - \bar{a}_{22}\alpha'_{12} = 1 - (1)(0) = 1\,.$$

From (4-33), we have for $i = 1, 2$,

$$-\pi'_1 = -\pi_1 - \bar{c}_2\alpha'_{11} = 0 - (1)(0.1) = -0.1\,,$$
$$-\pi'_2 = -\pi_2 - \bar{c}_2\alpha'_{12} = 0 - (1)(0) = 0\,.$$

From (4-44), we have for $i = 1, 2$,

$$-\sigma'_1 = -\sigma_1 - \bar{d}_2\alpha'_{11} = 0 - (-10)(0.1) = 1\,,$$
$$-\sigma'_2 = -\sigma_2 - \bar{d}_2\alpha'_{12} = 0 - (-10)(0) = 0\,.$$

Hence the inverse of the new basis is given by

$$\begin{bmatrix} x_2 \\ x_4 \\ -z \\ -w \end{bmatrix} = \begin{bmatrix} 0.1 & 0 & 0 & 0 \\ -0.1 & 1 & 0 & 0 \\ -0.1 & 0 & 1 & 0 \\ 1 & 0 & 0 & 1 \end{bmatrix} \begin{bmatrix} 8 \\ 1 \\ 0 \\ -8 \end{bmatrix} = \begin{bmatrix} 0.8 \\ 0.2 \\ -0.8 \\ 0 \end{bmatrix}.$$

We observe from the inverse of the new basis that $\sigma_1 = -1$ and $\sigma_2 = 0$. Using (4-38), we obtain for the nonbasic variables

$$\bar{d}_1 = d_1 - (a_{11}\sigma_1 + a_{21}\sigma_2) = -5 - (-1)(5) = 0\,,$$
$$\bar{d}_3 = d_3 - (a_{13}\sigma_1 + a_{23}\sigma_2) = 1 - (-1)(-1) = 0\,,$$
$$\bar{d}_5 = d_5 - (a_{15}\sigma_1 + a_{25}\sigma_2) = 0 - (-1)(1) = 1\,.$$

Since Min $\bar{d}_j$ is nonnegative, the procedure for Phase I terminates and $-w = 0$ or $w = 0$, as noted in the solution above. Hence we shall proceed to Phase II by deleting the last row and last column associated with Phase I in the inverse of the new basis. The remaining steps in Phase II are essentially the same as those in Example 4-5.

4-6 TABLEAU FOR REVISED SIMPLEX METHOD

The computation for the revised simplex method described in the previous section can also be carried out in tabulated form. Since the relative cost coefficients $\bar{c}_j$ or $\bar{d}_j$ are computed directly from the original data after the simplex multipliers π_i or σ_i are obtained at each cycle, their computed values together with the original data may be tabulated as shown in Table 4-3. The determination of inverse and simplex multipliers can be tabulated in a separate form, as shown in Table 4-4.

Table 4-3

INITIAL DATA AND COMPUTED RELATIVE COST COEFFICIENTS

Basis	Admissible variables	Artificial variables	Value
	$x_1 \quad x_2 \cdots x_q \cdots x_n$	$x_{n+1} \quad x_{n+2} \cdots x_{n+p} \cdots x_{n+m}$	
x_{n+1}	$a_{11} \quad a_{12} \cdots a_{1q} \cdots x_{1n}$	1	b_1
x_{n+2}	$a_{21} \quad a_{22} \cdots a_{2q} \cdots x_{2n}$	1	b_2
$\vdots$	$\vdots$	$\ddots$	$\vdots$
x_{n+p}	$a_{p1} \quad a_{p2} \cdots a_{pq} \cdots a_{pn}$	1	b_p
$\vdots$	$\vdots$	$\ddots$	$\vdots$
x_{n+m}	$a_{m1} \quad a_{m2} \cdots a_{mq} \cdots a_{mn}$	1	b_m
z-form	$c_1 \quad c_2 \cdots c_q \cdots c_n$		z
w-form	$d_1 \quad d_2 \cdots d_q \cdots d_n$		$w - w_0$
Cycle	Relative cost coefficients		
I	$\bar{d}_1 \quad \bar{d}_2 \cdots \bar{d}_n$	$\bar{d}_{n+1} \quad \bar{d}_{n+2} \cdots \bar{d}_{n+m}$	
0	$d_1 \quad d_2 \cdots d_n$	$0 \quad 0 \cdots 0$	
1	— — ⋯ —	— — ⋯ —	
$\vdots$	$\vdots$	$\vdots$	
k	— — ⋯ —	— — ⋯ —	
II	$\bar{c}_1 \quad \bar{c}_2 \cdots \bar{c}_n$		
k	— — ⋯ —		
$k+1$	— — ⋯ —		
$\vdots$	$\vdots$		

To start the iterative procedure, we take from the original data in Table 4-3 to form the Cycle 0 of Phase I in Table 4-4. We also note the relative cost coefficients in Table 4-3 and select $\bar{d}_s$; consequently, the corresponding $\bar{a}_{is}$ column will be included in that cycle. In all subsequent cycles, the new inverse and new simplex multipliers can be obtained from the preceding inverse and simplex multipliers in Table 4-4, while the relative cost coefficients are computed from that original data in Table 4-3; the results are tabulated in the bottom part of the same table.

The iterative procedure from a typical Cycle t to Cycle $t + 1$, which is applicable to both Phases I and II, is shown in Table 4-4. However, in Phase I, $\bar{\sigma}_i$ and $\bar{d}_s$ are required, and π_i and $\bar{c}_s$ are computed routinely so that they will be available upon the termination of Phase I. On the

Table 4-4

TABLEAU FOR REVISED SIMPLEX METHOD

Cycle and phase	Basis	Inverse	$-z$	$-w$	$\bar{b}_i$	$\bar{a}_{is}$
Cycle 0 (Phase I)	x_{n+1}	1			b_1	a_{1q}
	$\vdots$	$\ddots$			$\vdots$	$\vdots$
	x_{n+p}	1			b_p	a_{pq}
	$\vdots$	$\ddots$			$\vdots$	$\vdots$
	x_{n+m}	1			b_m	a_{mq}
	$-z$	$0 \cdots 0 \cdots 0$	1		0	c_q
	$-w$	$0 \cdots 0 \cdots 0$		1	$-w_0$	d_q
Cycle t (Phase I or II)†	x_1	$\alpha_{11} \cdots \alpha_{1s} \cdots \alpha_{1m}$			$\bar{b}_1$	$\bar{a}_{1s}$
	$\vdots$	$\vdots$			$\vdots$	
	x_r	$\alpha_{r1} \cdots \alpha_{rs} \cdots \alpha_{rm}$			$\bar{b}_r$	$\bar{a}_{rs}$
	$\vdots$	$\vdots$			$\vdots$	
	x_m	$\alpha_{m1} \cdots \alpha_{ms} \cdots \alpha_{mm}$			$\bar{b}_m$	$\bar{a}_{ms}$
	$-z$	$-\pi_1 \cdots -\pi_s \cdots -\pi_m$	1		$-\bar{z}$	$\bar{c}_s$
	$-w$	$-\sigma_1 \cdots -\sigma_s \cdots -\sigma_m$		1	$-\bar{w}$	$\bar{d}_s$
Cycle $t+1$ (Phase I or II)†	x_1	$\alpha_{11} - \bar{a}_{1s}\alpha'_{r1} \cdots \alpha_{1m} - \bar{a}_{1s}\alpha'_{rm}$			$\bar{b}_1 - \bar{a}_{1s}\bar{b}'_r$	
	$\vdots$	$\vdots$			$\vdots$	
	x_s	$\alpha'_{r1} \cdots \alpha'_{rm}$			$\bar{b}'_r$	
	$\vdots$	$\vdots$			$\vdots$	
	x_m	$\alpha_{m1} - \bar{a}_{ms}\alpha'_{r1} \cdots \alpha_{mm} - \bar{a}_{ms}\alpha'_{rm}$			$\bar{b}_m - \bar{a}_{ms}\bar{b}'_r$	
	$-z$	$-\pi'_1 \cdots -\pi'_m$	1		$-\bar{z}'$	
	$-w$	$-\sigma'_1 \cdots -\sigma'_m$		1	$-\bar{w}'$	

† For Phase II, delete the last row in the cycle involving w.

other hand, in Phase II only $\bar{\pi}_i$ and $\bar{c}_s$ are required, since $\bar{\sigma}_i$ and $\bar{d}_s$ are automatically eliminated at the end of Phase I.

Using the operation in Phase I as an example, the iterative procedure of the revised simplex method consists of the following steps:

1. After the inverse of basis and the simplex multipliers for Cycle t are determined, proceed to determine $\bar{d}_j$ from the relation in (4-38):

$$\bar{d}_j = d_j - \sum_{i=1}^{m} a_{ij}\sigma_i :$$

2. Determine s from the condition $\text{Min}\, \bar{d}_j = \bar{d}_s$.

3. If $\bar{d}_s < 0$, determine $\bar{a}_{is}$ according to (4-19)

$$\bar{a}_{is} = \sum_{k=1}^{m} \alpha_{ik} a_{ks} .$$

We can also compute from (4-20) and (4-38), respectively,

$$\bar{c}_s = c_s - \sum_{k=1}^{m} \pi_k a_{ks} ,$$

$$\bar{d}_s = d_s - \sum_{k=1}^{m} \sigma_k a_{ks} .$$

4. Determine r from the condition

$$\operatorname*{Min}_{\bar{a}_{is}>0} \frac{\bar{b}_i}{\bar{a}_{is}} = \frac{\bar{b}_r}{\bar{a}_{rs}} = \bar{b}'_r .$$

5. The entering variable is $x_s = \bar{b}'_r$, while the departing variable is $x_r = 0$. The new values of other basic variables ($i \neq r$) in Cycle $t + 1$ are obtained from (4-25) as follows:

$$x_i = \bar{b}_i - \bar{a}_{is}\bar{b}'_r = \bar{b}'_i .$$

Also, from (4-26) and (4-45), respectively, we have the new values of z and w.

$$-z = -\bar{z} - \bar{c}_s\bar{b}'_r ,$$

$$-w = -\bar{w} - \bar{d}_s\bar{b}'_r .$$

6. The new inverse for Cycle $t + 1$ may be obtained by computing its elements from (4-31) and (4-32):

$$\alpha'_{rk} = \frac{\alpha_{rk}}{\bar{a}_{rs}} \qquad \text{for } i = r ,$$

$$\alpha'_{ik} = \alpha_{ik} - \bar{a}_{is}\alpha'_{rk} \qquad \text{for } i \neq r .$$

7. The new simplex multipliers π'_i and σ'_i are computed from (4-33) and (4-44), respectively:

$$-\pi'_i = -\pi_i - \bar{c}_s\alpha'_{ri} ,$$

$$-\sigma'_i = -\sigma_i - \bar{d}_s\alpha'_{ri} .$$

8. Record the results obtained in steps 5, 6, and 7 for Cycle $t + 1$.

Example 4-9. Tabulate the solution of Example 4-5 by using both phases of the revised simplex method.

The results of computation for Phase I are already shown in Example 4-8, while those for Phase II are given in Example 4-5. Hence the complete solution may be as summarized in Tables 4-5 and 4-6.

Table 4-5

ORIGINAL DATA AND COST COEFFICIENTS FOR EXAMPLE 4-9

Basis	x_1	x_2	x_3	x_4	x_5	Value
x_5	5	10	-1		1	8
x_4	1	1		1		1
z	2	1				z
w	-5	-10	1			$w - 10$
I	$\bar{d}_1$	$\bar{d}_2$	$\bar{d}_3$	$\bar{d}_4$	$\bar{d}_5$	
0	-5	-10	1	0	0	
1	0	0	0	0	1	
II	$\bar{c}_1$	$\bar{c}_2$	$\bar{c}_3$	$\bar{c}_4$		
1	1.5	0	0.1	0		

Table 4-6

INVERSE AND SIMPLEX MULTIPLIERS FOR EXAMPLE 4-9

Basis	Inverse		$-z$	$-w$	$\bar{b}_i$	$\bar{a}_{is}$
x_5	1				8	$\boxed{10}$
x_4		1			1	1
$-z$			1		0	1
$-w$				1	-8	-10
0/I						$x_s = x_2$
x_2	0.1	0			0.8	
x_4	-0.1	1			0.2	
$-z$	-0.1	0	1		-0.8	
$-w$	1	0		1	0	
1/I–II						

Example 4-10. Solve the following linear programming problem by the revised simplex method.

$$
\begin{aligned}
5x_2 + 4x_3 + 4x_4 &= 5\,,\\
6x_1 + 4x_2 + 3x_3 + 2x_4 &= 3\,,\\
4x_1 + x_2 + 3x_3 + 4x_4 &= 2\,,\\
x_j \geq 0\,, \qquad j = 1, 2, 3, 4\,,&\\
8x_1 + 10x_2 + 12x_3 + 11x_4 &= z \quad \text{(Min)}\,.
\end{aligned}
$$

Let us introduce the artificial variables x_5, x_6, and x_7, one for each of the constraint equations. Thus the infeasibility form becomes

$$x_5 + x_6 + x_7 = w ,$$

from which we obtain

$$-10x_1 - 10x_2 - 10x_3 - 10x_4 = w - 10 \quad \text{(Min)} .$$

The detached coefficients for the problem are given in Table 4-7. Since the artificial variables form the initial basis, the inverse for the basis is a unit matrix and the simplex multipliers are zero, as indicated in the data for Cycle 0 (Phase I) in Table 4-8. The values of the basic variables are $\bar{b}_i = b_i$. Thus the basic variables, the inverse of the basis, and the simplex multipliers for the cycle are known, but column $\bar{a}_{is}$ is not known when the iteration begins.

The steps in the iterative procedure from Cycle 0 to Cycle 1 are carried out as follows:

1. Determine $\bar{d}_j$, which are identical to d_j in this case, since all $\sigma_i = 0$ in Cycle 0. The results are tabulated as relative cost coefficients for Cycle 0 in Table 4-7.

2. Since $\bar{d}_1 = \bar{d}_2 = \bar{d}_3 = \bar{d}_4 = -10$, we choose one of them at random, say $\bar{d}_s = \bar{d}_1$. Thus $x_s = x_1$ is recorded at the end of Cycle 0 in Table 4-8,

3. The column $\bar{a}_{is}$ of Cycle 0 in Table 4-8 then refers to $\bar{a}_{i1}$, which is computed from the data for a_{i1} in Table 4-7. Also, $\bar{c}_1$ and $\bar{d}_1$ in the $\bar{a}_{is}$ column are computed from the data for c_1 and d_1, respectively, in Table 4-7. Again, since the inverse of the basis is a unit matrix and the simplex

Table 4-7

ORIGINAL DATA AND COST COEFFICIENTS FOR EXAMPLE 4-10

Basis	x_1	x_2	x_3	x_4	x_5	x_6	x_7	Value
x_5	0	5	4	4	1			5
x_6	6	4	3	2		1		3
x_7	4	1	3	4			1	2
z	8	10	12	11				z
w	-10	-10	-10	-10				$w - 10$
I	$\bar{d}_1$	$\bar{d}_2$	$\bar{d}_3$	$\bar{d}_4$	$\bar{d}_5$	$\bar{d}_6$	$\bar{d}_7$	
0	-10	-10	-10	-10	0	0	0	
1	0	$-\frac{15}{2}$	$-\frac{5}{2}$	0	0	0	$\frac{5}{2}$	
2	0	0	-7	-12	0	3	-2	
3	$\frac{130}{7}$	0	$\frac{75}{7}$	0	0	$\frac{15}{7}$	$\frac{10}{7}$	

Table 4–8

INVERSE AND SIMPLEX MULTIPLIERS FOR EXAMPLE 4–10

Basis	Inverse			$-z$	$-w$	$\bar{b}_i$	$\bar{a}_{is}$
x_5	1					5	0
x_6		1				3	6
x_7			1			2	$\boxed{4}$
$-z$				1		0	8
$-w$					1	-10	-10
0/I							$x_s = x_1$
x_5	1		0			5	5
x_6		1	$-\frac{3}{2}$			0	$\boxed{\frac{5}{2}}$
x_1			$\frac{1}{4}$			$\frac{1}{2}$	$\frac{1}{4}$
$-z$			-2	1		-4	8
$-w$			$\frac{5}{2}$		1	-5	$-\frac{15}{2}$
1/I							$x_s = x_2$
x_5	1	-2	3			5	12
x_2		$\frac{2}{5}$	$-\frac{3}{5}$			0	$-\frac{8}{5}$
x_1		$-\frac{1}{10}$	$\frac{2}{5}$			$\frac{1}{2}$	$\boxed{\frac{7}{5}}$
$-z$		$-\frac{16}{5}$	$\frac{14}{5}$	1		-4	$\frac{79}{5}$
$-w$		3	-2		1	-5	-12
2/I							$x_s = x_4$
x_5	1	$-\frac{8}{7}$	$-\frac{3}{7}$			$\frac{5}{7}$	
x_2		$\frac{2}{7}$	$-\frac{1}{7}$			$\frac{4}{7}$	
x_4		$-\frac{1}{14}$	$\frac{2}{7}$			$\frac{5}{14}$	
$-z$		$-\frac{29}{14}$	$-\frac{12}{7}$	1		$-\frac{135}{14}$	
$-w$		$\frac{15}{7}$	$\frac{10}{7}$		1	$-\frac{5}{7}$	
3/I							

multipliers are zero in Cycle 0, we have

$$\bar{a}_{i1} = a_{i1}\,, \qquad \bar{c}_1 = c_1\,, \qquad \bar{d}_1 = d_1\,.$$

4. The ratio $\bar{b}_i/\bar{a}_{is}$ may be computed from the data now available for Cycle 0 in Table 4–8. Thus

$$\text{Min}\ \frac{\bar{b}_i}{\bar{a}_{is}} = \bar{b}'_r = \frac{1}{2}\,.$$

Since there is a tie between $i = 2$ and $i = 3$, we choose randomly $r = 3$. Then $\bar{a}_{rs} = \bar{a}_{31} = 4$, which is indicated by a block in Table 4–8. Hence x_7 becomes the departing variable.

5. With the entering variable $x_1 = \frac{1}{2}$ and the departing variable $x_7 = 0$, the new values of the remaining basic variables in Cycle 1 are found from (4–25), (4–26), and (4–45) to be

$$x_5 = 5\,, \qquad x_6 = 0\,, \qquad -z = -4\,, \qquad -w = -5\,,$$

6. The new inverse for Cycle 1 is determined from (4–31) and (4–32):

$$\begin{array}{llll} r = 3\,, & \alpha'_{31} = 0\,, & \alpha'_{32} = 0\,, & \alpha'_{33} = \frac{1}{4}\,, \\ i = 1\,, & \alpha'_{11} = 1\,, & \alpha'_{12} = 0\,, & \alpha'_{13} = 0\,, \\ i = 2\,, & \alpha'_{21} = 0\,, & \alpha'_{22} = 1\,, & \alpha'_{23} = -\frac{3}{2}\,. \end{array}$$

7. The new simplex multipliers are computed from (4–33) and (3–44) as follows:

$$\begin{array}{lll} -\pi_1 = 0\,, & -\pi_2 = 0\,, & -\pi_3 = -2\,, \\ -\sigma_1 = 0\,, & -\sigma_2 = 0\,, & -\sigma_3 = \frac{5}{2}\,. \end{array}$$

8. The results of steps 5, 6, and 7 are recorded in Cycle 1 of Table 4–8. Thus we have the basic variables, the inverse of the basis, and the simplex multipliers to start a new cycle all over again.

This problem terminates after three cycles in Phase I when Min $\bar{d}_j = \bar{d}_s > 0$ and Min $w \neq 0$. Hence, no solution is possible.

Example 4–11. Solve the following linear programming problem by the revised simplex method:

$$\begin{array}{rrrrrrrl} & 5x_2 + & 4x_3 + & 4x_4 - & 10x_5 & & & = 5 \\ 6x_1 + & 4x_2 + & 3x_3 + & 2x_4 & & - 10x_6 & & = 3 \\ 4x_1 + & x_2 + & 3x_3 + & 4x_4 & & & - 10x_7 & = 2 \end{array}$$

$$x_j \geq 0\,, \qquad j = 1, 2, \ldots, 7\,,$$

$$8x_1 + 10x_2 + 12x_3 + 11x_4 + 2x_5 + 2x_6 + 2x_7 = z \quad \text{(Min)}$$

Let us introduce the artificial variables x_8, x_9, and x_{10}, one for each of the constraint equations. Then the infeasibility form becomes

$$x_8 + x_9 + x_{10} = w\,,$$

from which we obtain

$$-10x_1 - 10x_2 - 10x_3 - 10x_4 + 10x_5 + 10x_6 + 10x_7 = w - 10 \quad \text{(Min)}\,.$$

Hence the detached coefficients for the problem are given in Table 4–9. The computed values of the relative cost coefficients at various cycles are also shown in Table 4–9, while the inverse and simplex multipliers

Table 4-9

ORIGINAL DATA AND COST COEFFICIENTS FOR EXAMPLE 4-11

Basis	x_1	x_2	x_3	x_4	x_5	x_6	x_7	x_8	x_9	x_{10}	Value
x_8	0	5	4	4	-10	0	0	1			5
x_9	6	4	3	2	0	-10	0		1		3
x_{10}	4	1	3	4	0	0	-10			1	2
z	8	10	12	11	2	2	2				z
w	-10	-10	-10	-10	10	10	10				$w-10$
I	$\bar{d}_1$	$\bar{d}_2$	$\bar{d}_3$	$\bar{d}_4$	$\bar{d}_5$	$\bar{d}_6$	$\bar{d}_7$	$\bar{d}_8$	$\bar{d}_9$	$\bar{d}_{10}$	
0	-10	-10	-10	-10	10	10	10	0	0	0	
1	0	$-\frac{15}{2}$	$-\frac{5}{2}$	0	0	0	-15	0	0	$\frac{5}{2}$	
2	0	-5	-4	-4	10	0	0	0	1	1	
3	0	0	-7	-12	0	-20	30	0	3	-2	
4	0	0	0	0	0	0	0	1	1	1	
II	$\bar{c}_1$	$\bar{c}_2$	$\bar{c}_3$	$\bar{c}_4$	$\bar{c}_5$	$\bar{c}_6$	$\bar{c}_7$				
4	0	0	$-\frac{11}{10}$	$-\frac{23}{5}$	19	0	25				
5	$\frac{15}{4}$	0	$\frac{65}{16}$	0	$\frac{175}{8}$	0	$\frac{85}{8}$				

for these cycles are tabulated in Table 4-10. It is interesting to note that in Table 4-10 the entering variable in Cycle 2 based on $\text{Min}\, d_j = d_s$ is x_7, but from Table 4-10, this variable is seen to enter the basis in Cycle 2 and leave the basis at the end of Cycle 3. In other words, if we had chosen x_2 instead of x_7 as the entering variable at Cycle 2, we would have saved the computation for one cycle. This illustrates the fact that the rule of selection $\text{Min}\, d_j = d_s$ does not guarantee a most efficient algorithm for either the simplex method or the revised simplex method. Furthermore, because of a tie in selecting a departing variable at Cycle 0, the basis in Cycle 1 is a degenerate case. It can be seen from Table 4-10 that the iterative procedure does not improve the value of w from Cycle 1 through Cycle 3 inclusive. However, this circling without improvement to w is finally broken up in Cycle 4, at which Phase I terminates. It takes one more cycle to complete the solution of the problem in Phase II.

4-7 PRIMAL-DUAL RELATIONSHIP

We shall now introduce the concept of *duality* in linear programming problems. Let us consider the minimization problem represented by

Table 4-10

INVERSE AND SIMPLEX MULTIPLIERS FOR EXAMPLE 4-11

Basis	Inverse			$-z$	$-w$	$\bar{b}_i$	$\bar{a}_{is}$
x_8	1					5	0
x_9		1				3	6
x_{10}			1			2	$\boxed{4}$
$-z$				1		0	8
$-w$					1	-10	10
0/I							$x_s = x_1$
x_8	1		0			5	0
x_9		1	$-\frac{3}{2}$			0	$\boxed{15}$
x_1			$\frac{1}{4}$			$\frac{1}{2}$	$-\frac{5}{2}$
$-z$			-2	1		-4	22
$-w$			$\frac{5}{2}$		1	-5	-15
1/I							$x_s = x_7$
x_8	1	0	0			5	5
x_7		$\frac{1}{15}$	$-\frac{1}{10}$			0	$\boxed{\frac{1}{6}}$
x_1		$\frac{1}{6}$	0			$\frac{1}{2}$	$\frac{2}{3}$
$-z$		$-\frac{22}{15}$	$\frac{1}{5}$	1		-4	$\frac{13}{3}$
$-w$		1	1		1	-5	-5
2/I							$x_s = x_2$
x_8	1	-2	3			5	$\boxed{20}$
x_2		$\frac{2}{5}$	$-\frac{3}{5}$			0	-4
x_1		$-\frac{1}{10}$	$\frac{2}{5}$			$\frac{1}{2}$	1
$-z$		$-\frac{16}{5}$	$\frac{14}{5}$	1		-4	34
$-w$		3	-2		1	-5	-20
3/I							$x_s = x_6$
x_6	$\frac{1}{20}$	$-\frac{1}{10}$	$\frac{3}{20}$			$\frac{1}{4}$	$\frac{3}{5}$
x_2	$\frac{1}{5}$	0	0			1	$\frac{4}{5}$
x_1	$-\frac{1}{20}$	0	$\frac{1}{4}$			$\frac{1}{4}$	$\boxed{\frac{4}{5}}$
$-z$	$-\frac{17}{10}$	$\frac{1}{5}$	$-\frac{23}{10}$	1		$-\frac{25}{2}$	$-\frac{6}{5}$
$-w$	1	1	1		1	0	0
4/I–II							$x_s = x_4$
x_6	$\frac{7}{80}$	$-\frac{1}{10}$	$-\frac{3}{80}$			$\frac{1}{16}$	
x_2	$\frac{1}{4}$	0	$-\frac{1}{4}$			$\frac{3}{4}$	
x_4	$-\frac{1}{16}$	0	$\frac{5}{16}$			$\frac{5}{16}$	
$-z$	$-\frac{159}{80}$	$\frac{1}{5}$	$-\frac{69}{80}$	1		$-\frac{177}{16}$	
5/II							

(3-4), (3-5), and (3-6) in the inequality form:

$$x_j \geq 0\,, \qquad j = 1, 2, \ldots, q\,; \tag{4-46}$$

$$\begin{aligned} a_{11}x_1 + a_{12}x_2 + \cdots + a_{1q}x_q &\geq b_1\,, \\ a_{21}x_1 + a_{22}x_2 + \cdots + a_{2q}x_q &\geq b_2\,, \\ &\vdots \\ a_{p1}x_1 + a_{p2}x_2 + \cdots + a_{pq}x_q &\geq b_p\,; \end{aligned} \tag{4-47}$$

$$c_1x_1 + c_2x_2 + \cdots + c_qx_q = \text{Min } z\,. \tag{4-48}$$

If this problem is called a *primal problem*, which has a set of *primal variables* $x_j \geq 0$ $(j = 1, 2, \ldots, q)$, we can construct a new linear programming problem called the *dual problem*, which has a set of dual variables $y_i \geq 0$ $(i = 1, 2, \ldots, p)$, and has an inequality form as follows:

$$y_i \geq 0\,, \qquad i = 1, 2, \ldots, p\,; \tag{4-49}$$

$$\begin{aligned} a_{11}y_1 + a_{21}y_2 + \cdots + a_{p1}y_p &\leq c_1\,, \\ a_{12}y_1 + a_{22}y_2 + \cdots + a_{p2}y_p &\leq c_2\,, \\ &\vdots \\ a_{1q}y_1 + a_{2q}y_2 + \cdots + a_{pq}y_p &\leq c_q\,; \end{aligned} \tag{4-50}$$

$$b_1y_1 + b_2y_2 + \cdots + b_py_p = \text{Max } A\,. \tag{4-51}$$

By definition, a dual problem is constructed from a primal problem as follows:

1. Change the primal variables x_j in (4-46) to dual variable y_i in (4-49), noting that i refers to the constraint number in the primal problem and that j refers to the constraint number in the dual problem.

2. Transpose the rows and columns of the coefficients a_{ij} in (4-47) to a_{ji} in (4-50).

3. Interchange b_i in (4-47) with c_j in (4-48) to form the new coefficients c_j in (4-50) and b_i in (4-51), respectively.

4. Reverse the greater-than or equal-to signs of the inequalities in (4-47) to less-than or equal-to signs in (4-50).

5. Change the minimization of z in (4-48) to the maximization of A in (4-51).

It can be shown that if the maximization problem represented by (4-49), (4-50), and (4-51) is regarded as a *primal problem*, then by definition the *dual problem* proves to be the minimization problem represented by (4-46), (4-47), and (4-48). Let us multiply (4-50) and (4-51) through by -1 to obtain

$$\begin{aligned} -a_{11}y_1 - a_{21}y_2 - \cdots - a_{p1}y_p &\geq -c_1\,, \\ -a_{12}y_1 - a_{22}y_2 - \cdots - a_{p2}y_p &\geq -c_2\,, \\ &\vdots \\ -a_{1p}y_1 - a_{2p}y_2 - \cdots - a_{pq}y_p &\geq -c_q\,; \end{aligned} \tag{4-52}$$

$$-b_1y_1 - b_2y_2 - \cdots - b_py_r = \text{Min}\,(-A)\,. \tag{4-53}$$

By regarding $-a_{ji} = a'_{ji}$, $-c_j = c'_j$, $-b_i = b'_i$, and $-A = z'$, then (4-52) and (4-53), together with $y_i \geq 0$ $(i = 1, 2, \ldots, p)$ in (4-49), represent a minimization problem which has the form of a primal problem. We now construct a dual of the this problem in accordance with the definition. Let x_j $(j = 1, 2, \ldots, q)$ be the dual variable. By transposing the rows and columns of a_{ji} in (4-52), reversing the signs of inequalities in (4-52), interchanging $-c_j$ in (4-52) with $-b_i$ in (4-53), and maximizing $-z$ instead of minimizing $-A$ in (4-53), then, for $x_j \geq 0$ $(j = 1, 2, \ldots, q)$, we have

$$\begin{aligned} -a_{11}x_1 - a_{12}x_2 - \cdots - a_{1q}x_q &\leq -b_1\,, \\ -a_{21}x_1 - a_{22}x_2 - \cdots - a_{2q}x_q &\leq -b_2\,, \\ &\vdots \\ -a_{p1}x_1 - a_{p2}x_2 - \cdots - a_{pq}x_q &\leq -b_p\,; \end{aligned} \tag{4-54}$$

$$-c_1x_1 - c_2x_2 - \cdots - c_qx_q = \text{Max}\,(-z)\,. \tag{4-55}$$

Again, by multiplying both (4-54) and (4-55) by -1, the results will be identical to (4-47) and (4-48), respectively. Hence, the *dual* of the *dual problem* turns out to be the *primal problem*. In other words, it does not matter whether the minimization or the maximization problem is identified as the primal problem since one is the dual of the other.

We shall now consider some important duality relations which may be derived from the definition of primal and dual problems. Let each of the inequalities in (4-47) be multiplied by $y_i \geq 0$, which corresponds to the ith inequality $(i = 1, 2, \ldots, p)$. Then the sum of the products becomes

$$\begin{aligned} \sum_{i=1}^{p} (a_{i1}y_i)x_1 + \sum_{i=1}^{p} (a_{i2}y_i)x_2 + \cdots \\ + \sum_{i=1}^{p} (a_{iq}y_i)x_q \geq \sum_{i=1}^{p} b_iy_i\,. \end{aligned}$$

By multiplying each of the inequalities in (4-50) by $x_j \geq 0$ $(j = 1, 2, \ldots, q)$, we obtain the sum of the products:

$$\sum_{i=1}^{p} (a_{i1}y_i)x_1 + \sum_{i=1}^{p} (a_{i2}y_i)x_2 + \cdots + \sum_{i=1}^{p} (a_{iq}y_i)x_q \leq \sum_{j=1}^{q} c_j x_j \, .$$

Hence

$$\sum_{i=1}^{p} b_i y_i \leq \sum_{j=1}^{q} c_j x_j \, .$$

In view of (4–48) and (4–51), we can conclude that regardless of the values of $y_i \geq 0$ and $x_j \geq 0$,

$$A \leq z \, . \tag{4–56}$$

That is, any feasible solution of the primal is greater than or equal to any feasible solution of the dual. *If a feasible solution to the primal exists but z approaches minus infinity, then no feasible solution to the dual exists, since $A \leq z$; similarly, if a feasible solution to the dual exists, but A approaches infinity, then no feasible solution to the primal exists.* In either case, there is no finite optimum for the primal or the dual. If feasible solutions to both the primal and the dual exist, the optimal solution for the primal, $\min z = z^*$, cannot be less than any feasible solution A of the dual, and the optimal solution for the dual $\max A = A^*$, must be less than any feasible solution z of the primal. If there exists a feasible solution A whose value is equal to z^*, or a feasible z whose value is equal to A^*, it can be concluded that an optimal solution exists for both problems, and that the optimal value is $A^* = z^*$. The existence of such solutions can be demonstrated by considering the simplex multipliers to be associated with the optimal solution if it exists. For the primal problem, we obtain from (4–16),

$$\min z = \bar{z} = \sum_{i=1}^{m} \pi_i b_i \, .$$

Replacing π_i by y_i, we have $\min z = A$. Hence

$$\max A = \min z \, . \tag{4–57}$$

This property is known as the *duality theorem*.

An example of primal-dual relationship in linear programming has been given in Examples 4–1 and 4–2.

4–8 COMPLEMENTARY SLACKNESS

The primal and dual problems in inequalities may also be expressed in the standard form. For the case of the primal, let $x_{q+1}, x_{q+2}, \ldots, x_{q+p}$ be nonnegative slack variables. Then from (4-47),

$$\begin{aligned} a_{11}x_1 + a_{12}x_2 + \cdots + a_{1q}x_q - x_{q+1} \qquad\qquad &= b_1 , \\ a_{21}x_1 + a_{22}x_2 + \cdots + a_{2q}x_q \qquad - x_{q+2} \qquad &= b_2 , \\ \vdots \qquad\qquad\qquad\qquad & \\ a_{p1}x_1 + a_{p2}x_2 + \cdots + a_{pq}x_q \qquad\qquad - x_{q+p} &= b_p ; \end{aligned} \tag{4-58}$$

$$\begin{aligned} x_j &\geq 0 , \quad j = 1, 2, \ldots, q ; \\ x_{q+i} &\geq 0 , \quad i = 1, 2, \ldots, p ; \end{aligned} \tag{4-59}$$

$$c_1x_1 + c_2x_2 + \cdots + c_qx_q = \text{Min } z . \tag{4-60}$$

Similarly, for the case of the dual, let $y_{p+1}, y_{p+2}, \ldots, y_{p+q}$ be nonnegative slack variables. Then from (4-50),

$$\begin{aligned} a_{11}y_1 + a_{21}y_2 + \cdots + a_{p1}y_p + y_{p+1} \qquad\qquad &= c_1 , \\ a_{12}y_1 + a_{22}y_2 + \cdots + a_{p2}y_p \qquad + y_{p+2} \qquad &= c_2 , \\ \vdots \qquad\qquad\qquad\qquad & \\ a_{1q}y_1 + a_{2q}y_2 + \cdots + a_{pq}y_p \qquad\qquad + y_{p+q} &= c_q ; \end{aligned} \tag{4-61}$$

$$\begin{aligned} y_i &\geq 0 , \quad i = 1, 2, \ldots, p ; \\ y_{p+j} &\geq 0 , \quad j = 1, 2, \ldots, q ; \end{aligned} \tag{4-62}$$

$$b_1y_1 + b_2y_2 + \cdots + b_py_p = \text{Max } A . \tag{4-63}$$

Again, multiplying the ith equation in (4-58) by y_i for each of the equations ($i = 1, 2, \ldots, p$), and subtracting the sum of these products from (4-60), we get

$$\begin{aligned} \left(c_1 - \sum_{i=1}^{p} a_{i1}y_i\right) x_1 + \left(c_2 - \sum_{i=1}^{p} a_{i2}y_i\right) x_2 + \cdots \\ + \left(c_q - \sum_{i=1}^{p} a_{iq}y_i\right) x_q + y_1x_{q+1} + y_2x_{q+2} + \cdots \\ + y_px_{q+p} = z - \sum_{i=1}^{p} b_iy_i . \end{aligned} \tag{4-64}$$

In view of (4-61), we have

$$\begin{aligned} (y_{p+1}x_1 + y_{p+2}x_2 + \cdots + y_{p+q}x_q) \\ + (y_1x_{q+1} + y_2x_{q+2} + \cdots + y_qx_{q+p}) = z - A . \end{aligned} \tag{4-65}$$

Since all slack variables in the primal and dual are nonnegative, it is seen that $z - A \geq 0$. If an optimal solution exists to both the primal and dual, $\max A = \min z$. Let $x_j = x_j^* \geq 0$ and $\min z = z^*$ be the values associated with the optimal solution to the primal, and $y_i = y_i^* \geq 0$ and $\max A = A^*$

be the values associated with the optimal solution to the dual. Then

$$(y^*_{p+1}x^*_1 + y^*_{p+2}x^*_2 + \cdots + y^*_{p+q}x^*_q) + (y^*_1x^*_{q+1} + y^*_2x^*_{q+2} + \cdots + y^*_qx^*_{q+p}) = 0\,. \qquad (4\text{–}66)$$

In other words, each product term in (4–66) must be zero. For

$$y^*_{p+j}x^*_j = 0\,, \qquad j = 1, 2, \ldots, q\,,$$

we have

$$x^*_j = 0 \quad \text{if} \quad y^*_{p+j} > 0 \qquad \text{and} \qquad y^*_{p+j} = 0 \quad \text{if} \quad x^*_j > 0\,. \qquad (4\text{–}67)$$

Similarly, for

$$y^*_ix^*_{q+i} = 0 \qquad i = 1, 2, \ldots, p\,,$$

we have

$$y^*_i = 0 \quad \text{if} \quad x^*_{q+i} > 0 \qquad \text{and} \qquad x^*_{q+i} = 0 \quad \text{if} \quad y^*_i > 0\,. \qquad (4\text{–}68)$$

Thus for an optimal solution to both the primal and dual, if a slack variable occurs in the kth constraint of either system of equations in the standard form, then the kth variable of its dual vanishes; furthermore, if the kth variable is positive in either system, the kth constraint of its dual is an equation. This is known as the *complementary slackness theorem.*

In examining the equation in (4–66), we conclude that the restrictions on x^*_j and y^*_i can be relaxed if $y^*_{p+j} = 0$ and $x^*_{q+i} = 0$, respectively. For $y^*_{p+j} = 0$, x^*_j is unrestricted in sign; and for $x^*_{q+i} = 0$, y^*_i is unrestricted in sign. In other words, if the kth relation in either system is an equation,

Table 4–11

SUMMARY OF PRIMAL-DUAL RELATIONSHIPS

Item	Primal	Dual
1. Constraints	$i = 1, 2, \ldots, p$	$j = 1, 2, \ldots, q$
2. Variables	x_j and x_{q+i}	x_i and x_{p+j}
3. Structural coefficients	a_{ij}	a_{ji}
4. Stipulations	b_i	c_j
5. Cost coefficients	c_j	b_i
6. Objective	Min z	Max A
7. Relation between ith constraint in primal and ith variable in dual:		
a) inequality constraint	ith inequality ($\geq$)	$y_i \geq 0$
b) equation constraint	ith equation ($=$)	$-\infty < y_i < +\infty$
8. Relation between jth variable in primal and jth constraint in dual:		
a) nonnegative variable	$x_j \geq 0$	jth inequality ($\leq$)
b) unrestricted variable	$-\infty < x_j < +\infty$	jth equation ($=$)

the kth variable in its dual is unrestricted in sign. Thus a mixed system of inequality and equation constraints in the primal results in a mixed system of nonnegative and restricted variables in the dual, and conversely. The primal-dual relationship in the most general form is summarized in Table 4-11.

Example 4-12. Construct the dual of the following linear problem:

$$x_1 \geq 0\,, \qquad x_2 \geq 0\,, \qquad x_3 \geq 0\,,$$

$$\begin{aligned} x_1 + x_2 - x_3 &\geq 1\,, \\ -2x_1 \qquad + x_3 &\geq 2\,, \end{aligned}$$

$$2x_1 + 9x_2 + 5x_3 = \text{Min } z\,.$$

Since all constraints are inequalities, the dual problem may be constructed according to the definition as follows:

$$y_1 \geq 0\,, \qquad y_2 \geq 0\,,$$

$$\begin{aligned} y_1 - 2y_2 &\leq 2\,, \\ y_1 \qquad &\leq 9\,, \\ -y_1 + y_2 &\leq 5\,, \end{aligned}$$

$$y_1 + 2y_2 = \text{Max } A\,.$$

However, if the primal problem is expressed in the standard form, we can obtain the same dual problem by the use of the complementary slackness theorem. Consider

$$x_j \geq 0\,, \qquad j = 1, 2, 3, 4, 5\,,$$

$$\begin{aligned} x_1 + x_2 - x_3 - x_4 \qquad &= 1\,, \\ -2x_1 \qquad + x_3 \qquad - x_5 &= 2\,, \end{aligned}$$

$$2x_1 + 9x_2 + 5x_3 = \text{Min } z\,.$$

Then the dual problem is seen to be

$$-\infty \leq y_1 \leq +\infty\,, \qquad -\infty \leq y_2 \leq +\infty\,,$$

$$\begin{aligned} y_1 - 2y_2 &\leq 2\,, \\ y_1 \qquad &\leq 9\,, \\ -y_1 + y_2 &\leq 5\,, \\ -y_1 \qquad &\leq 0\,, \\ - y_2 &\leq 0\,, \end{aligned}$$

$$y_1 + 2y_2 = \text{Max } A\,.$$

Note that although the dual variables y_1 and y_2 appear to be unconstrained because the constraint relations in the primal problem are equations, they are in reality constrained to be nonnegative by the last two constraint relations in the dual problem, i.e.,

$$-y_1 \leq 0\,, \qquad y_1 \geq 0\,,$$
$$-y_2 \leq 0\,, \qquad y_2 \geq 0\,.$$

Thus when some or all constraint relations in the primal problem are equations, we can construct the dual problem by use of the complementary slackness theorem without first expressing the primal problem in the inequality form.

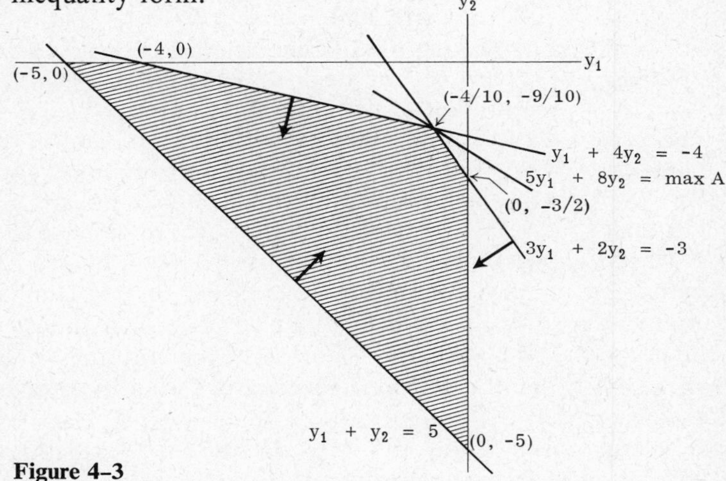

Figure 4-3

Example 4-13. Construct the dual of the following linear programming problem:

$$x_1 \geq 0\,, \qquad x_2 \geq 0\,, \qquad x_3 \geq 0\,,$$

$$3x_1 + x_2 - x_3 = 5\,,$$
$$2x_1 + 4x_2 - x_3 = 8\,;$$

$$-3x_1 - 4x_2 + 5x_3 = \text{Min } z\,.$$

The dual problem is seen to be

$$-\infty \leq y_1 \leq +\infty\,, \qquad -\infty \leq y_1 \leq +\infty\,,$$

$$3y_1 + 2y_2 \leq -3\,,$$
$$y_1 + 4y_2 \leq -4\,,$$
$$-y_1 - y_2 \leq 5\,;$$

$$5y_1 + 8y_2 = \text{Max } A\,.$$

In this case, the dual variables y_1 and y_2 are unconstrained. The graphical solution of the dual problem in Fig. 4-3 indeed indicates that both y_1 and y_2 are negative when z reaches the maximum value.

4-9 DUAL SIMPLEX METHOD

In the discussion of the simplex method, it has been stated that the stipulations b_i $(i = 1, 2, \ldots, m)$ in (4-1) must be nonnegative or made nonnegative before the introduction of artificial variables and/or slack variables in order to obtain a feasible solution. While this requirement of nonnegative b_i is not a limitation on the application of the simplex method, it does increase the computational steps. However, this handicap may be removed by making use of the complementary relationship between the bases of the primal and dual problems observed by Lemke. If $x_1, x_2, \ldots, x_m$ represent a set of basic variables which provides a basic, but not necessarily feasible, solution to the primal problem, it can be shown that $\pi_1, \pi_2, \ldots, \pi_m$ and $\bar{c}_{m+1}, \bar{c}_{m+2}, \ldots, \bar{c}_n$ constitute a basic feasible solution for the dual problem where π_i $(i = 1, 2, \ldots, m)$ are simplex multipliers and $\bar{c}_j$ $(j = m + 1, m + 2, \ldots, n)$ are modified cost coefficients of the primal problem, as indicated in (4-6). Thus the simplex algorithm may be modified for the direct solution of problems having the requirement $c_j \geq 0$ instead of $b_i \geq 0$ without actually constructing a dual. If it also happens that $b_i \geq 0$ for all i, the solution is optimal as well as feasible. This modified algorithm is called the *dual simplex method*.

Without going through the complete derivation, the dual simplex method for the problem represented by (4-1), (4-2), and (4-3) is simply stated as follows:

1. Express the equations in canonical form, say with $x_1, x_2, \ldots, x_m$ as basic variables, in which not all $\bar{b}_i$ $(i = 1, 2, \ldots, m)$ are nonnegative. If all b_i are nonnegative, the existing solution is optimal. If one or more $\bar{b}_i$ are negative, select $\bar{b}_r$ in the rth row which is the smallest of $\bar{b}_i$:

$$\bar{b}_r = \text{Min}\, \bar{b}_i < 0\,. \tag{4-69}$$

2. Examine coefficients $\bar{a}_{rj}$ in row r. If all $\bar{a}_{rj}$ are nonnegative, then the solution is unbounded. If one or more $\bar{a}_{rj}$ are negative, select the smallest ratio of $\bar{c}_j/(-\bar{a}_{rj})$ for all negative $\bar{a}_{rj}$. Then the coefficient $\bar{a}_{rs}$ in the sth column, which contains the smallest ratio, is the pivotal element in changing the basis, i.e.,

$$\frac{\bar{c}_s}{-\bar{a}_{rs}} = \underset{\bar{a}_{rj}<0}{\text{Min}} \frac{\bar{c}_j}{-\bar{a}_{rj}}\,. \tag{4-70}$$

In case of a tie, the choice is arbitrary.

Table 4–12

SOLUTION OF EXAMPLE 4–14

Basis	x_1	x_2	x_3	x_4	x_5	Value
x_3	-1	-1	1			-1
x_4	-2	$\boxed{-4}$		1		-3
x_5	3	7			1	6
	4	5				z
0/II		↑		↓		
x_3	$\boxed{-\frac{1}{2}}$		1	$-\frac{1}{4}$		$-\frac{1}{4}$
x_2	$\frac{1}{2}$	1		$-\frac{1}{4}$		$\frac{3}{4}$
x_5	$-\frac{1}{2}$			$\frac{7}{4}$	1	$\frac{3}{4}$
	$\frac{3}{2}$			$\frac{5}{4}$		$z - \frac{15}{4}$
1/II	↑		↓			
x_1	1		-2	$\frac{1}{2}$		$\frac{1}{2}$
x_2		1	1	$-\frac{1}{2}$		$\frac{1}{2}$
x_5			-1	2		1
			3	$\frac{1}{2}$	1	$z - \frac{9}{2}$
2/II						

3. Using $\bar{a}_{rs}$ as a pivotal element, obtain a new basic feasible solution by having x_s as an entering variable and x_r as a departing variable.

4. The procedure is repeated as many times as needed until all $\bar{b}_i$ in the new solution become positive. Then, Min $(z - \bar{z}) = 0$, or

$$\min z = \bar{z} . \tag{4–71}$$

Example 4–14. Solve the linear programming problem in Example 4–3 by the dual simplex method.

The solution of a linear programming problem by the dual simplex method can also be carried out in the simplex tableau, since the steps in this procedure are very similar to those in the standard simplex method. The solution for this particular example is shown in Table 4–12.

The coefficients for this problem are first tabulated for Cycle 0. Since the equations are already expressed in canonical form, we need not introduce any artificial variables and use Phase I. In examining b_i in Cycle 0, we select $b_r = -3$ in the second row, which is the smallest b_i and is negative. Since both $\bar{a}_{rj}$ in this row are negative, we compute the ratios $4/[-(-2)] = 2$ and $5/[-(-4)] = 5/4$. Since the latter ratio is the smallest of the pair, select a_{rs} in the second column as the pivotal element. Thus x_2 becomes the entering variable and x_4 is the departing variable. The same procedure is repeated for Cycle 1 and Cycle 2, until at the end of Cycle 2, all $\bar{b}_i \geq 0$. Then from the objective form $z - \frac{9}{2} = 0$ we obtain $\min z = \frac{9}{2}$.

If artificial variables are needed in order to get an initial solution in canonical form, the problem can be solved in the same manner by going through Phase I as well as Phase II, except that the artificial variables can be discarded once they depart from the basis in Phase I. However, the artificial variables are seldom needed in the dual simplex method since all slack variables can be used as basic variables regardless of the signs preceding them. For example, consider the set of constraint equations

$$\begin{aligned} x_1 + x_2 - x_3 &= 1\,, \\ 2x_1 + 4x_2 - x_4 &= 3\,, \\ 3x_1 + 7x_2 + x_5 &= 6\,. \end{aligned}$$

If the standard simplex method is used, we must introduce two artificial variables in order to get an initial solution in canonical form. Using the dual simplex method, however, we can simply multiply the first two equations by -1, since b_i can be negative. Thus the set of equations become

$$\begin{aligned} -x_1 - x_2 + x_3 &= -1\,, \\ -2x_1 - 4x_2 + x_4 &= -3\,, \\ 3x_1 + 7x_2 + x_5 &= 6\,. \end{aligned}$$

This is the set of equations used for Cycle 0 in Table 4-12.

4-10 SENSITIVITY ANALYSIS

In the formulation of practical problems, the coefficients a_{ij}, b_i and/or c_j are selected judiciously to reflect the real situation, but they seldom can be exact. Therefore we are interested not only in the optimal solution of a linear programming problem, but also how sensitive this solution is to variations in the coefficients. We often want to determine the effects of these variations without reworking the problems completely many times.

Referring again to the linear programming problem in the standard form represented by (4-1), (4-2), and (4-3), we find that the sensitivity analysis can be facilitated by the use of the inverse of basis x_i in (4-5), the relations of the modified cost coefficients $\bar{c}_j$ in (4-6) and (4-9), the expression of $\bar{z}$ in (4-7), the simplex multipliers in (4-11), and, in general, the operations in the revised simplex method. However, we shall confine our discussion to the study of the effects of only those changes brought about when the optimal basis is not changed. Let the original basis be the set of variables $x_1, x_2, \ldots, x_m$. Then the original values of the basic variables are given by (4-5) for $x_j = x_i$:

$$x_i = \sum_{k=1}^{m} \alpha_{ik} b_k = \bar{b}_i\,, \qquad i = 1, 2, \ldots, m\,. \tag{4-72}$$

The original optimal value of z is given by (4-7),

$$z = \bar{z} = \sum_{i=1}^{m} b_i \pi_i\,. \tag{4-73}$$

Also, from (4-9) and (4-6), respectively,

$$\bar{c}_j = c_j - \sum_{i=1}^{m} a_{ij} \pi_i = 0\,, \qquad j = 1, 2, \ldots, m\,; \tag{4-74}$$

$$\bar{c}_j = c_j - \sum_{i=1}^{m} a_{ij} \pi_i \geq 0\,, \qquad j = m+1, m+2, \ldots, n\,; \tag{4-75}$$

and from (4-11),

$$\pi_i = \sum_{k=1}^{m} \alpha_{ki} c_k\,. \tag{4-76}$$

Let us consider first the changes in coefficients a_{ij}. If the changes are confined to the a_{ij}-coefficients of the nonbasic variables in the original system, that is $j = m + 1, m + 2, \ldots, n$, we can see by examining the inequalities in (4-75) whether the changes will affect the original system. If the relations in (4-75) hold for the new coefficients a_{ij}, then the changes will affect neither the original optimal basis, nor the original optimal value of z; otherwise the original optimal basis will be changed. The changes in the a_{ij}-coefficients of each basic variable in the original optimal basis must be examined separately in light of the revised simplex method. For example, if the a_{is}-coefficients for x_s in the original optimal basis are changed, the $\bar{a}_{is}$ for the new a_{is}-coefficients can be determined from (4-19), and the reoptimization can be carried out by the use of the revised simplex procedure described in Section 4-6. The details of computation will not be described here.

The changes in coefficients b_i affect the values of x_i in the original optimal basis. Let Δb_i be the amounts of changes in b_i, and let Δx_i be the corresponding changes in the values of the basic variables. Then from (4-72), we can deduce that

$$x_i + \Delta x_i = \sum_{k=1}^{m} \alpha_{ik}(b_k + \Delta b_k)\,, \qquad i = 1, 2, \ldots, m\,,$$

or

$$\Delta x_i = \sum_{k=1}^{m} \alpha_{ik} \Delta b_k\,, \qquad i = 1, 2, \ldots, m\,. \tag{4-77}$$

The original optimal basis will remain optimal provided that

$$x_i + \Delta x_i \geq 0 .$$

The changes Δb_i must therefore satisfy the conditions

$$x_i + \sum_{k=1}^{m} \alpha_{ik} \, \Delta b_k \geq 0 , \qquad i = 1, 2, \ldots, m , \tag{4–78}$$

if the original optimal basis is to be maintained. The corresponding change Δz for the value of z can be deduced from (4–73) as follows:

$$\Delta z = \sum_{i=1}^{m} \pi_i \, \Delta b_i . \tag{4–79}$$

The effects of changes in coefficients c_j may be examined if we first substitute the values of π_i from (4–76) into (4–75):

$$\bar{c}_j = c_j - \sum_{i=1}^{m} a_{ij} \sum_{k=1}^{m} \alpha_{ki} c_k \geq 0 , \qquad j = m + 1, m + 2, \ldots, n .$$

By reversing the order of summation, we obtain

$$\bar{c}_j = c_j - \sum_{k=1}^{m} c_k \sum_{i=1}^{m} a_{ij} \alpha_{ki} \geq 0 , \qquad j = m + 1, m + 2, \ldots, n .$$

Let Δc_j be the amounts of changes in c_j. We can then deduce that

$$\Delta \bar{c}_j = \Delta c_j - \sum_{k=1}^{m} \Delta c_k \sum_{i=1}^{m} a_{ij} \alpha_{ki} \geq 0 , \qquad j = m + 1, m + 2, \ldots, n .$$

Hence the original optimal basis will remain provided that $\bar{c}_j + \Delta \bar{c}_j \geq 0$. Therefore the changes Δc_j must satisfy the conditions

$$\bar{c}_j + \Delta c_j - \sum_{k=1}^{m} \Delta c_k \sum_{i=1}^{m} a_{ij} \alpha_{ki} \geq 0 , \qquad j = m + 1, m + 2, \ldots, n , \tag{4–80}$$

if the original basis is to be maintained. The corresponding change Δz in the optimal value of z can be deduced from

$$z = \sum_{j=1}^{m} c_j x_j ,$$

such that

$$\Delta z = \sum_{j=1}^{m} x_j \, \Delta c_j , \tag{4–81}$$

in which x_j ($j = 1, 2, \ldots, m$) are the original basic variables. Note that if changes in c_j are made only in coefficients associated with the nonbasic variables in the original optimal basis, the relations in (4–80) reduce to

$$\bar{c}_j + \Delta c_j \geq 0 \qquad j = m + 1, m + 2, \ldots, n ,$$

and the change Δz in (4–81) becomes zero.

Example 4–15. Solve Example 4–4 on the basis of changing the b_i-coefficients in Example 3–1.

Example 3–1 can be expressed in the standard form as follows:

$$\begin{aligned} x_1 + x_2 - x_3 \quad\quad &= 1\,, \\ 2x_1 + 4x_2 \quad - x_4 \quad &= 3\,, \\ 3x_1 + 7x_2 \quad\quad + x_5 &= 6\,, \\ x_j \geq 0\,, \qquad j = 1, 2, 3, 4, 5\,; \\ 4x_1 + 5x_2 = \text{Min } z\,. \end{aligned}$$

The solution of this problem by the dual simplex method has been given in Example 4–14 (Table 4–12), in which $x_1 = \frac{1}{2}$, $x_2 = \frac{1}{2}$, $x_5 = 1$, $x_3 = x_4 = 0$, and $\min z = \frac{9}{2}$. Thus the original optimal basis consists of x_1, x_2, and x_5. The changes in b_i for $i = 1, 2, 5$ are indicated by

$$\Delta b_1 = 1 - 1 = 0\,, \qquad \Delta b_2 = 2.5 - 3 = -0.5\,,$$

and

$$\Delta b_5 = 5 - 6 = -1\,.$$

For the original optimal basis, we have for $i = 1, 2, 5$ and $j = 1, 2, 5$,

$$[a_{ij}] = \begin{bmatrix} 1 & 1 & 0 \\ 2 & 4 & 0 \\ 3 & 7 & 1 \end{bmatrix}, \qquad [\alpha_{ij}] = [a_{ij}]^{-1} = \begin{bmatrix} 2 & -0.5 & 0 \\ -1 & 0.5 & 0 \\ 1 & -2 & 1 \end{bmatrix}.$$

Thus the relations in (4–78) become

$$\begin{aligned} i = 1\,, \quad & x_1 + (\alpha_{11}\,\Delta b_1 + \alpha_{12}\,\Delta b_2 + \alpha_{15}\,\Delta b_5) \\ & = 0.5 + [(2)(0) + (-0.5)(-0.5) + (0)(-1)] = 0.75 > 0\,; \\ i = 2\,, \quad & x_2 + (\alpha_{21}\,\Delta b_1 + \alpha_{22}\,\Delta b_2 + \alpha_{25}\,\Delta b_5) \\ & = 0.5 + [(-1)(0) + (0.5)(-0.5) + (0)(-1)] = 0.25 > 0\,; \\ i = 5\,, \quad & x_5 + (\alpha_{51}\,\Delta b_1 + \alpha_{52}\,\Delta b_2 + \alpha_{55}\,\Delta b_5) \\ & = 1 + [(1)(0) + (-2)(-0.5) + (1)(-1)] = 1 > 0\,. \end{aligned}$$

Since all these conditions are satisfied, the original optimal basis remains. From (4–76), we have

$$\begin{aligned} \pi_1 &= \alpha_{11}c_1 + \alpha_{21}c_2 + \alpha_{51}c_5 \\ &= (2)(4) + (-1)(5) + (1)(0) = 3\,, \\ \pi_2 &= \alpha_{12}c_1 + \alpha_{22}c_2 + \alpha_{52}c_5 \\ &= (-0.5)(4) + (0.5)(5) + (-2)(0) = 0.5\,, \\ \pi_5 &= \alpha_{15}c_1 + \alpha_{25}c_2 + \alpha_{55}c_5 \\ &= (0)(4) + (0)(5) + (1)(0) = 0\,. \end{aligned}$$

Hence from (4–79), it follows that

$$\begin{aligned}\Delta z &= \pi_1 \Delta b_1 + \pi_2 \Delta b_2 + \pi_5 \Delta b_5 \\ &= (3)(0) + (0.5)(-0.5) + (0)(-1) = -0.25\,,\end{aligned}$$

and the new value of z is

$$z + \Delta z = 4.5 - 0.25 = 4.25\,.$$

Example 4–16. Given that the coefficient $c_2 = 5$ in Example 3–1 is changed to $c_2 = 4$, while all other coefficients remain, solve the problem on the basis of changing the c_j-coefficients.

In this example, $\Delta c_2 = 4 - 5 = -1$ and $\Delta c_1 = \Delta c_3 = \Delta c_4 = \Delta c_5 = 0$. Also, from the solution of the original problem by the dual simplex method in Example 4–12 (Table 4–12), we have $\bar{c}_1 = \bar{c}_2 = \bar{c}_5 = 0$, $\bar{c}_3 = 3$, and $\bar{c}_4 = \frac{1}{2}$. All other pertinent data are given previously in Example 4–15. Thus from (4–80), for j referring to each of the nonbasic variables, we have

$$\begin{aligned} j = 3\,, \quad & \bar{c}_3 - (\Delta c_2)[a_{13}\alpha_{21} + a_{23}\alpha_{22} + a_{53}\alpha_{25}] \\ & = 3 - (-1)[(-1)(-1) + (0)(0.5) + (0)(0)] = 4 > 0\,; \\ j = 4\,, \quad & \bar{c}_4 - (\Delta c_2)[a_{14}\alpha_{21} + a_{24}\alpha_{22} + a_{54}\alpha_{25}] \\ & = 0.5 - (-1)[(0)(-1) + (-1)(0.5) + (0)(0)] = 0\,. \end{aligned}$$

Since all conditions in (4–80) are satisfied, the original optimal basis remains. Then from (4–81), we can write

$$\Delta z = x_2 \Delta c_2 = (0.5)(-1) = -0.5\,.$$

Hence the new value of z is

$$z + \Delta z = 4.5 - 0.5 = 4\,.$$

REFERENCES

4–1. HADLEY, G., *Linear Programming*, Addison Wesley, Reading, Mass., 1961.

4–2. GASS, S. I., *Linear Programming*, McGraw-Hill, New York, 1958.

4–3. LEMKE, C. E., "The Dual Method of Solving Linear Programming Problems," *Naval Research Logistics Quarterly*, **1** (1954), 36–47.

4–4. DANTZIG, G. B., "Recent Advances in Linear Programming," *Management Science*, **2**, No. 2 (1956), 131–144.

4–5. SCHMIT, L. A., JR., G. G. GOBLE, R. L. FOX, L. LASDON, F. MOSES, and R. RAZANI, *Structural Synthesis* (Summer course notes), Case Institute of Technology, Cleveland, Ohio, July, 1965.

4–6. STARK, R. M., *Unbalanced Bidding Models*, Dept. of Civil Engineering, University of Delaware, Newark, Del. 1966.

PROBLEMS

P4–1 to P4–4. Verify the solution of Problems P3–1 to P3–4 in Chapter 3, respectively, according to the following steps:

a) Express the problem in the standard form.
b) Select the a_{ij}-coefficients associated with the basic variables in the optimal basis to form the $[a_{ij}]$-matrix.
c) Find the inverse $[\alpha_{ij}]$ of the optimal basis.
d) Determine the simplex multipliers π_i.
e) Find the optimal solution using the simplex multipliers.

P4–5 to P4–8. Solve Problems P3–1 to P3–4 in Chapter 3, respectively, by the use of the revised simplex method.

P4–9. Show that Problems P3–1 and P3–2 in Chapter 3 are primal-dual problems by first expressing the primal problem in

a) the inequality form, and
b) the standard form.

P4–10. Construct the dual of Problem P3–3 in Chapter 3 and express it in the standard form.

P4–11. Solve Problem P3–5 in Chapter 3 by the dual simplex method.

P4–12. Solve the following linear programming problem by the dual simplex method:

$$x_j \geq 0\,,$$

$$\begin{aligned} -x_1 - x_2 - 2x_3 \qquad\qquad + x_5 + x_6 \qquad &= -1\,, \\ 2x_1 \qquad\qquad - x_3 - x_4 - x_5 \qquad + x_7 &= -2\,, \\ 2x_1 + 9x_2 + 24x_3 + 8x_4 + 5x_5 \qquad &= z \quad \text{(Min)} \end{aligned}$$

P4–13. Investigate the change in the optimal solution of Problem P3–2 in Chapter 3 given that the following changes are made one at a time:

a) $b_1 = 7$, $b_2 = 3$,
b) $c_1 = 1.5$, $c_2 = 2.5$.

P4–14. Investigate the change in the optimal solution of Problem P3–1 in Chapter 3 given that the following changes are made one at a time:

a) $b_2 = 3.5$,
b) $c_1 = 9$.

CHAPTER 5

TRANSPORTATION AND ASSIGNMENT PROBLEMS

5-1 ELEMENTARY EXAMPLES

Among the many problems for distribution and allocation of products and materials which can be solved by linear programming, a special class commonly known as the *transportation problem*, or its variations, has certain characteristics which permit the simplification of the method of solution to a great extent, and hence merits separate treatment. In this section, several elementary examples are given to illustrate the nature of *transportation and assignment problems*.

Example 5–1. A rapid-transit system serving a metropolitan area has five terminals in the suburbs. Each morning, cars are dispatched to these terminals from three carbarns located in the central city. The cost of dispatching a car from each carbarn to a terminal is given in the cost matrix in Table 5–1, and the number of cars available in each carbarn

Table 5–1

COST MATRIX FOR EXAMPLE 5–1

Cost matrix		Terminal number 1	2	3	4	5	Number of cars available
Carbarn number	1	4	2	1	3	6	30
	2	3	5	2	1	4	20
	3	2	1	3	4	5	25
Number of cars required		25	20	15	5	10	75

and that required at each terminal are also given. How should the cars be allocated so as to minimize the total cost?

In Table 5-1, there are $3 \times 5 = 15$ unit cost coefficients in the cost matrix. For example, the unit cost of dispatching a car from carbarn 1 to terminal 1 is given as 4, that from carbarn 1 to terminal 2 is given as 2, etc. Corresponding to the unit cost coefficients, there are 15 variables in the problem, each representing the number of cars dispatched from a carbarn to a terminal. The values of these variables must be integers. Table 5-1 is called a standard transportation array for the given transportation problem. Since the total number of cars available in all carbarns is equal to the total number of cars required at all terminals, as indicated in the stubs of Table 5-1, every car will be dispatched to meet the demand. However, from every carbarn, cars will be dispatched to, at the most, five terminals; and at every terminal, cars can come from, at the most, three carbarns. Hence the total number of variables in each constraint equation resulting from the balance of demand and supply will not be greater than five (out of a total of 15). If the simplex method is used in the solution of this problem, the computational procedure is very inefficient, since many structural coefficients are zero. A special algorithm is therefore desirable for this class of problems.

Example 5-2. In Example 5-1, how should the cars be allocated to minimize the total cost if some of the routes from carbarns to terminals are closed for repair?

When a route is closed for repair, the cost coefficients for that route may be regarded as infinity so that no car will be dispatched through it. If only a few routes are inaccessible, we can bypass them in dispatching cars. However, when many routes become inaccessible, it may be impossible to allocate sufficient cars as requested through the remaining routes. Modifications of the standard procedure are necessary in order to determine whether a feasible solution exists for such problems.

Example 5-3. A contractor has six pieces of special excavation equipment of different categories, and he is engaged in six excavation jobs with different site conditions. He has rated the performance of the equipment in each category for each site condition and uses the rating of 1, 2, 3, . . . , 9, as shown in Table 5-2, in which 1 represents the most desirable and 9 the least desirable. How should he assign one piece of equipment to each job so that the total rating scores will be a minimum?

This problem is obviously a special case of the transportation problem in which the amount of supply in each origin is one, and the amount of demand in each destination is also one. Hence no stubs need be added to the effectiveness coefficient matrix to remind us of the amounts of supply and demand. Such a special case is called an *assignment problem*.

Table 5-2

EFFECTIVENESS MATRIX FOR EXAMPLE 5-3

		Site number					
		1	2	3	4	5	6
Equipment number	1	4	5	3	2	7	1
	2	5	4	2	3	4	6
	3	6	8	1	4	3	7
	4	3	2	9	7	6	5
	5	1	3	5	1	8	4
	6	2	6	7	5	2	3

Example 5-4. Suppose that the contractor in the previous example has only two types of equipment, I and II, but he has three of each. Assume also that the site conditions for the jobs can be classified as types *A*, *B*, and *C*, with the number of jobs corresponding to these types being 2, 3, and 1, respectively. If a new rating of performance is as given in Table 5-3, how should he assign his equipment to each job?

In this assignment problem, the rating for equipment type I for soil condition *A* is 4, regardless which of the three pieces of equipment is chosen or which of the two jobs with soil condition *A* the equipment is assigned to. The same argument is applied to the rating for I-*B*, I-*C*, II-*A*, II-*B*, and II-*C*. Thus the assignment problem may be reduced to the form of the transportation problem, as indicated in Table 5-4.

We may therefore regard the transportation problem as a special case of the assignment problem in which some of the origins and destinations can be classified into subgroups. This is an interesting conclusion, since we have observed in Example 5-3 that the assignment problem is also a special case of the transportation problem in which the amount of supply at each origin or the amount of demand at each destination is always one.

If we consider that equipment types I and II are located in Cleveland and New York, respectively, and that site conditions *A*, *B*, and *C* are those for Philadelphia, Chicago, and Pittsburgh, respectively, this problem becomes identical to Example 1-3, which has been formulated previously in Chapter 1.

Table 5–3

EFFECTIVENESS MATRIX FOR EXAMPLE 5–4

		Site condition					
		A	*A*	*B*	*B*	*B*	*C*
Equipment type	I	4	4	3	3	3	1
	I	4	4	3	3	3	1
	I	4	4	3	3	3	1
	II	2	2	7	7	7	5
	II	2	2	7	7	7	5
	II	2	2	7	7	7	5

Table 5–4

EQUIVALENT TRANSPORTATION PROBLEM

		Site condition			Supply
		A	*B*	*C*	
Equipment type	I	4	3	1	3
	II	2	7	5	3
Demand		2	3	1	6

5–2 CLASSICAL TRANSPORTATION PROBLEM

We shall consider first the classical transportation problem of balanced supply and demand, which is sometimes referred to as the *Hitchcock problem*. In general, the number of origins of the supply may be denoted by m and the number of destinations for the demand by n, as shown in

Fig. 5-1. The level of supply at origin i ($i = 1, 2, \ldots, m$) is a_i and the level of demand at destination j ($j = 1, 2, \ldots, n$) is b_j. Assuming that the unit cost of shipping from origin i to destination j is c_{ij}, we wish to determine the shipping pattern which minimizes the total transportation cost.

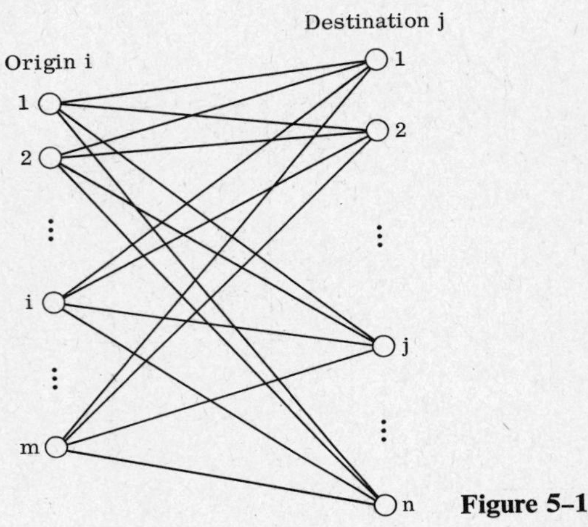

Figure 5-1

Let x_{ij} be the amount of goods shipped from origin i to destination j. Then the total shipment from each origin to all destinations is the amount available at the origin; similarly, the total shipment received by each destination from all origins is the amount required at the destination. Mathematically, we have

$$x_{ij} \geq 0 \,; \tag{5-1}$$

$$\sum_{j=1}^{n} x_{ij} = a_i \,, \qquad i = 1, 2, \ldots, m \,; \tag{5-2}$$

$$\sum_{i=1}^{m} x_{ij} = b_j \,, \qquad j = 1, 2, \ldots, n \,; \tag{5-3}$$

$$\sum_{i=1}^{m} \sum_{j=1}^{n} c_{ij} x_{ij} = \text{Min } z \,. \tag{5-4}$$

It may be observed that the condition of balanced supply and demand is implicit in (5-2) and (5-3), since it can be obtained by summing (5-2) over all i and summing (5-3) over all j. The sum of all m equations in (5-2) is given by

$$\sum_{i=1}^{m} \sum_{j=1}^{n} x_{ij} = \sum_{i=1}^{m} a_i \,,$$

and the sum of all n equations in (5-3) is

$$\sum_{i=1}^{m}\sum_{j=1}^{n} x_{ij} = \sum_{j=1}^{n} b_j \, .$$

Hence

$$\sum_{i=1}^{m} a_i = \sum_{j=1}^{n} b_j \, . \tag{5-5}$$

Although (5-2) and (5-3) together represent a system of $m + n$ constraint equations in $m \times n$ variables, one of the constraint equations is not independent because of the known condition of balanced supply and demand. Thus *the classical transportation problem may be regarded as a linear programming problem having $m + n - 1$ constraint equations in $m \times n$ variables.* There will be only $(m + n - 1)$ basic variables in the optimal solution; the remaining nonbasic variables must be zero. In the process of the solution, however, we may preserve the appearance of $m + n$ equations for easy identification and checking, and set them in a simplified form such as Table 5-5, in which the equations of (5-2) are represented by the rows and those of (5-3) are represented by the columns. Hence the equations (5-2) and (5-3) are referred to as *row equations* and *column*

Table 5-5

TRANSPORTATION ARRAY

		Destination j				Supply
		1	2	...	n	
Origin i	1	x_{11} c_{11}	x_{12} c_{12}	... ...	x_{1n} c_{1n}	a_1 u_1
	2	x_{21} c_{21}	x_{22} c_{22}	... ...	x_{2n} c_{2n}	a_2 u_2
	...	... ...	... ...	... ...	... ...	... ...
	m	x_{m1} c_{m1}	x_{m2} c_{m2}	... ...	x_{mn} c_{mn}	a_m u_m
Demand		b_1 v_1	b_2 v_2	... ...	b_n v_n	

equations, respectively. The unit cost c_{ij} corresponding to each x_{ij} in (5-4) is shown in the lower right-hand corner of each cell. The condition of balanced supply and demand in (5-5) is indicated by the equality of the sum of all supplies in the summation column for a_i and the sum of all demands in the summation row for b_j. The simplex multipliers u_i and v_j in the stubs will be explained later.

Any feasible solution of the problem must satisfy the nonnegative constraints of (5-1) and the constraint conditions of (5-2) and (5-3). Since the number of nonbasic variables is the difference between the total number of variables $(m \times n)$ and the number of basic variables $(m + n - 1)$, namely,

$$m \times n - (m + n - 1) = (m - 1)(n - 1),$$

at least $(m - 1)(n - 1)$ variables in Table 5-5 should be zero if a basic feasible solution is obtained. The optimal solution is a basic feasible solution which also satisfies the objective function in (5-4).

Like the solution of all linear programming problems, the classical transportation problem can be solved by the determination of an initial basic feasible solution, the test of optimality of the solution, and the improvement of the solution by changing the basis, if necessary, until an optimal solution is reached; but unlike the solution of the general linear programming problem, the classical transportation problem has certain mathematical properties which permit considerable simplification in its solution. We shall examine these properties in the development of a procedure for the systematic solution of the classical transportation problem.

5-3 DETERMINATION OF AN INITIAL SOLUTION

The determination of an initial solution for the classical transportation problem may be illustrated by a simple example (Example 1-3 in Chapter 1), which can be written as follows:

$$x_{ij} \geq 0, \qquad i = 1, 2; \quad j = 1, 2, 3;$$

$$\begin{aligned}
x_{11} + x_{12} + x_{13} \qquad\qquad\qquad\qquad &= 3, \\
x_{21} + x_{22} + x_{23} &= 3, \\
x_{11} \qquad\qquad + x_{21} \qquad\qquad &= 2, \\
x_{12} \qquad\qquad + x_{22} \qquad &= 3, \\
x_{13} \qquad\qquad + x_{23} &= 1, \\
4x_{11} + 3x_{12} + x_{13} + 2x_{21} + 7x_{22} + 5x_{23} &= \text{Min } z.
\end{aligned}$$

In this problem there are $m + n = 5$ constraint equations in $m \times n = 6$ variables, but only $m + n - 1 = 4$ of them are independent. Because the coefficients of variables in the constraint equations are either one or zero, the task of obtaining an initial basic feasible solution is greatly simplified. For example, we can take the first four of the five constraint equations and rearrange them in the following manner:

$$\begin{aligned}
x_{13} + x_{12} \qquad\qquad\qquad + x_{11} \qquad &= 3\,, \\
x_{12} + x_{22} \qquad\qquad\qquad\qquad &= 3\,, \\
x_{22} + x_{21} \qquad + x_{23} &= 3\,, \\
x_{21} + x_{11} \qquad\quad &= 2\,,
\end{aligned}$$

in which x_{21}, x_{22}, x_{12}, and x_{13} are regarded as basic variables, and x_{11} and x_{23} are treated as nonbasic variables. The initial selection of the set of basic variables can be arbitrary, but a reasonable guide is the minimum-cost rule, which is to select variables with lowest cost coefficients in the objective function that are compatible with the number of basic variables allowed in the constraint equations. This set of equations is seen to have a triangular basis. When the nonbasic variables are set equal to zero, that is, $x_{11} = 0$ and $x_{23} = 0$, the basic variables can be obtained by back substitution, starting from the last equation; thus we get $x_{21} = 2$, $x_{22} = 1$, $x_{12} = 2$, and $x_{13} = 1$. This solution together with the given data can be tabulated in a transportation array like that in Table 5-6.

Table 5-6

INITIAL BASIC FEASIBLE SOLUTION

 4	2 3	1 1	3
2 2	1 7	 5	3
2	3	1	6

Note that the nonbasic variables $x_{11} = 0$ and $x_{23} = 0$ do not appear in the array, so that they will not be confused with the degenerate cases of basic variables if the latter should occur. Thus the array provides a convenient form to select an initial basic feasible solution according to the minimum-cost rule. For example, we may assign $x_{13} = 1$, $x_{21} = 2$, and $x_{12} = 2$, in that order, corresponding to the cost coefficients 1, 2, and 3, respectively, but we must choose x_{22} as the fourth basic variable in order to satisfy all constraint equations (rows and columns in the array). Furthermore, we are not completely free in assigning the values of the basic variables, since the number of basic variables must be $m + n - 1 = 4$ in this case. For example, if for any reason we choose a solution with only three basic variables, say $x_{12} = 3$, $x_{21} = 2$ and $x_{23} = 1$, the solution is not basic, although the constraint equations are satisfied, unless a fourth variable, say $x_{22} = 0$, is added as a degenerate basic variable. On the other hand, if we assign $x_{13} = 1$ and $x_{21} = 1$, then we must have $x_{11} = 1$, $x_{12} = 1$ and $x_{22} = 2$ in order to satisfy all constraint equations. This solution with five nonzero variables violates the allowed number of basic variables and is therefore not acceptable.

Returning to the general problem of an $m \times n$ transportation array in Table 5-5, we wish to generalize two properties observed in the simple example, that is, (*1*) *the classical transportation problem has a triangular basis, and* (*2*) *the values of the basic variables can be so chosen that the number of basic variables is* $m + n - 1$. If these properties are established, an initial solution for the general problem can readily be obtained.

In examining the rows and columns in the $m \times n$ transportation array, we conclude that at least one basic variable appears in each row and in each column, since the supplies at the origins and the demands at the destinations are nonzero. On the other hand, it is not possible to have two or more basic variables in every row and in every column because it would contradict the total number of basic variables, $k = m + n - 1$, allowed for the system of constraint equations. For example, if all rows and all columns contain two or more basic variables, then

$$k \geq 2m \qquad \text{and} \qquad k \geq 2n\,,$$

in which m or n may be the larger number. Consequently, it follows that

$$k \geq m + n\,.$$

This is an obvious contradiction to $k = m + n - 1$, and is not acceptable. Hence we conclude that at least one row or column has only one basic variable which can readily be computed. After this computation is completed, then that row or column can be eliminated from the array. The same argument may be applied to the reduced array in the computation of the next basic variable, and the procedure can be repeated until all basic variables are computed. This is essentially a process of back substitution, starting from the row or column which has been assigned only one basic variable. Hence all bases of the transportation array are triangular.

In computing each basic variable from the row or column that contains only one basic variable, we find that the value of the basic variable must equal either the supply value corresponding to the row or the demand value corresponding to the column, whichever is smaller. If the basic variable takes a larger value, at least one condition in the row or the column will be violated; if it takes a value smaller than the one thus specified, the difference must be assigned to another variable which will then violate the condition of only one basic variable in the row or column. Hence, in choosing the first basic variable x_{rs} in row r and column s, let

$$x_{rs} = \text{Min}\,(a_r, b_s)\,.$$

That is, x_{rs} equals the smaller of the two numbers, a_r (the supply in the rth row) and b_s (the demand in the sth column). At an intermediate step, however, a basic variable x_{rs} should be assigned the value

$$x_{rs} = \text{Min}\,(\bar{a}_r, \bar{b}_s)\,,$$

where $\bar{a}_r$ and $\bar{b}_s$ represent, respectively, the available supply in the rth row and the remaining demand in the sth column in the reduced array as a result of the elimination of some rows and columns in the previous steps. This method of selection will ensure that there will be one row or one column left in the reduced array after the last step. If, at any stage of computation, we encounter the case of $\bar{a}_r = \bar{b}_s$, we can choose to eliminate either the row or the column, but not both, unless a zero is also assigned to another variable either in the row or the column under consideration to represent the degenerate case. Hence the total number of basic variables will always correspond to the total number of rows and columns eliminated, which is $m + n - 1$.

5-4 TEST OF OPTIMALITY

After an initial basic feasible solution is obtained, it can be tested for optimality by use of simplex multipliers. We can illustrate this procedure by a simple example without losing generality. Consider again Example 1-3 written in the following form:

$$x_{ij} \geq 0\,, \qquad i = 1, 2\,; \quad j = 1, 2, 3\,;$$

$$\begin{aligned}
x_{11} + x_{12} + x_{13} \qquad\qquad\qquad\qquad &= a_1\,, \\
x_{21} + x_{22} + x_{23} &= a_2\,, \\
x_{11} \qquad\qquad + x_{21} \qquad\qquad &= b_1\,, \\
x_{12} \qquad\qquad + x_{22} \qquad &= b_2\,, \\
x_{13} \qquad\qquad + x_{23} &= b_3\,; \\
c_{11}x_{11} + c_{12}x_{12} + c_{13}x_{13} + c_{21}x_{21} + c_{22}x_{22} + c_{23}x_{23} &= \text{Min}\, z\,.
\end{aligned}$$

By introducing the simplex multipliers, u_i $(i = 1, 2)$ and v_j $(j = 1, 2, 3)$, for the row equations and column equations, respectively, we multiply the first of the row constraint equations by u_1, and the second by u_2; we also multiply the first of the column constraint equations by v_1, the second by v_2, and the third by v_3. Then the sum of these products becomes

$$\begin{aligned}(u_1 + v_1)x_{11} + (u_1 + v_2)x_{12} + (u_1 + v_3)x_{13} + (u_2 + v_1)x_{21} \\ + (u_2 + v_2)x_{22} + (u_2 + v_3)x_{23} \\ = (u_1a_1 + u_2a_2) + (v_1b_1 + v_2b_2 + v_3b_3) .\end{aligned}$$

Note that since each variable x_{ij} appears only twice in the constraint equations (once in the row equations and once in the column equations), each x_{ij} in the resulting equation has a coefficient of $(u_i + v_j)$. If this equation is subtracted from the objective function, we obtain

$$\bar{c}_{11}x_{11} + \bar{c}_{12}x_{12} + \bar{c}_{13}x_{13} + \bar{c}_{21}x_{21} + \bar{c}_{22}x_{22} + \bar{c}_{23}x_{23} = \text{Min}\,(z - \bar{z}) ,$$

where

$$\begin{aligned}\bar{c}_{ij} &= c_{ij} - (u_i + v_j) , \qquad i = 1, 2 ; \quad j = 1, 2, 3 ; \\ \bar{z} &= (u_1a_1 + u_2a_2) + (v_1b_1 + v_2b_2 + v_3b_3) .\end{aligned}$$

In order to test the optimality of a basic solution by use of this modified objective function, it is necessary that the coefficients $\bar{c}_{ij}$ for the basic variables be zero and the coefficients $\bar{c}_{ij}$ for the nonbasic variables be examined to determine whether the criterion of optimality is satisfied. Hence for the basic variables

$$\bar{c}_{ij} = c_{ij} - (u_i + v_j) = 0 \qquad \text{or} \qquad u_i + v_j = c_{ij} .$$

Although there are five constraint equations in this problem, only four of them are independent, and the remaining one should be eliminated. This can be accomplished by letting any one of the five simplex multipliers be zero, since any one of the five equations can be considered as redundant. Thus the remaining four simplex multipliers can be determined from the conditions of $\bar{c}_{ij} = 0$ for the four basic variables. For example, if x_{12}, x_{13}, x_{21}, and x_{22} are the basic variables of the initial solution, we have

$$\begin{aligned}u_1 + v_2 &= c_{12} , \\ u_1 + v_3 &= c_{13} , \\ u_2 + v_1 &= c_{21} , \\ u_2 + v_2 &= c_{22} .\end{aligned}$$

By assuming that $u_1 = 0$, then the other simplex multipliers, u_2, v_1, v_2, and v_3, can be determined in terms of the known cost coefficients. After the simplex multipliers are determined, the coefficients $\bar{c}_{ij}$ for the nonbasic variables can be obtained by the relation

$$\bar{c}_{ij} = c_{ij} - (u_i + v_j) \, .$$

In this case, the nonbasic variables are x_{11} and x_{23}. Hence

$$\bar{c}_{11} = c_{11} - (u_1 + v_1) \, ,$$
$$\bar{c}_{23} = c_{23} - (u_2 + v_3) \, .$$

If all $\bar{c}_{ij}$ for the nonbasic variables are nonnegative, the current solution cannot be further improved and min $z = \bar{z}$ *is optimal; otherwise improvement can be made by changing the basis.*

Table 5-7

TEST OF OPTIMALITY

<table>
<tr><td>
4</td><td>2
3</td><td>1
1</td><td>3
0</td></tr>
<tr><td>2
2</td><td>1
7</td><td>
5</td><td>3
4</td></tr>
<tr><td>2
−2</td><td>3
3</td><td>1
1</td><td>6
</td></tr>
</table>

The test of optimality for this example can be carried out numerically in the transportation array in Table 5-7. Note that the simplex multipliers u_1 and u_2 are shown at the lower right-hand corner of the cells for a_1 and a_2, respectively, and that v_1, v_2, and v_3 are shown at the lower right-hand corner of the cells for b_1, b_2, and b_3, respectively. By assuming that $u_1 = 0$, we can compute $v_2 = 3$ and $v_3 = 1$ by noting that

$$u_1 + v_2 = 3 \qquad \text{and} \qquad u_1 + v_3 = 1 \, ,$$

since x_{12} and x_{13} are basic variables. Also, from

$$u_2 + v_2 = 7 \qquad \text{and} \qquad u_2 + v_1 = 2 \, ,$$

we can obtain successively $u_2 = 4$ and $v_1 = -2$. After all the simplex multipliers are determined, we can proceed to find

$$\bar{c}_{11} = 4 - (0 - 2) = 6 \, ,$$
$$\bar{c}_{23} = 5 - (4 + 1) = 0 \, .$$

Since $\bar{c}_{ij}$ for all nonbasic variables are nonnegative, the current solution is optimal. The minimum cost may be obtained either from the original objective function,

$$\begin{aligned}\min z &= (4)(0) + (3)(2) + (1)(1) + (2)(2) + (7)(1) + (5)(0)\\ &= 6 + 1 + 4 + 7 = 18\,,\end{aligned}$$

or from the modified objective function, which leads to

$$\begin{aligned}\min z = \bar{z} &= (0)(3) + (4)(3) + (-2)(2) + (3)(3) + (1)(1)\\ &= 12 - 4 + 9 + 1 = 18\,.\end{aligned}$$

The numerical calculation can easily be carried out in the transportation array.

Returning to the general problem of an $m \times n$ transportation array, we note that the same criterion of optimality holds true. That is, in the modified objective function

$$\sum_i \sum_j \bar{c}_{ij} x_{ij} = \mathrm{Min}\,(z - \bar{z})\,, \tag{5-6}$$

where

$$\bar{z} = \sum_i a_i u_i + \sum_j b_j v_j\,, \tag{5-7}$$

$$\bar{c}_{ij} = c_{ij} - (u_i + v_j)\,. \tag{5-8}$$

Choose u_i and v_j such that for all basic variables,

$$c_{ij} = u_i + v_j\,. \tag{5-9}$$

If, for all nonbasic variables, the resulting coefficients

$$\bar{c}_{ij} = c_{ij} - (u_i + v_j) \geq 0\,,$$

then the current solution is optimal.

5-5 SEEKING AN IMPROVED BASIC SOLUTION

If the initial solution of a classical transportation problem is not optimal, an improved basic solution may be obtained by changing the basis. We must first determine the entering variable for the new basis and the departing variable from the current basis, and then test the new basic solution for optimality.

Again, we shall consider the numerical example used in the previous section. If we had chosen a basis consisting of x_{11}, x_{12}, x_{22}, and x_{23}, the initial solution would not have been optimal. This can be seen from four of the five constraint equations arranged in a triangular form, together with the modified objective function obtained according to the equations in (5-6) through (5-9):

$$\begin{aligned}
x_{23} + x_{22} \qquad\qquad\qquad + x_{21} &= 3\,,\\
x_{22} + x_{12} \qquad\qquad\qquad &= 3\,,\\
x_{12} + x_{11} + x_{13} \qquad &= 3\,,\\
x_{11} \qquad + x_{21} &= 2\,,\\
-\,6x_{21} &= \text{Min}\,(z - 30)\,.
\end{aligned}$$

Since the coefficient of x_{21} in the objective function is negative, the initial solution is not optimal. If x_{21} is selected as the entering variable and x_{11} as the departing variable, a new triangular basis can easily be obtained by interchanging the positions of x_{11} and x_{21} in the constraint equations. A new modified objective function corresponding to the new basis may also be obtained:

$$\begin{aligned}
x_{23} + x_{22} \qquad + x_{21} \qquad\qquad &= 3\,,\\
x_{22} + x_{12} \qquad\qquad\qquad &= 3\,,\\
x_{12} \qquad + x_{13} + x_{11} &= 3\,,\\
x_{21} \qquad + x_{11} &= 2\,,\\
4x_{13} + 6x_{11} &= \text{Min}\,(z - 18)\,.
\end{aligned}$$

This latter basic solution is seen to be optimal, since the coefficients for both nonbasic variables are nonnegative.

The above example only serves to illustrate that, *because of the properties of the transportation problem, we can change the basis very easily, and that the change affects only the variables which are directly or indirectly related to the entering and departing variables by the row and column equations.* The computation of the basic variables for the new basis may be carried out in the transportation array by letting the entering variable equal θ, which will disturb the balance in the row and column equations unless necessary changes, including the deletion of the departing variable, are made to restore the balance. By observing the nonnegative constraints for all variables, the value of θ can be determined from the adjusted basic variables in the new basis. The modified cost coefficients $\bar{c}_{ij}$ may also be computed from the transportation array.

This procedure is illustrated in the transportation array in Table 5-8 by the same numerical example. The initial solution is obtained by assuming that $x_{11} = 2$, $x_{12} = 1$, $x_{22} = 2$, and $x_{23} = 1$. After the simplex multipliers are computed, it is found that $\bar{c}_{13} = 0$ and $\bar{c}_{21} = -6$. Thus x_{21} becomes the entering variable and a parameter θ is added to the upper left-hand corner of the square for x_{21}. Since the balance of the row and column equations is disturbed by this addition, an attempt is made to

Table 5–8

CHANGE OF BASIS

$2-\theta$ 4	$1+\theta$ 3	 1	3 0
θ 2	$2-\theta$ 7	1 5	3 4
2 4	3 3	1 1	6

rebalance these equations. We therefore subtract θ from x_{11} which, in turn, causes θ to be added to x_{12}; finally, θ is also subtracted from x_{22} to complete the adjustment. When all rows and columns are again balanced, we inspect all the basic variables of the form $x_{ij} = f - \theta$, in which f is the current value of the variable. Since all basic variables must remain nonnegative, the variable in the form of $x_{ij} = f - \theta$ that has the smallest f will become the departing variable as θ takes the value of the smallest f. In the case of a tie for the smallest f, only one variable should depart from the basis, and the other takes the value of zero as a degenerate basic variable, since $m + n - 1 = 4$ basic variables must be maintained. Thus in the example we have $\theta = 2$, and $x_{11} = x_{22} = 2 - \theta = 0$. If we select x_{11} as the departing variable, then $x_{22} = 0$ stays in the basis as a degenerate basic variable. Table 5–9 shows the improved solution with a new basis which upon test is found to be optimal.

Table 5–9

OPTIMAL SOLUTION

 4	3 3	 1	3 −4
2 2	0 7	1 5	3 0
 2	 7	 5	

For the general problem of an $m \times n$ transportation array, the same procedure may be used to obtain an improved solution by changing the basis. Although more basic variables in the current basis may be affected after the entering variable is assigned a value θ, the rows and columns can always be rebalanced by alternately subtracting and adding θ to the basic variables in the closed path leading back to the entering variable. Hence we can always seek the smallest f among basic variables with $x_{ij} = f - \theta$, and set θ equal to the smallest f. The variable which vanishes under such conditions becomes the departing variable.

5–6 PROCEDURE FOR SOLVING TRANSPORTATION PROBLEM

The computational procedure for the classical transportation problem is seen to be an iteration procedure which can be carried out in a transportation array. The procedure can be summarized as follows:

1. Find an initial basic feasible solution by having a set of basic variables whose values are chosen to be as large as possible, but which are consistent with the sums of rows and columns. In general, basic variable x_{rs} should be set equal to

$$x_{rs} = \text{Min}\,(\bar{a}_r, \bar{b}_s)\,,$$

where $\bar{a}_r = a_r$ and $\bar{b}_s = b_s$ for the first basic variable selected, but, in general, $\bar{a}_r$ and $\bar{b}_s$ represent, respectively, the available supply in the rth row and the remaining demand in the sth column in the reduced array. If $\bar{a}_r < \bar{b}_s$, the rth row is to be eliminated; if $\bar{a}_r > \bar{b}_s$, the sth column is to be eliminated. If $\bar{a}_r = \bar{b}_s$, another variable in the rth row or the sth column must be given a value of zero before both are eliminated.

2. A basic feasible solution can be tested for optimality by first evaluating the simplex multipliers, with the exception of one which is set equal to zero, from the condition

$$c_{ij} = u_i + v_j$$

for all basic variables. Then the criterion for optimality is given by

$$\bar{c}_{ij} = c_{ij} - (u_i + v_j) \geq 0$$

for all nonbasic variables. If this criterion is satisfied, the solution is optimal.

3. If a solution does not satisfy the criterion for optimality, an improved solution may be obtained by changing the basis. First, let the entering variable equal θ and then rebalance rows and columns by alternately subtracting θ from, and adding θ to, the basic variables in the closed path leading back to the entering variable. The basic variable in the

Table 5-10

CLOSED PATH WITH HORIZONTAL
AND VERTICAL LINKS

(a)

	1		3		4
	5			4	9
7		2			9
			6	2	8
7	6	2	9	6	

current basis which first becomes zero as the value of θ is gradually increased will be the departing variable. In the case of a tie, only one will be selected as the departing variable, and the other remains as a degenerate basic variable (of zero value).

4. It should be kept in mind that a set of basic variables, including the degenerate basic variables in the set, must not form a closed path with horizontal and vertical links if the solution is unique. A counter example is shown in Table 5-10, in which both sets of variables in (a) and (b) satisfy the constraint conditions, but neither is unique. On the other hand, in the process of changing the basis, θ must be added to and subtracted from cells such that a closed path is finally formed. After a new basic feasible solution is obtained, the entire procedure is repeated until the optimal solution is reached.

5. The optimal value of z is obtained by substituting into the objective function (5-4) the set of x_{ij}^* representing an optimal basic feasible solution. Thus

$$\min z = z^* = \sum_{i=1}^{m} \sum_{j=1}^{n} c_{ij} x_{ij}^* .$$

In the case of balanced supply and demand, (5-6) may also be used.

Table 5-10

(Continued)

(b)

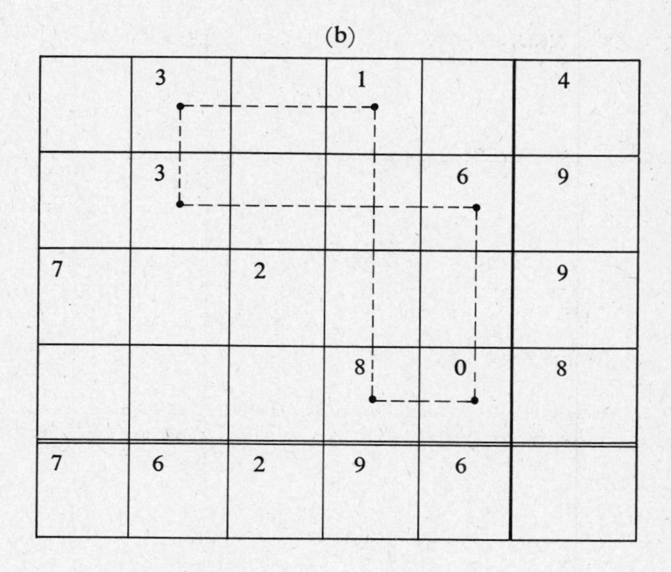

	3		1		4
	3			6	9
7		2			9
			8	0	8
7	6	2	9	6	

Example 5-5. Find the optimal solution for the problem in Example 5-1.

The problem, which is already expressed in a transportation array, may be solved in a tabulated form, as shown in Table 5-11. Since $m = 3$ and $n = 5$ in the cost matrix, there exist $3 + 5 - 1 = 7$ basic variables. Using the minimum cost rule, we screen the cost coefficients in Table 5-11(a) one by one. Since $c_{13} = 1$ is one of the lowest costs, we let $x_{13} = 15$, which is the smaller of $a_1 = 30$ and $b_3 = 15$, and check off column 3 as having been satisfied. Next we let $x_{24} = 5$ and $x_{32} = 20$, and check off columns 2 and 4 accordingly. In examining column 5, we let $x_{25} = 10$, which is the smaller of 10 and $20 - 5 = 15$, and check off column 5. Next we check off the rows after letting $x_{11} = 15$, $x_{21} = 5$, and $x_{31} = 5$, which are the only admissible values for rows 1, 2 and 3, respectively. Finally, we note that column 1 is also satisfied but that it has not been checked off because it is regarded as the redundant equation. The number of check marks, which is $3 + 5 - 1 = 7$, equals the number of basic variables assigned to the transportation array. Hence an initial basic solution has been obtained.

To test the optimality of the solution, we introduce a set of simplex multipliers u_i and v_j. Since any one of the constraint equations may be considered as redundant, we can set any one of these multipliers equal

Table 5-11

SOLUTION FOR EXAMPLE 5-5

(a)

15 4	 2	15 1	 3	 6	30 1	✓
5 3	 5	 2	5 1	10 4	20 0	✓
5 2	20 1	 3	 4	 5	25 −1	✓
25 3	20 2	15 0	5 1	10 4	75	
	✓	✓	✓	✓		

(b)

$15-\theta$	θ −1	15	 1	 1	30
5	 3	 2	5	10	20
$5+\theta$	$20-\theta$	 4	 4	 2	25
25	20	15	5	10	75

Table 5–11

(Continued)

(c)

 4	15 2	15 1	 3	 6	30 0
5 3	 5	 2	5 1	10 4	20 0
20 2	5 1	 3	 4	 5	25 −1
25 3	20 2	15 1	5 1	10 4	75

to zero. In order to reduce the amount of numerical computation, however, it is advisable to select that multiplier corresponding to the row or column which contains the largest number of basic variables. Hence in this case let $u_2 = 0$. Then from the relation in (5–9) for all basic variables, we can compute directly from Table 5–11(a) the following:

$$v_1 = 3 - 0 = 3\,, \qquad v_4 = 1 - 0 = 1\,,$$
$$v_5 = 4 - 0 = 4\,, \qquad u_1 = 4 - 3 = 1\,,$$
$$u_3 = 2 - 3 = -1\,, \qquad v_2 = 1 - (-1) = 2\,,$$
$$v_3 = 1 - 1 = 0\,.$$

For the nonbasic variables, the $\bar{c}_{ij}$ are computed according to (5–8). Each of the $\bar{c}_{ij}$ is recorded at the lower left corner of the cell corresponding to that nonbasic variable in Table 5–2(b). For example,

$$\bar{c}_{12} = 2 - (1 + 2) = -1\,, \qquad \bar{c}_{14} = 3 - (1 + 1) = 1\,,$$
$$\bar{c}_{15} = 6 - (1 + 4) = 1\,, \qquad \bar{c}_{22} = 5 - (0 + 2) = 3\,,$$
$$\bar{c}_{23} = 2 - (0 + 0) = 2\,, \qquad \bar{c}_{33} = 3 - (-1 + 0) = 4\,,$$
$$\bar{c}_{34} = 4 - (-1 + 1) = 4\,, \qquad \bar{c}_{35} = 5 - (-1 + 4) = 2\,.$$

Since $\bar{c}_{12} < 0$, the solution is not optimal. Therefore we let $x_{12} = \theta$ be the entering variable, and rebalance the rows and columns by completing the closed path involving x_{12}, x_{32}, x_{31}, and x_{11}, as indicated by the dashed line in Table 5-11(b). It may be observed that, if we had tried to modify other basic variables in the balancing process, we would not have been able to obtain a closed path. By increasing the value of θ, it is seen that $x_{11} = 15 - \theta$ will first reach zero, and hence it becomes the departing variable as $\theta = 15$. By substituting $\theta = 15$ into the basic variables in Table 5-11(b), we obtain a new basis, as shown in part 11(c) of the table. The new solution can be tested for optimality and is found to be optimal.

Finally, the minimum cost of dispatching the cars is found to be

$$z^* = (15)(2) + (15)(1) + (5)(3) + (5)(1) + (10)(4) + (20)(2) + (5)(1) = 150\,.$$

This value can also be checked by using the simplex multipliers as follows:

$$z^* = (0)(30) + (0)(20) + (-1)(25) + (3)(25) + (2)(20) + (1)(15) + (1)(5) + (4)(10) = 150\,.$$

An alternative solution of this problem is also given in Table 5-12 in which a different initial basic solution has been selected. The interme-

Table 5-12

ALTERNATIVE SOLUTION FOR EXAMPLE 5-5

Cycle 0 $\theta = 5$

$5-\theta$ 4	θ 2	15 1	 3	10 6	30 0
15 3	 5	 2	5 1	 4	20 −1
$5+\theta$ 2	$20-\theta$ 1	 3	 4	 5	25 −2
25 4	20 3	15 1	5 2	10 6	75

Table 5–12

(Continued)

Cycle 1 $\theta = 10$

4	$5 + \theta$ 2	15 1	3	$10 - \theta$ 6	0
$15 - \theta$ 3	5	2	5 1	θ 4	0
$10 + \theta$ 2	$15 - \theta$ 1	3	4	5	-1
3	2	1	1	6	

Cycle 2 Optimal

4	15 2	15 1	3	6	0
5 3	5	2	5 1	10 4	0
20 2	5 1	3	4	5	-1
3	2	1	1	4	

diate step of computing $\bar{c}_{ij}$ is not shown; thus all operations in each cycle are indicated in a separate table. Note that the alternative solution involves one more cycle than the first solution in Table 5-11. In each cycle, the basic variables are first evaluated, the simplex multipliers are then introduced, and the optimality test is carried out directly by using the table in checking the $\bar{c}_{ij}$ for the nonbasic variables. If all $\bar{c}_{ij}$ are nonnegative, the solution is optimal; if not, determine the entering and departing variables for the new basis, and finally, evaluate θ.

Example 5-6. Rework Example 1-3 by selecting an initial solution, as shown in Table 5-13.

In Table 5-13, we first let $x_{21} = 2$ and check off column 1. Since we let $x_{12} = 3$, both column 2 and row 1 are satisfied. However, we shall check off only one row or column, say column 2, because only one basic variable has been added in this process. Similarly, we let $x_{23} = 1$, and check off only column 3. When we let $x_{22} = 0$ and check off row 2, the initial solution is complete. Again, the number of check marks equals the number of basic variables. This initial basic solution is found to be optimal.

Table 5-13

SOLUTION FOR EXAMPLE 5-6

Cycle 0 Optimal

	3		3	
4	3	1	−4	
2	0	1	3	√
2	7	5	0	
2	3	1	6	
2	7	5		
√	√	√		

If $x_{11} = 0$ is assumed instead of $x_{22} = 0$ in the initial basic solution, as shown in cycle 0 of Table 5-14, not all $\bar{c}_{ij}$ are nonnegative and the solution is not optimal. After changing the basis, however, we see in cycle 1 that the solution is optimal although only the degenerate basic variable has changed its position. This indicates that, while the location of zero representing the degenerate basic variable need not be assigned to a specific cell in the initial solution, its location may affect the number of cycles in reaching an optimal solution.

Table 5–14

ALTERNATIVE SOLUTION
FOR EXAMPLE 5–6

Cycle 0 $\theta = 0$

$0 - \theta$ 4	3 3	θ 1	3 0
$2 + \theta$ 2	 7	$1 - \theta$ 5	3 −2
2 4	3 3	1 7	6

Cycle 1 Optimal

 4	3 3	0 1	3 0
2 2	 7	1 5	3 4
 −2	 3	 1	

5–7 INACCESSIBLE ROUTES

In a transportation problem, some of the routes may be inaccessible. In order to avoid shipment over such routes, the cost coefficient c_{ij} of such a route should have an infinite value. Furthermore, if a fixed shipment is assigned to a certain route, regardless of the quantities shipped through other routes, this fixed shipment for the route between i and j can first be taken out from the supply at i and the demand at j. The remaining problem then becomes one which prohibits further shipment from i to j, and it can be treated as a transportation problem with an inaccessible route between i and j.

If only a few routes of a large transportation system are inaccessible, the problem can be solved by simply noting that the inaccessible routes have extremely high (say, infinite) cost coefficients, and they should be bypassed in selecting basic variables for an initial solution. On the other hand, if many routes in a system are inaccessible, not only the selection of an initial solution may be difficult, but the problem may not even have a feasible solution. Hence a special procedure is needed to test the feasibility of the solution and, if the solution is feasible, to provide an initial feasible solution.

The procedure for testing the feasibility of a solution of a transportation problem is analogous to Phase I of the simplex method for solving general linear programming problems. Since all x_{ij} in the inaccessible routes must be zero, the infeasible form may be given by

$$w = \sum_i \sum_j x_{ij} ,$$

in which the summation covers only those x_{ij} in the inaccessible routes. The infeasible form may also be written as follows:

$$w = \sum_{i=1}^{m} \sum_{j=1}^{n} d_{ij} x_{ij} , \tag{5-11}$$

in which

$$d_{ij} = \begin{cases} 1 , & \text{for all inaccessible routes,} \\ 0 , & \text{for all accessible routes.} \end{cases}$$

Thus the problem is temporarily changed to the optimization of w represented by (5-11). If min $w = 0$, the problem has a feasible solution; if not, it has no feasible solution.

It may be observed that Phase I in the solution of a transportation problem can also be carried out in a standard transportation array in which the coefficients c_{ij} are replaced by coefficients d_{ij}, as defined by (5-11). If min $w = 0$ at the end of Phase I, an initial solution will also be provided for entering into Phase II.

Example 5-7. Solve Example 5-1 given that (a) $c_{13} = \infty$ because that route is inaccessible, and (b) $c_{13} = 1$ but only five cars (no more and no less) are dispatched through that route.

In case (a), we have $c_{13} = \infty$ for the inaccessible route, and we shall avoid selecting x_{13} as a basic variable. However, it is not difficult to find an initial solution, as shown in Table 5-15(a), which also happens to be an optimal solution upon verification. Thus

$$\begin{aligned} z^* &= (20)(2) + (10)(6) + (15)(2) + (5)(1) + (25)(2) \\ &= 185 . \end{aligned}$$

Table 5–15

SOLUTION OF EXAMPLE 5–7

(a)

 4	20 2	 ∞	 3	10 6	30 2
 3	 5	15 2	5 1	0 4	20 0
25 2	 1	 3	 4	0 5	25 1
25 1	20 0	15 2	5 1	10 4	75

(b)

 4	20 2	 ∞	 3	5 6	25 2
 3	 5	10 2	5 1	5 4	20 0
25 2	 1	 3	 4	0 5	25 1
25 1	20 0	10 2	5 1	10 4	70

In case (b), we shall assign $x_{13} = 5$, as required, before entering $c_{13} = \infty$ into the transportation array, as shown in Table 5–15(b). Note also in the table that

$$a_1 = 30 - 5 = 25\,,$$
$$b_3 = 15 - 5 = 10\,.$$

Thus the problem can now be solved as if the route between 1 and 3 were inaccessible. The initial solution in the table is found to be an optimal solution. Thus according to the table,

$$z^* = (20)(2) + (5)(6) + (10)(2) + (5)(1) + (5)(4) + (25)(2)$$
$$= 165\,.$$

Table 5-16

SOLUTION FOR EXAMPLE 5-8

Given data

∞	1	6	∞	5	6
7	5	4	3	∞	5
∞	2	∞	1	∞	8
3	∞	∞	6	4	3
3	2	5	4	8	22

Cycle 0/Phase I $\theta = 1$

1	$2 - \theta$ 0	0	1	$4 + \theta$ 0	6 0
3 0	0	2 0	0	1	5 0
1	θ 0	3 1	4 0	$1 - \theta$ 1	8 1
0	1	1	0	3 0	3 0
3 0	2 0	5 0	4 −1	8 0	22

Table 5–16

(Continued)

Cycle 1/Phase I $\theta = 1$

 1	$1-\theta$ 0	θ 0	 1	5 0	6 0
3 0	 0	2 0	 0	 1	5 −1
 1	$1+\theta$ 0	$3-\theta$ 1	4 0	 1	8 0
 0	 1	 1	 0	3 0	3 0
3 1	2 0	5 1	4 0	8 0	22

Cycle 2/Phase I

 1	 0	1 0	 1	5 0	6 0
3 0	 0	2 0	 0	 1	5 0
 1	2 0	2 1	4 0	 1	8 1
 0	 1	 1	 0	3 0	3 0
3 0	2 −1	5 0	4 −1	8 0	22

However, the final result shall also include the cost of shipping $x_{13} = 5$, which has been excluded from the table. Hence the minimum total cost is

$$\min z = (5)(1) + 165 = 170 .$$

Example 5-8. Given the transportation problem with inaccessible routes shown in Table 5-16, determine if a feasible solution can be obtained.

Phase I is introduced by changing the given cost coefficients c_{ij} to

$$d_{ij} = \begin{cases} 1 , & \text{for all inaccessible routes,} \\ 0 , & \text{for all accessible routes.} \end{cases}$$

After Cycle 2, all $\bar{d}_{ij}$ are nonnegative, and it is found that

$$\text{Min } w = (2)(1) = 2 .$$

This indicates that the problem has no feasible solution. The operation is therefore terminated at the end of Phase I.

Example 5-9. Given the problem shown in Table 5-17, find the optimal solution, using Phase I and Phase II operations.

Since all basic variables can be placed in accessible routes in Phase I, it is obvious that min $w = 0$. Proceeding to Phase II, an optimal solution is obtained after Cycle 1, as shown in Table 5-13. Thus the optimal solution is given by

$$z^* = (6)(5) + (5)(4) + (2)(2) + (4)(1) + (2)(7) + (3)(3) = 81 .$$

5-8 MODIFICATION OF THE COST MATRIX

If a constant is added to every element of a row or a column of the cost matrix, a set of basic variables which optimizes the total cost in the original matrix is found to optimize also the total cost of the modified cost matrix. However, the total cost itself will be affected by such modifications. A modified matrix with a constant added to some rows and/or some columns of the original matrix is sometimes used to simplify numerical computations.

For example, consider, the transportation array in Table 5-18 in which the cost coefficients in the kth column ($j = k$) are given by

$$c'_{ik} = c_{ik} + c_k , \qquad i = 1, 2, \ldots, m ,$$

in which c_{ik} ($i = 1, 2, \ldots, m$) represent the cost coefficients in the kth column of the original matrix before modification, and c_k is the constant added to every element of the column. Let u_i and v_k be the simplex multipliers which, along with other v_j ($j \neq k$), are used in the optimality test for a set of basic variables that optimizes the total cost in the original

Table 5-17

SOLUTION FOR EXAMPLE 5-9

Given data

∞	1	∞	∞	5	6
7	5	4	∞	∞	5
∞	2	∞	1	7	8
3	∞	∞	6	4	3
3	2	5	4	8	22

Cycle 0/Phase I

 1	2 0	 1	 1	4 0	6 0
0 0	 0	5 0	 1	 1	5 0
 1	 0	 1	4 0	4 0	8 0
3 0	 1	 1	 0	0 0	3 0
3 0	2 0	5 0	4 0	8 0	22

(Cont.)

Table 5-17

(Continued)

Cycle 0/Phase II $\theta = 2$

 ∞	$2-\theta$ 1	 ∞	 ∞	$4+\theta$ 5	6 0
0 7	 5	5 4	 ∞	 ∞	5 3
 ∞	θ 2	 ∞	4 1	$4-\theta$ 7	8 2
3 3	 ∞	 ∞	 6	0 4	3 -1
3 4	2 1	5 1	4 -1	8 5	22

Cycle 1/Phase II

 ∞	 1	 ∞	 ∞	6 5	6 -2
0 7	 5	5 4	 ∞	 ∞	5 1
 ∞	2 2	 ∞	4 1	2 7	8 0
3 3	 ∞	 ∞	 6	0 4	3 -3
3 6	2 2	5 3	4 1	8 7	22

Table 5-18

MODIFICATION OF A COLUMN IN THE COST MATRIX

c_{11}	...	c'_{1k}	...	c_{1n}	
c_{21}	...	c'_{2k}	...	c_{2n}	u_2
...	...	...	...	...	
c_{m1}	...	c'_{mk}	...	c_{mn}	
		v'_k			

matrix before modification. Since at least one basic variable, say x_{2k}, exists in the kth column, we can compute v_k from u_2, which is assumed to be determined first; this value of v_k will, in turn, be used to compute other u_i corresponding to other basic variables in the column, or to compute the $\bar{c}_{ik}$ associated with the nonbasic variables in the column. In other words, before modification,

$$v_k = c_{ik} - u_i \,, \qquad \text{for basic variables,}$$

$$\bar{c}_{ik} = c_{ik} - (u_i + v_k) \,, \qquad \text{for nonbasic variables.}$$

In the modified matrix, these relations become

$$v'_k = c'_{ik} - u_i \,, \qquad \text{for basic variables,}$$

$$\bar{c}'_{ik} = c'_{ik} - (u_i + v'_k) \,, \qquad \text{for nonbasic variables.}$$

From the stated relationship between c'_{ik} and c_{ik}, we have

$$v'_k = c_{ik} + c_k - u_i = v_k + c_k \,,$$

$$\bar{c}'_{ik} = (c_{ik} + c_k) - [u_i + v_k + c_k] = c_{ik} - (u_i + v_k) \,.$$

Hence it is seen that the set of basic variables which optimizes the total cost in the original matrix also optimizes the modified cost matrix, and conversely, because the conditions for $\bar{c}_{ik} \geq 0$ and for $\bar{c}'_{ij} \geq 0$ are the same.

The minimum total cost for the problem involving c'_{ik} $(i = 1, 2, \ldots, m)$ is then given by

$$\text{Min } z' = \sum_{i=1}^{m} \sum_{j=1}^{n} c_{ij} x_{ij} + \sum_{i=1}^{m} c_k x_{ik}, \tag{5-12}$$

in which the last term in the equation represents the contribution of the constant c_k in every element of the kth column.

Example 5-10. In Table 5-19(a), each coefficient c_{ij} in the transportation array represents the profit for shipment of one unit from origin i to destination j. The units of supply at the origins and the units of demand at the destinations are also given. Determine the shipping pattern which optimizes the total profit.

This is an optimization problem for which the objective function is

$$\text{Max } A = \sum_{i=1}^{m} \sum_{j=1}^{n} c_{ij} x_{ij}.$$

It can be converted to a minimization problem with an objective function

$$\text{Min } z = \sum_{i=1}^{m} \sum_{j=1}^{n} (-c_{ij}) x_{ij}.$$

Since the constraint equations for balanced supply and demand remain unchanged, it is easy to solve the equivalent minimization problem by changing the signs of all c_{ij} coefficients in the transportation array.

It is desirable, though not necessary, to add a constant to each row or column of the matrix with negative coefficients, such that all coefficients can be made positive again. Let us add 9 to each element in all rows (or all columns) of the 3×5 matrix in Table 5-19(a) after all the coefficients have undergone sign changes. We have, in effect, added 9 to each of the $-c_{ij}$, and the result is shown by the coefficients in part (b) of the table. The initial basic solution leads to $\bar{c}_{25} = -2$. Consequently, x_{25} is selected as the entering variable and $x_{15} = 10 - \theta = 0$ as the departing variable. Upon the substitution of $\theta = 10$ into other basic variables, an optimal shipping pattern is obtained.

5-9 ASSIGNMENT PROBLEM

The assignment problem is a type of allocation problem in which n positions are to be filled by n candidates, and each position can be filled by only one candidate. Let the candidates be designated by $i = 1, 2, \ldots, n$, and the positions be designated by $j = 1, 2, \ldots, n$. The effectiveness of candidate i in performing the task specified for position j is denoted by c_{ij}. The problem is to assign all candidates to different positions in order to optimize the overall effectiveness, which is measured by factors such

Table 5-19

SOLUTION FOR EXAMPLE 5-10

(a)

6	8	9	7	4	25
7	5	8	9	6	25
8	9	7	6	5	25
25	20	15	5	10	75

(b)

3	$0 + \theta$ 1	15 0	2	$10 - \theta$ 5	25 0
$20 - \theta$ 2	4	1	5 0	θ 3	25 0
$5 + \theta$ 1	$20 - \theta$ 0	2	3	4	25 -1
25 2	20 1	15 0	5 0	10 5	75

as the number of man-hours required to perform the tasks, or the scores obtained by the candidates in a set of tests. Although the problem may involve maximization or minimization, we shall formulate the problem as a minimization problem with no loss in generality.

Let x_{ij} be the fraction of the time that candidate i should fill the position j ($x_{ij} \geq 0$). Then for each candidate

$$\sum_{j=1}^{n} x_{ij} = 1 , \qquad i = 1, 2, \ldots, n , \tag{5-13}$$

and for each position

$$\sum_{i=1}^{n} x_{ij} = 1 , \qquad j = 1, 2, \ldots, n . \tag{5-14}$$

The objective function is given by

$$\sum_{i=1}^{n} \sum_{j=1}^{n} c_{ij} x_{ij} = \text{Min } z . \tag{5-15}$$

Since the assignment problem is shown to be a special case of the transportation problem with balanced damand and supply in which $m = n$, $a_i = 1$ ($i = 1, 2, \ldots, n$) and $b_j = 1$ ($j = 1, 2, \ldots, n$), there should be $(2n - 1)$ basic variables in the solution. Because a basic variable will be assigned as large a value as possible in obtaining an initial solution, n of the $(2n - 1)$ basic variables must be 1, and the remaining $(n - 1)$ variables must be zero. Since any basic solution will be highly degenerate, it is important that the zeros in the initial solution be placed in locations which do not form a closed path for the basic variables. One way to avoid this situation is to place all zeros in the same row or the same column. This is always possible, though not necessarily efficient for obtaining an optimal solution, because in each row and each column, there is only one basic variable with a value of 1, and the remaining $(n - 1)$ elements may be assigned the value of 0.

Example 5-11. Find an optimal solution for the assignment problem in Example 5-3.

An initial solution can easily be obtained by the minimum cost rule, as shown in Cycle 0 of Table 5-20. After two cycles, an optimal solution is obtained. Thus

$$z^* = 1 + 3 + 1 + 2 + 1 + 2 = 10 .$$

Example 5-12. Five persons are assigned to perform five different tasks. The efficiency of each person in performing each task is based on his scores on a set of tests, in which the highest score represents the best efficiency. In the given data of Table 5-21, each c_{ij} represents the efficiency of the ith person in performing the jth task. Determine the assignment which will maximize the overall efficiency.

The problem may be changed to one of minimization if we first multiply all coefficients by -1. Furthermore, the coefficients may be modified

Table 5-20

SOLUTION OF EXAMPLE 5-11

Cycle 0 $\theta = 0$

4	5	3	0 2	7	1 1	1 1
5	θ 4	0 2	3	$1-\theta$ 4	6	1 3
6	8	1 1	4	3	7	1 2
3	1 2	9	7	6	5	1 -1
$0+\theta$ 1	$0-\theta$ 3	5	1 1	8	4	1 0
$1-\theta$ 2	6	7	5	$0+\theta$ 2	3	1 1
1 1	1 3	1 -1	1 1	1 1	1 0	

Cycle 1 $\theta = 1$

4	5	3	0 2	7	1 1	1 -2
5	0 4	0 2	θ 3	$1-\theta$ 4	6	1 0
6	8	1 1	4	3	7	1 -1
3	1 2	9	7	6	5	1 -2
$0+\theta$ 1	3	5	$1-\theta$ 1	8	4	1 -3
$1-\theta$ 2	6	7	5	$0+\theta$ 2	3	1 -2
1 4	1 4	1 2	1 4	1 4	1 3	

(Cont.)

Table 5-20 (Continued)

Cycle 2 Optimal

4	5	3	0 / 2	7	1 / 1	1 / −1
5	0 / 4	0 / 2	1 / 3	4	6	1 / 0
6	8	1 / 1	4	3	7	1 / −1
3	1 / 2	9	7	6	5	1 / −2
1 / 1	3	5	0 / 1	8	4	1 / −2
0 / 2	6	7	5	1 / 2	3	1 / −1
1 / 3	1 / 4	1 / 2	1 / 3	1 / 3	1 / 2	

by adding 10 to each element in the cost matrix. Thus we have new coefficients, as shown in Cycle 0 of the problem. In the initial solution, all basic variables with zero values have been assigned to the fifth column, as indicated. After two cycles, an optimal solution for assignment is obtained.

5-10 AN ALTERNATIVE PROCEDURE FOR OPTIMAL ASSIGNMENT

The general algorithm for the transportation problem may prove to be rather tedious for the solution of assignment problems if many cycles are required to drive the zeros from one location to another. On the other hand, an optimal solution can be identified directly from the cost coefficients if they contain the same minimum cost for all assignments. For example, if for all $x_{ij} = 1$, the corresponding $c_{ij} = 0$ while all other $c_{ij} \geq 0$, then the basic solution containing this set of x_{ij} leads to $z^* = 0$, which cannot be made smaller. Thus, *it is desirable to develop a procedure which will generate as many zeros as possible in the cost coefficients and, more specifically, will produce at least one such coefficient in every row and every column.* If this is accomplished by repeatedly adding a constant to every element in a row or a column, the set of basic variables which optimizes the modified problem will also optimize the original problem.

Table 5–21

SOLUTION OF EXAMPLE 5–12

Given data

5	3	7	4	5	1
6	4	6	3	6	1
3	6	5	6	3	1
6	8	4	7	8	1
2	5	3	8	6	1
1	1	1	1	1	

Cycle 0 $\theta = 1$

 5	 7	1 3	 6	0 5	1 5
1 4	 6	 4	 7	0 4	1 4
 7	θ 4	 5	 4	$1-\theta$ 7	1 7
 4	$1-\theta$ 2	 6	 3	$0+\theta$ 2	1 2
 8	 5	 7	1 2	0 4	1 4
1 0	1 0	1 −2	1 −2	1 0	

(Cont.)

Table 5–21

(Continued)

Cycle 1 $\theta = 1$

 5	 7	1 3	 6	0 5	1 5
1 4	 6	 4	 7	0 4	1 4
 7	1 4	 5	θ 4	$0-\theta$ 7	1 7
 4	 2	 6	 3	1 2	1 2
 8	 5	 7	$1-\theta$ 2	$0+\theta$ 4	1 4
1 0	1 −3	1 −2	1 −2	1 0	

Cycle 2 Optimal

 5	 7	1 3	 6	0 5	1 5
1 4	 6	 4	 7	0 4	1 4
 7	1 4	 5	0 4	 7	1 6
 4	 2	 6	 3	1 2	1 2
 8	 5	 7	1 2	0 4	1 4
1 0	1 −2	1 −2	1 −2	1 0	

We need to examine only the cost matrix, which consists of nonnegative coefficients in a minimization problem. We first subtract the smallest element in each row from all elements in its row, and then, where necessary, the smallest element in each column from all elements in its column. If a solution can be obtained by assigning all nondegenerate basic variables to positions of zero cost coefficients, the sloution is optimal, since the total cost thus obtained is zero.

In general, if the minimum number of lines that can be constructed to pass through all the zeros of the matrix equals n, an optimal solution has been found; if not, more zeros must be introduced at appropriate positions by further adding and subtracting a constant to every element in a row or a column such that all elements of the matrix will remain nonnegative. This can be done by following certain simple rules which will be explained in a later example.

Example 5–13. Solve the assignment problem in Example 5–12 by using the method of selecting an optimal solution from the reduced cost matrix.

The original cost matrix with negative coefficients for the minimization problem can be used to obtain a reduced matrix with nonnegative coefficients. In Table 5–22, a constant $c_1 = 7$ is added to elements of the first row of the given matrix, $c_2 = 6$ to the second row, $c_3 = 6$ to the third row, $c_4 = 8$ to the fourth row, and $c_5 = 8$ to the fifth row. The reduced matrix thus obtained cannot be further reduced by adding a negative constant to all elements in any column without causing some coefficients to become negative. Hence we start checking the positions of zeros without further reduction.

We can proceed to make tentative assignments by screening the positions of zeros row by row, or column by column. For example, starting with the first row, we can select a zero (c_{13} in this case) and cross out other zeros in row 1 and in column 3 (c_{23} in this case). In the second row, we can select from those zeros which have not previously been crossed out (say, select c_{21} from c_{21} and c_{25} in this case), and cross out other zeros in row 2 and in column 1 (c_{25} in this case). In the third row, we select the zero at c_{32} and cross out the zeros c_{34} and c_{42} in row 3 and column 2, respectively. Finally, we can pick c_{45} in row 4 and c_{54} in row 5 to complete the assignment. Thus the optimal assignment is given by

$$x_{13} = 1\,, \qquad x_{21} = 1\,, \qquad x_{32} = 1\,, \qquad x_{45} = 1\,, \qquad \text{and} \qquad x_{54} = 1\,.$$

As a check, we can construct a minimum number of lines passing through all zeros. This is accomplished by drawing the lines in such an order that a line containing the most zeros will be drawn first. Thus we first draw a line passing through the elements in row 2, and then draw lines passing through the elements in columns 2, 3, 4, and 5. Of course,

Table 5-22

ALTERNATIVE SOLUTION OF EXAMPLE 5-12

Given matrix

−5	−3	−7	−4	−5
−6	−4	−6	−3	−6
−3	−6	−5	−6	−3
−6	−8	−4	−7	−8
−2	−5	−3	−8	−6

Optimal solution

		[0]		
[0]		×0		×0
	[0]		×0	
	×0			[0]
			[0]	

Reduced matrix

2	4	0	3	2
0	2	0	3	0
3	0	1	0	3
2	0	4	1	0
6	3	5	0	2

Incomplete

		[0]		
×0		×0		[0]
	×0		[0]	
	[0]			×0
			×0	

we can also draw lines through rows 2, 3, and 4, and through columns 3 and 4. In either case, the minimum number of lines is 5, and a complete assignment is possible. This check is a safeguard against possible initial assignments which may lead to the erroneous conclusion that a complete assignment is impossible. One such case is shown in the last attempt in Table 5-22.

Example 5-14. Solve the assignment problem in Example 5-3 by using the method of selecting an optimal solution from the reduced cost matrix.

Again, the original cost matrix is reduced by eliminating the smallest element in each row from all elements in its row, as shown in Table 5-23. As we become more familiar with the procedure, we need not single out the zeros in a separate table. Thus we proceed to make tentative assignments by selecting $c_{16} = 0$ in row 1, $c_{23} = 0$ in row 2, $c_{42} = 0$ in row 4, $c_{51} = 0$ in row 5, and $c_{65} = 0$ in row 6; but we cannot select $c_{33} = 0$ in

Table 5-23

ALTERNATIVE SOLUTION OF EXAMPLE 5-3

Given matrix

4	5	3	2	7	1
5	4	2	3	4	6
6	8	1	4	3	7
3	2	9	7	6	5
1	3	5	1	8	4
2	6	7	5	2	3

Cycle 0

3	4	2	1	6	[0]
3	2	[0]	1	2	4
5	7	~~0~~	3	2	6
1	[0]	7	5	4	3
[0]	2	4	~~0~~	7	3
~~0~~	4	5	3	[0]	1

Reduced matrix

3	4	2	1	6	0
3	2	0	1	2	4
5	7	0	3	2	6
1	0	7	5	4	3
0	2	4	0	7	3
0	4	5	3	0	1

Cycle 1

2	4	2	~~0~~	5	[0]
2	2	~~0~~	[0]	1	4
4	7	[0]	2	1	6
~~0~~	[0]	7	4	3	3
[0]	3	5	~~0~~	7	4
~~0~~	5	6	3	[0]	2

row 3 because we have already chosen $c_{23} = 0$ in the same column. As a check, the minimum number of lines passing through all zero values may consist of a set of five lines passing through rows 5 and 6, and columns 3, 2, and 6, constructed in that order. Since the minimum number of lines covering all zeros is less than n, no complete assignment can be made in the reduced matrix in Cycle 0.

As we attempt to reduce the cost coefficients further in order to produce more zeros in the proper positions for a complete assignment, we are confronted with the prospect of introducing negative cost coefficients. However, this can be avoided if, for any constant subtracted from all elements in a row containing zeros, we add the same constant to all elements in each column corresponding to each zero in that row. This same procedure may be applied to the modifying of all elements in a row containing zero. Since we are particularly interested in reducing to zero the smallest element in the reduced matrix which is not covered by any line, we subtract this element from every element in all rows which are not covered by a line, and add the same element to every element in all columns which are covered by a line. Specifically for this problem, we first subtract 1 from all elements in rows 1, 2, 3, and 4, and then add 1 to all elements in columns 2, 4, and 5 in the reduced matrix at Cycle 0. The result of this modification is shown in the reduced matrix at Cycle 1. The same result can be obtained if we simply subtract this smallest element from every element in the reduced matrix at Cycle 0, which is not covered by a line, and add it to every element which lies at an intersection of two lines.

We proceed to make assignments in the reduced matrix at Cycle 1, and an optimal solution is indicated by the zeros in c_{16}, c_{24}, c_{33}, c_{42}, c_{51}, and c_{65}. Hence the optimal assignment is given by

$$x_{16} = 1\,, \quad x_{24} = 1\,, \quad x_{33} = 1\,, \quad x_{42} = 1\,, \quad x_{51} = 1 \quad \text{and} \quad x_{65} = 1\,.$$

5-11 TRANSPORTATION WITH SURPLUS AND SHORTAGE

In real situations for transportation, the total in supplies at the origins and the total in demands at the destinations are seldom equal to each other. Nevertheless, it is often possible to isolate the situation in such a way that the problem becomes one of balanced supply and demand. On the other hand, such an idealization is not always necessary, since the conditions of surplus and/or shortage can be included in the problem.

In the case of surplus at the origins, we have the following relationship for total supply and demand:

$$\sum_{i=1}^{m} a_i > \sum_{j=1}^{n} b_j\,.$$

At each origin, we have

$$\sum_{j=1}^{n} x_{ij} \leq a_i\,, \qquad i = 1, 2, \ldots, m\,.$$

Let

$$\sum_{i=1}^{m} a_i - \sum_{j=1}^{n} b_j = b_0\,. \tag{5-16}$$

Also, let x_{i0} be the surplus at the ith origin. Then

$$\sum_{j=1}^{n} x_{ij} + x_{i0} = a_i\,, \qquad i = 1, 2, \ldots, m\,, \tag{5-17}$$

in which $x_{i0} \geq 0$ are slack variables introduced to balance the supply and demand. If all demands at the destinations are to be met, the following conditions remain unchanged:

$$\sum_{i=1}^{m} x_{ij} = b_j\,, \qquad j = 1, 2, \ldots, n\,.$$

If we can assign appropriate values to coefficients c_{i0} $(i = 1, 2, \ldots, m)$ corresponding to x_{i0} $(i = 1, 2, \ldots, m)$, then the problem is reduced to that of the classical transportation problem by introducing a fictitious destination 0 (and hence a fictitious column 0 in the transportation array) to which the total surplus b_0 will be dumped. The objective is to optimize the total cost of transportation and disposal of the surplus.

On the other hand, in the case of shortage for distribution at the destinations, we have the following relationship:

$$\sum_{i=1}^{m} a_i < \sum_{j=1}^{n} b_j\,.$$

At each destination, we have

$$\sum_{i=1}^{m} x_{ij} \leq b_j\,, \qquad j = 1, 2, \ldots, n\,.$$

Let

$$\sum_{j=1}^{n} b_j - \sum_{i=1}^{m} a_i = a_0\,. \tag{5-18}$$

Also, let x_{0j} be the storage at the jth destination. Then

$$\sum_{i=1}^{m} x_{ij} + x_{0j} = b_j\,, \qquad j = 1, 2, \ldots, n\,, \tag{5-19}$$

in which $x_{0j} \geq 0$ are slack variables introduced to balance the supply and demand. If all supplies at the origins are to be shipped out to meet the demands until exhausted, the following conditions remained unchanged:

$$\sum_{j=1}^{n} x_{ij} = a_i\,, \qquad i = 1, 2, \ldots, m\,.$$

Similarly, if we can assign appropriate values to coefficients c_{0j} $(j = 1, 2, \ldots, n)$ corresponding to x_{0j} $(j = 1, 2, \ldots, n)$, then the problem is reduced to that of the classical transportation problem by introducing a fictitious origin 0 (and hence a fictitious row 0 in the transportation array), from which the total shortage a_0 will come. The objective is to optimize the total cost of transportation and penalty for not meeting the demand.

If the demands are to be met solely on the basis of profits, regardless of the amount of supply available, a problem may involve both surplus and shortage, since not all supplies at the origins are shipped even if they are in demand, and not all demands at the destinations are to be met even if such demands can be met. Thus

$$\sum_{j=1}^{n} x_{ij} \leq a_i\,, \qquad i = 1, 2, \ldots, m\,,$$

and

$$\sum_{i=1}^{m} x_{ij} \leq b_j\,, \qquad j = 1, 2, \ldots, n\,.$$

Again, introducing slack variables x_{i0} and x_{0j}, we have

$$\sum_{j=1}^{n} x_{ij} + x_{i0} = a_i\,, \qquad i = 1, 2, \ldots, m\,; \tag{5-20}$$

$$\sum_{i=1}^{m} x_{ij} + x_{0j} = b_j\,, \qquad j = 1, 2, \ldots, n\,. \tag{5-21}$$

Summing Eqs. (5-20) over all values of i and summing Eqs. (5-21) over all values of j, we have, respectively,

$$\sum_{i=1}^{m}\sum_{j=1}^{n} x_{ij} + \sum_{i=1}^{m} x_{i0} = \sum_{i=1}^{m} a_i\,,$$

$$\sum_{j=1}^{n}\sum_{i=1}^{m} x_{ij} + \sum_{j=1}^{n} x_{0j} = \sum_{j=1}^{n} b_j\,.$$

Let

$$x_{00} = \sum_{i=1}^{m}\sum_{j=1}^{n} x_{ij}\,, \tag{5-22}$$

which represents the total amount actually shipped from all origins to all destinations. Then

$$x_{00} + \sum_{i=1}^{m} x_{i0} = \sum_{i=1}^{m} a_i\,, \tag{5-23}$$

$$x_{00} + \sum_{j=1}^{n} x_{0j} = \sum_{j=1}^{n} b_j\,. \tag{5-24}$$

Hence a fictitious origin 0 and fictitious destination 0 may be added to the transportation array, as shown in Table 5-24, provided that the c_{i0} $(i = 1, 2, \ldots, m)$ and c_{0j} $(j = 1, 2, \ldots, n)$ assigned to the array are meaningful. Since x_{00} is not an independent variable but is the sum of all x_{ij} actually shipped, as represented by (5-22), the cost coefficient c_{00} must be zero so that the shipping cost will not be counted twice. Furthermore, (5-23) and (5-24) are not independent constraint equations, and we can conveniently assume $u_0 = 0$ and $v_0 = 0$. Consequently, x_{00} may be treated as if it were a basic variable in the enlarged transportation array, since

$$c_{00} = u_0 + v_0\,.$$

Thus the criterion for optimality also holds true for $i = 0$ and $j = 0$.

Table 5-24

TRANSPORTATION ARRAY WITH SURPLUS AND SHORTAGE

$i \backslash j$	0	1	2	...	n	
0	x_{00} c_{00}	x_{01} c_{01}	x_{02} c_{02}	... ...	x_{0n} c_{0n}	$\sum b_j$ u_0
1	x_{10} c_{10}	x_{11} c_{11}	x_{12} c_{12}	... ...	x_{1n} c_{1n}	a_1 u_1
2	x_{20} c_{20}	x_{21} c_{21}	x_{22} c_{22}	... ...	x_{2n} c_{2n}	a_2 u_2
	... ...	... ...	... ...	... ...	... ...	... ...
m	x_{m0} c_{m0}	x_{m1} c_{m1}	x_{m2} c_{m2}	... ...	x_{mn} c_{mn}	a_m u_m
	$\sum a_i$ v_0	b_1 v_1	b_2 v_2	... ...	b_n v_n	

In the case of surplus only, from the row for $i = 0$ in the array, we have

$$\sum_{j=1}^{n} x_{0j} = 0$$

and

$$x_{00} = \sum b_j .$$

Hence the row for $i = 0$ can be eliminated from the array in Table 5-24 if the sum of the column for $j = 0$ is replaced by

$$\sum a_i - x_{00} = \sum a_i - \sum b_j = b_0 .$$

Similarly, in the case of shortage only, from the column for $j = 0$ in the array, we have

$$\sum_{i=1}^{m} x_{i0} = 0$$

and

$$x_{00} = \sum a_i .$$

Hence the column for $j = 0$ can be eliminated from the array in Table 5-24 if the sum of the row for $i = 0$ is replaced by

$$\sum b_j - x_{00} = \sum b_j - \sum a_i = a_0 .$$

In either case, the cell for x_{00} is eliminated from the array, and we need not be concerned with the coefficient c_{00}. Although it is logical to assume that $v_0 = 0$ for a transportation problem with surplus and $u_0 = 0$ for a transportation problem with shortage, such assumptions are not necessary. Instead, any row or column may be treated as redundant in the enlarged problem; hence the simplex multiplier corresponding to that row or column can be assumed to be zero.

Example 5-15. The rapid transit system in Example 5-1 is under study for one of the two possible changes in the conditions of demand and supply: (A) The numbers of cars available in carbarns 1, 2, and 3 are changed to 30, 40, and 30, respectively, and no cost is involved in storing the surplus cars at the origins after all demands have been met; and (B) the numbers of cars required at terminals 1, 2, 3, 4, and 5 are changed to 25, 30, 15, 10, and 20, respectively, and no penalty is involved for shortage of cars at the destinations after all supplies have been exhausted. In each case, how should the cars be allocated in order to minimize the transportation cost?

In formulating transportation problems with surplus and/or shortage, we must decide whether we want to minimize the transportation cost alone, or to minimize the total cost of transportation plus cost of disposing surplus and/or the penalty of shortage. In this particular example, it is clearly stated that the system encompasses only transportation itself, and other factors in supply and demand are considered as outside the influence of the system.

In case (A) of the example, the surplus column ($j = 0$) is shown dashed in Table 5-25(a). The cost coefficients c_{i0} ($i = 1, 2, \ldots, m$) are assigned zero values. Since all demands must be met first before surplus is allowed to accumulate, it makes no difference if c_{i0} is assigned a constant other than 0. In other words, so long as all demands must be met first, the coefficients c_{i0} ($i = 1, 2, \ldots, m$) represent only the relative costs of storing the surplus at origins $1, 2, \ldots, m$; they may be quite independent of the cost coefficients of transportation, since they are outside the influences of the system. Thus the coefficients $c_{i0} = 0$ ($i = 1, 2, \ldots, m$) are also applicable to the case when the storage costs for surplus at various origins are equal but not necessarily zero. If the storage costs at various origins are not equal, then the cost coefficients should reflect the difference.

Similarly, in case (B) of the example, the shortage row ($i = 0$) is shown dashed in Table 5-25(a). The cost coefficients c_{0j} ($j = 1, 2, \ldots, n$) are

Table 5–25

SOLUTION OF EXAMPLE 5–15

(a)

15 0	 4	 2	15 1	 3	 6	30 0
10 0	15 3	 5	 2	5 1	10 4	40 0
 0	10 2	20 1	 3	 4	 5	30 −1
25 0	25 3	20 2	15 1	5 1	10 4	

(b)

5 0	 0	 0	 0	20 0	25 0
$10-\theta$ 4	$5+\theta$ 2	15 1	 3	 6	30 4
10 3	 5	 2	10 1	 4	20 3
θ 2	$25-\theta$ 1	 3	 4	 5	25 3
25 0	30 −2	15 −3	10 −2	20 0	

assigned zero values. Again, the coefficients c_{0j} are assumed to be outside of influences of the system, since all demands must first be met until the total supply is exhausted. Thus c_{0j} need not be zero or equal for all j if the given conditions indicate otherwise.

The solution of the problem is carried out in Table 5-25 which is self-explanatory. It is sufficient to add that because the demand and supply are no longer in balance, there will be $m + n$ basic variables in both cases. However, we may treat Table 5-25(a) as a standard transportation array of $m \times n'$ in which $n' = n + 1$. Then, the number of basic variables will be

$$m + n' - 1 = m + n .$$

Similarly, Table 5-24(b) may be treated as a standard transportation array of $m \times n$ in which $m' = m + 1$. Again, the number of basic variables is

$$m' + n - 1 = m + n .$$

Thus the method of solving the classical transportation problem can be applied to the solution of transportation problems with surplus or shortage.

In Table 5-25(a), an initial solution is obtained by first assigning values to the variables inside the system according to minimum cost rule, and then assigning the surplus to column $j = 0$. The procedure is terminated after one cycle, since the initial basic solution is optimal. In Table 5-25(b), an initial solution is obtained in a similar manner. However, the initial solution is not optimal, and a change of basis is necessary, as indicated by the entering variable θ. Upon the substitution of $\theta = 10$ into the basic variables, we will obtain a new solution, which is found to be optimal.

Example 5-16. The rapid transit system in Example 5-1 has suddenly received new demands in terminals 2, 4, and 5 one morning because a major highway repair near these terminals has slowed down the traffic and some motorists have decided to use public transportation. The total demands of all terminals including new demands are shown in Table 5-26. Before this crisis happened unexpectedly, a private organization had made a very attractive offer to charter as many cars as can be spared in carbarn 1 on that same day. Thus the administrator of the system faces the problem of assigning values to the cost coefficients in the surplus column and in the shortage row in Table 5-26. This is not an easy task, but he has to decide the best he can. He would like to keep some cars in carbarn 1 in anticipation of extra income from charter arrangements, but he is quite willing, with different degrees of eagerness, to deplete the cars in carbarns 2 and 3. Thus the cost coefficients c_{i0} ($i = 1, 2, 3$) in the surplus column are assigned values of -1, 6, and 2, respectively; the negative sign in a cost coefficient represents a net profit for not supplying. On the other hand, he also attaches different values or penalties for not meeting demands at various terminals since, according to his experience, the riders at terminal 5 have been most faithful but the patronage at terminals 2 and 4 has been poor and there has been talk of moving these terminals elsewhere. Thus the cost coefficients c_{0j} ($j = 1, 2, 3, 4, 5$) in the shortage

Table 5-26

SOLUTION OF EXAMPLE 5-16

60 0	 5	30 −1	 3	10 −2	 10	100 0
15 −1	 4	 2	15 1	0 3	30 6	 −1
 6	 3	 5	 2	20 1	20 4	 −3
 2	25 2	 1	 3	0 4	25 5	 −2
75 0	25 4	30 −1	15 2	10 −2	20 7	

row are assigned values of 5, −1, 3, −2, and 10, respectively; the negative sign in a cost coefficient means actually a net gain for not meeting the demand. (For example, if the patrons are unhappy enough about the transit system, it will be easier to discontinue the service at those terminals.) The cost coefficient $c_{00} = 0$ is also noted in Table 5-26. Determine the minimum cost of supplying the demands under such a value system.

The solution of this problem may be carried out in a manner similar to the classical transportation problem with balanced demand and supply, except that x_{00} is always the last variable to be selected, since it is not independent of other basic variables. Thus for the enlarged array of $m' \times n'$, where $m' = m + 1$ and $n' = n + 1$, the number of basic variables, including x_{00}, appears to be

$$m' + n' - 1 = m + n + 1\,.$$

However, in reality, there are only $m + n$ independent basic variables in the original problem, as anticipated, since all $m + n$ constraint equations are independent when supply and demand are not in balance.

The minimum cost of supplying the demands under the prescribed circumstances is

$$z^* = (15)(1) + (20)(4) + (25)(2) = 145\,.$$

Note that this is different from the minimum cost of dispatching all available cars to meet the demand. In fact, we would be paying a price if 15 cars in carbarn 1 are dispatched to terminal 2 in order to meet part of the demand there.

REFERENCES

5-1. HITCHCOCK, F. L., "The Distribution of a Product from Several Sources to Numerous Localities," *Journal of Mathematics and Physics*, **20** (1941), 224–230.

5-2. PRAGER, W., "A Generalization of Hitchcock's Transportation Problem," *Journal of Mathematics and Physics*, **36** (1957), 99–106.

5-3. KOOPMANS, T. C., and S. REITER, "A Model of Transportation," *Activity Analysis of Production and Allocation* (T. C. Koopmans, editor), Wiley, New York (1951), 222–259.

PROBLEMS

P5-1. Solve for the minimum cost of the transportation problem shown in Table P5-1.

Table P5-1

2	1	6	4	5	9
7	5	4	3	2	5
4	2	3	1	7	8
3	5	2	6	4	3
6	2	5	4	8	

P5-2. Determine the minimum cost of the transportation problem shown in Table P5-2.

Table P5-2

9	7	8	5	6	11
5	6	7	8	9	7
4	5	9	6	7	15
10	7	5	8	3	

P5-3. Rework Problem P5-1 given that $c_{31} = 1$ and $c_{34} = 8$, where other cost coefficients remain unchanged.

P5-4. Rework Problem P5-2 given that $c_{32} = 9$ and $c_{33} = 5$, while other cost coefficients remain unchanged.

P5-5. Rework Example 5-9 by using the initial solution for Phase I in Table P5-5.

Table P5-5

(0) 1	(2) 0	1	1	(4) 0	6
0	0	(5) 0	(0) 1	1	5
1	0	1	(4) 0	(4) 0	8
(3) 0	1	1	0	0	3
3	2	5	4	8	

P5-6. Rework Example 5-7 for both (a) and (b) given that $c_{12} = \infty$ and $c_{22} = \infty$, while other cost coefficients remain unchanged.

P5-7. Find the optimal solution (minimum cost) for the assignment problem shown in Table P5-7.

Table P5–7

60	42	25	31	1
36	81	62	70	1
47	38	74	65	1
56	62	33	52	1
1	1	1	1	

P5–8. The central office of a car-truck rental service company has five available vehicles: (1) a two-door sedan, (2) a four-door sedan, (3) a station wagon, (4) a small truck, and (5) a large truck. There are five district offices which have demands for all of these vehicles, and the central office has decided to supply only one to each district office. The cost of dispatching each vehicle to each destination is given in Table P5–8. Determine the assignment for which the cost is minimum.

Table P5–8

		Destination				
		1	2	3	4	5
Vehicles	1	9	11	14	7	8
	2	7	9	11	7	10
	3	10	9	15	9	11
	4	9	10	12	11	9
	5	13	11	14	10	14

P5–9. The cost matrix for an assignment problem is shown in Table P5–9. Find the minimum cost solution for the problem.

Table P5–9

60	41	82	54	33	72
58	44	77	60	30	75
70	52	90	71	42	83
66	58	83	67	46	80
75	70	88	73	51	85
77	68	95	80	60	90

Table P5–10

	Job				
Man	6	7	9	5	2
	3	4	8	6	6
	7	2	5	8	9
	4	5	6	3	4
	2	5	8	9	7

P5–10. The matrix is given in Table P5–10 for an assignment problem of maximizing the effectiveness of performance. The elements of the matrix represent the ranking scores, with the highest number representing the best performance. Find the assignment leading to the best possible total score by any formal method. What is the score?

P5–11. A contractor plans to purchase five types of construction equipment. There are manufacturers which have large supplies of all types of equipment in their district warehouses. The relative costs for various types of equipment by different manufacturers are given in the cost matrix in Table P5–11. If the contractor immediately needs 10, 7, 5, 8, and 3 pieces of Types 1, 2, 3, 4, and 5, respectively, and the manufacturers 1, 2, and 3 can supply not more than 20, 25, and 15 pieces, respectively, in any combination, how should orders be placed in order to minimize the cost? (Assume that the manufacturers are willing to get the order and are not charging storage cost for surplus.)

Table P5–11

		Equipment type				
		1	2	3	4	5
Manufacturers	1	9	7	8	5	6
	2	5	6	7	8	9
	3	4	9	5	6	7

P5–12. Rework Problem 5–1 for the case when the supplies at origins 1, 2, 3, and 4 are 12, 10, 15, and 5, respectively, but the demands in destinations remain the same. Assume that the costs of storing surplus at origins 1, 2, 3, and 4 are 3, 1, 4, and -2, respectively, and that all demands must be met. Determine the minimum total cost including the cost of storage.

P5–13. Rework Problem 5–1 for the case when the demands at destinations 1, 2, 3, 4, and 5 are 10, 5, 5, 6, and 10, respectively, but the supplies remain the same. Assume that penalties of not meeting the demands at destinations 1, 2, 3, 4, and 5 are 2, 4, -1, 3, and -5, respectively, and that the demands must be met until the supplies are exhausted. Determine the minimum total cost including the cost of penalties.

P5–14. Five students are registered for a project course, and are assigned to three faculty members. The areas of interest of the students S_i $(i = 1, 2, \ldots, 5)$ are materials, transportation, water resources, foundation engineering, and structures, respectively. The fields of specialization of the faculty members F_j $(j = 1, 2, 3)$ are fluid mechanics, solid mechanics, and systems engineering, respectively. They represent eight different subjects altogether. The desirability of matching student interest to faculty specialization according to subjects is given by a score (with the smallest number representing the most desirable) in Table P5–14. Given that each faculty member must supervise at least one student but not more than two students, determine the combination of faculty and students which minimizes the total score.

Table P5–14

	F_1	F_2	F_3	S_1	S_2	S_3	S_4	S_5
F_1	0	3	3	2	5	1	2	4
F_2	3	0	3	2	5	4	3	1
F_3	3	3	0	5	1	3	4	2
S_1	2	2	5	0	5	3	2	1
S_2	5	5	1	4	0	1	3	2
S_3	1	4	3	3	1	0	4	5
S_4	2	3	4	2	3	4	0	2
S_5	4	1	2	1	2	5	2	0

CHAPTER 6

NETWORK ANALYSIS

6–1 ELEMENTARY EXAMPLES

In a broad class of distribution problems, the throughputs in a distribution system may be regarded as flows in a network, such as traffic in a highway network or communication messages in a transmission network. In general, a network is depicted by a graph consisting of a set of nodes interconnected by branches and it can be analyzed according to the theory of finite graphs. In this section, examples are given to illustrate the determination of the flow patterns in general, and the maximum flows in particular, in networks with capacity constraints.

Example 6–1. Figure 6–1 shows a network of one-way streets between junctions 0 and 6. The capacities of these streets in terms of cars per minute in the directions indicated by the arrowheads are given in the figure. Determine the maximum capacity of the network.

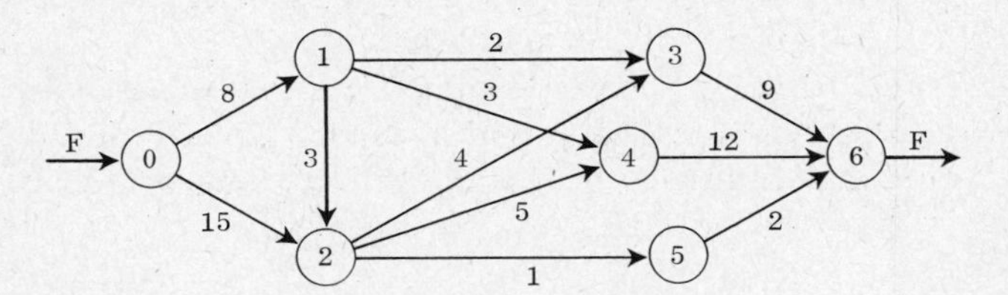

Figure 6–1

At junction 0, cars can enter at a rate of $8 + 15 = 23$ cars/min, and at junction 6, cars can leave at a rate of $9 + 12 + 2 = 23$ cars/min. However, cars at junctions 1 and 2 can move to junctions 3, 4, and 5 only through five routes, the total capacity of which is $2 + 3 + 4 + 5 + 1 = 15$ cars/min. Since all cars entering junction 0 may pass through junctions 1 and 2 and then through junctions 3, 4, and 5 before reaching junction 6, the bottleneck limits the maximum capacity of the entire network to 15 cars/min.

Example 6–2. A network of city streets shown in Fig. 6–2 has a number of one-way streets. All except one (the route between nodes 5 and 6) have been designated a direction for traffic flow. The capacity of each street (cars/min) in the direction of the flow is shown at the tail of the arrow, and the capacity of route 5–6 is 6 cars/min when it is used as a one-way street. What direction should route 5–6 be assigned in order to permit the maximum amount of traffic flow in the network?

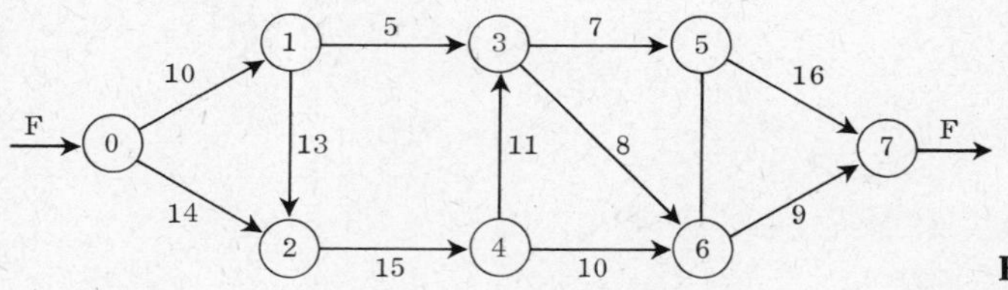

Figure 6–2

The network in this problem can be replaced by two networks, one in which the direction of route 5–6 points toward node 6 and the other in which the direction of route 5–6 points toward node 5. We can solve each network separately and determine which has a greater flow capacity. However, it would be more convenient if we could replace route 5–6 by two directed routes (one from 5 to 6 and the other from 6 to 5) on the condition that, if a flow exists in one direction, it must be zero in the other direction.

Example 6–3. In the previous example, let it be required that route 5–6 is kept as a two-way street with capacity of 3 cars/min in each direction, while all other streets remain one-way, as directed. Determine the maximum flow capacity of the network.

In this case, the directional flows in opposite directions may be represented by two directed routes, each with a capacity of 3. Simultaneous flows from 5 to 6 and from 6 to 5 are allowed so long as the capacity of the network is not exceeded.

6–2 NETWORK CONCEPT

When a physical system is modeled as a network problem, two aspects of the system can be distinctly separated: (1) the physical relationships of each component in the system, and (2) the interconnection of components for the system. The former depends on the properties of a particular system, such as electrical network, hydraulic pipeline network, trussed structures, etc., while the latter is concerned only with the topology of the network, i.e., the form of interconnection between components which can be treated as an abstract graph. Given the physical properties of the components and the form of interconnections between components, the performance of the system under any imposed effects can be determined.

The interconnection of components or elements in a network is specified by an abstract graph which is formed by joining each point in a set of points by means of one or more lines to some other points. Two types of problems that may arise naturally in many physical systems are characterized by Kirchhoff's laws in electrical network as follows:

1. *Kirchhoff's node law.* The algebraic sum of the currents meeting at any junction in a network of conductors is zero.
2. *Kirchhoff's loop law.* The algebraic sum of the potential differences around any closed loop in a network of conductors is zero.

The variables (currents) in Kirchhoff's node law are called *through* variables, while the variables (potential differences) in Kirchhoff's loop law are known as *across* variables. The topology of the abstract graph imposes specific algebraic relationships on these variables describing the properties of components or elements. Thus the algebraic relationships either in the system of node equations or in the system of loop equations are associated only with the graph, and are independent of the physical properties of the variables. Instead of considering the current-voltage relationships in an electrical network as originally intended in Kirchhoff's laws, the two systems of equations may, for example, represent the force equations and displacement equations in the analysis of trussed structures.

The type of problem involving through variables is applicable to the determination of throughputs in a network, and will therefore be the primary concern of this chapter. However, the interrelation of the two types of problems arising from Kirchhoff's laws will also be discussed briefly.

6-3 FINITE GRAPHS

We shall now consider the topology of networks in terms of abstract graphs. The points in a graph are called *nodes*, or *junctions*, or *vertices*; the lines are called *branches*, or *arcs*, or *edges*. A branch has two distinct endpoints, each of which is a node. A node and a branch are said to be *incident* with each other if the node is an endpoint of the branch.

A *linear graph* is a collection of branches, each of which can have a common point with another branch only at a node. If the set of branches in a linear graph is finite, it is called a *finite graph*. The *topology* of a graph describes the interconnection of the nodes, and not their relative space locations. A graph is said to be *planar* if it can be represented on a plane such that the nodes are distinct points and the branches are distinct lines, and such that no two branches intersect except at their endpoints. A *subgraph* is a subset of branches of a graph, and thus is a graph itself.

If the branches of a graph can be ordered such that each branch has one node in common with the preceding branch and the other node in

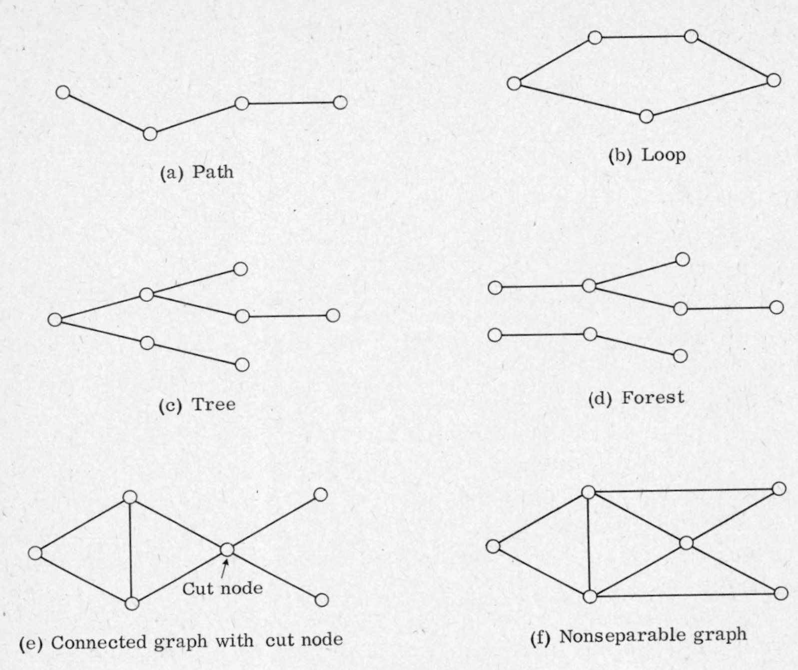

(a) Path

(b) Loop

(c) Tree

(d) Forest

(e) Connected graph with cut node

(f) Nonseparable graph

Figure 6-3

common with the succeeding branch, the graph is called a *path*. If the starting and terminal nodes of a path are one and the same node, the path is called a *loop*. A graph whose nodes are completely connected to each other by paths but one which has no loops is called a *tree*. The path, loop or tree can be, and often is, the subgraph of a graph. A graph is *connected* if there exists a path between any two nodes of a graph. A graph which contains a set of unconnected subgraphs is called a *forest*. If two subgraphs have only one common node, the node is called a *cut-node* or *articulate point* of the graph. A connected graph is said to be *nonseparable* if it contains no cut-node. These definisions are illustrated by examples in Fig. 6-3.

A branch connecting an unordered pair of nodes is called an *undirected* branch, and a branch connecting an ordered pair is called a *directed* branch. Although multiple branches may be introduced between two nodes, we shall allow at most one branch leading from any node to another without losing generality. For an unordered pair of nodes i and j, a *forward* branch (i, j) and a *backward* or reverse branch (j, i) will be permitted. For an ordered pair, only one directed branch is allowed. A graph is said to be *directed* if every branch in the network is directed, or *undirected* if none of its branches is directed, or *mixed*, if only some of its branches are directed. However, each undirected branch can often be replaced by a pair of directed branches, according to the statement of the problem. We shall therefore devote our discussion primarily to directed graphs.

6-4 BASIC LOOPS AND BASIC CUTS

In general, a connected graph of n nodes and b branches contains at least one subgraph which is a tree with n nodes. Since a tree is connected and contains no loops, any given tree with n nodes has $(n - 1)$ branches which allow one and only one path between any two nodes. The remaining $b - (n - 1)$ branches in the graph constitute the *complement* of the tree, and each of these branches is called a *chord* or *link*. If we add one chord to a tree, the resulting graph constitutes a *basic loop* or *mesh*. The *basic loops* of a connected graph for a tree are the $m = b - n + 1$ loops formed by each chord and its unique tree. For example, consider the network in Fig. 6-4(a) in which there are four nodes and five branches. The subgraph containing branches A, B, and C is a tree connecting all four nodes in the graph, and branch D and E are chords or links, as indicated by the dashed lines in Fig. 6-4(b). If the chords (D and E) are added to the tree one at a time, each will form a basic loop of the graph. The number of basic loops will be $m = 5 - 4 + 1 = 2$.

If the nodes of a connected graph are partitioned into two sets S_1 and S_2 such that any two nodes in the same set can be connected by paths not containing a node of the other set, then the branches of the graph

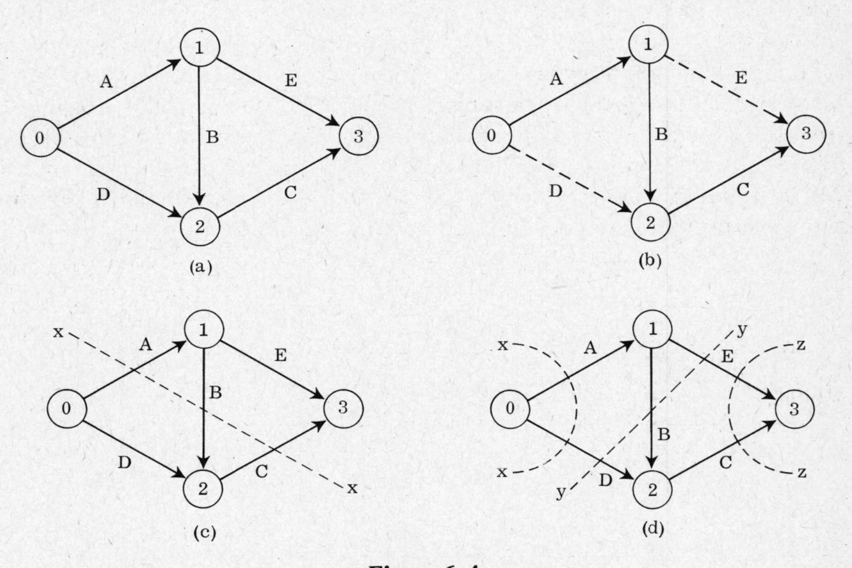

Figure 6-4

which have one node in S_1 and the other in S_2, constitute a *cut*. Thus the set of branches intersecting the line of separation of a connected graph is a cut; the set of branches incident at a node is also a cut, provided that the node is not a cut-node. As shown by the examples in Fig. 6-4, the set of branches A, B, and C intersecting the dotted line in case (c) is a cut; so is the set of branches at the intersection line x–x, y–y, or z–z in case (d).

The *cut of a tree* is defined as the branches which separate all nodes in a tree; hence the cut consists of all $(n - 1)$ branches for a tree with n nodes. For a given tree which is a subgraph of a connected graph of n nodes and b branches, a *basic cut* is the set of branches at a cut which contain only one branch of the tree. The *basic cuts* with respect to a tree in a connected graph are the $(n - 1)$ cuts, each containing only one branch of the unique tree. For example, in Fig. 6-4(c), the cut of the tree consisting of branches A, B, and C consists of these branches themselves; and the basic cuts with respect to this tree in the graph are given by sets of branches intersected by x–x, y–y and z–z in Fig. 6-4(d). For this graph of four nodes, the number of basic cuts is $4 - 1 = 3$.

6-5 INCIDENCE MATRICES

The incidence relationships between nodes and branches of a directed graph are completely specified by a node incidence matrix in which the incidence is indicated by 1, −1, or 0 representing respectively the incidence at the initial node (tail end) of the branch, the incidence at the terminal node (arrow head) of the branch, or no incidence. A node incidence matrix for the network shown in Fig. 6-5(a) is given in Table 6-1. Note that for each branch represented by a column in the table, there can be only one set of coefficients (1, −1) corresponding to the initial and terminal nodes of the branch; the remaining coefficients for the column must be zero. Then each row in the table represents the coef-

Table 6-1

NODE INCIDENCE MATRIX

Node \ Branch	A (0, 1)	B (1, 2)	C (2, 3)	D (0, 2)	E (1, 3)
0	1	0	0	1	0
1	−1	1	0	0	1
2	0	−1	1	−1	0
3	0	0	−1	0	−1

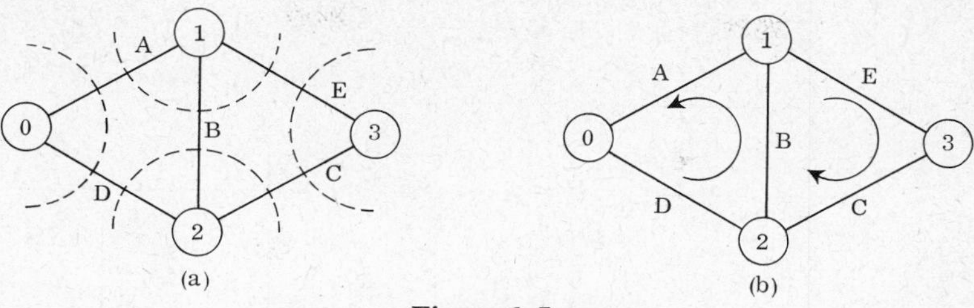

Figure 6-5

ficients of the node equation based on Kirchhoff's node law. This relation can easily be seen by letting the currents through branches A, B, C, D, and E be I_a, I_b, I_c, I_d, and I_e, respectively, and by summing up algebraically the currents in branches meeting at each node as follows:

$$\begin{array}{lrcl}
\text{node } 0\,, & I_a & + I_d & = 0\,, \\
\text{node } 1\,, & -I_a + I_b & + I_e & = 0\,, \\
\text{node } 2\,, & - I_b + I_c - I_d & & = 0\,, \\
\text{node } 3\,, & - I_c & - I_e & = 0\,.
\end{array}$$

Generally, n rows corresponding to n node equations are present in the node incidence matrix. However, only $(n - 1)$ of them are independent, since any one of the n node equations can be obtained by the linear combination of the other $(n - 1)$ equations.

If a branch is a part of a loop in a finite graph, the branch is said to be *incident* with the loop. The incidence relationships between loops and branches can be specified by a loop incidence matrix in which the incidence is indicated by 1, -1, or 0, denoting respectively the incident branch in the same direction of the loop, the incident branch in the opposite direction of the loop, or no incidence. Although the direction of the loop can be specified as clockwise or counter-clockwise, it is convenient to specify that the direction of a basic loop follow the direction of the chord. A loop incidence matrix for the network in Fig. 6-5 is shown in Table 6-2. Note that the loop containing branches A, D, B and that containing branches B, E, C are basic loops if branches A, B, C are regarded as a tree. Thus the directions of these two loops follow those of D and E, respectively, as shown in Fig. 6-5(b). The loop containing branches A, D, C, E is not a basic loop, and the direction is therefore arbitrarily assigned. Then the incident branches for each loop can be entered into a row in the table, and each row gives the coefficients of the loop equation based on Kirchhoff's loop law. This

Table 6-2

LOOP INCIDENCE MATRIX

Loop \ Branch	A (0, 1)	B (1, 2)	C (2, 3)	D (0, 2)	E (1, 3)
ADB (1, 0, 2, 3)	−1	−1	0	1	0
BEC (2, 1, 3, 2)	0	−1	−1	0	1
$ADCE$ (1, 0, 2, 3, 1)	−1	0	1	1	−1

relation can also be seen by letting the potential differences across branches, A, B, C, D, and E be P_a, P_b, P_c, P_d, and P_e, respectively, and by summing up algebraically the potential differences across the branches in each loop as follows:

$$\begin{aligned}
&\text{loop } ADB\,, \qquad -P_a - P_b \qquad + P_d \qquad = 0\,,\\
&\text{loop } BEC\,, \qquad \quad\; - P_b - P_c \qquad + P_e = 0\,,\\
&\text{loop } ADCE\,, \quad -P_a \qquad + P_c + P_d - P_e = 0\,.
\end{aligned}$$

It is obvious that only two of these three loops are independent, since the third equation can be obtained by subtracting the second equation from the first. In general, the number of independent loops in a network of n nodes and b branches is equal to the number of basic loops, $m = b - n + 1$. In the case of this example, $m = 5 - 4 + 1 = 2$. Hence it is necessary to include only the basic loops in the loop incidence matrix once a tree has been specified.

6-6 PREDECESSOR-SUCCESSOR RELATIONSHIPS

In the analysis of directed networks, it is convenient to denote a node by a number i, and a branch by a pair of numbers (i, j) denoting the initial and terminal nodes (tail end and arrowheads) of the branch. It is also desirable, though not necessary, to denote the nodes in such a way that the number i of the initial node is smaller than the number j of the terminal node in every branch. To provide proper topological ordering of the node numbers, a systematic sorting procedure may be followed. In other words, the node numbers are first arbitrarily assigned, and then arranged in the topological order by specifying that no branches be allowed to start from a node until all branches merging into that node have been accounted for in the sorting process. For example, consider the network in Fig. 6-6(a), in which the node numbers have been arbitrarily assigned. The corresponding sorting table is shown in case (a) of Table 6-3. We start with node 0, which has no prodecessor. Two branches, (0, 1) and (0, 3), come out from node 0; hence nodes 1 and 3 are

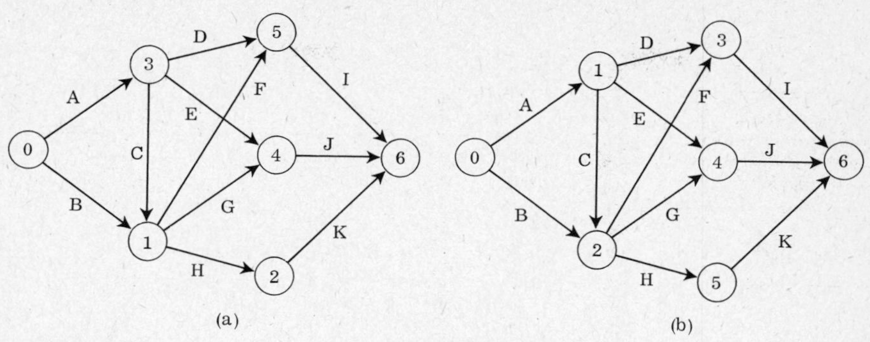

Figure 6–6

considered to be $-$nodes and node 0 to be a $+$node. Next, either node 1 or node 3 may be treated as a $+$node, provided that all branches merging into that node are sorted first. At this point, however, node 1 does not satisfy this requirement, but node 3 does. We must therefore consider node 3 first, and then go back to node 1 after all branches, i.e., (0, 1) and (3, 1), merging into it have been sorted. The order of sorting nodes 5, 4, and 2 follows that of node 1, but each does not depend on the other. Node 6 is the last node which has no successor. We can therefore reshuffle the node numbers in the graph according to the order obtained in the $+$node column of the sorting table. The new order of node numbers is shown in Fig. 6–6(b), for which the corresponding sorting table is given in case (b) of Table 6–3.

The incidence relationships between nodes and branches may therefore be indicated indirectly by a node predecessor-successor relationship in which the nodes in a network are listed in both the rows and columns in the increasing order. The coefficient in the ith row and the jth column of the matrix is denoted by x if a branch (i, j) is incident to an initial node i and a terminal node j; and the coefficient is 0 if no branch exists between i and j. The node predecessor-successor matrix corresponding to the graph in Fig. 6–6(a) is shown in Table 6–4(a). Since node 0 has no predecessor and node 6 has no successor, the first column and the last row in the matrix have no incidence, and can therefore be omitted. The node predecessor-successor matrix corresponding to the graph in Fig. 6–6(b) is shown in Table 6–4(b). Note that in this table, all coefficients below the main diagonal of the matrix are zero. This is the characteristic of the node predecessor-successor matrix when the nodes of a directed graph are numbered according to the suggested topological order.

The incidence relationship between nodes and branches may also be indicated indirectly by a branch predecessor-successor relationship matrix in

Table 6–3

SORTING TABLE FOR NETWORKS IN FIG. 6–6

(a)		(b)	
+Node (tail end)	−Node (arrowhead)	+Node (tail end)	−Node (arrowhead)
0	1 3	0	1 2
3	1 4 5	1	2 3 4
1	2 4 5	2	3 4 5
5	6	3	6
4	6	4	6
2	6	5	6

Table 6–4

NODE PREDECESSOR-SUCCESSOR RELATIONSHIP

(a)

i \ j	0	1	2	3	4	5	6
0	0	x	0	x	0	0	0
1	0	0	x	0	x	x	0
2	0	0	0	0	0	0	x
3	0	x	0	0	x	x	0
4	0	0	0	0	0	0	x
5	0	0	0	0	0	0	x
6	0	0	0	0	0	0	0

(b)

i \ j	0	1	2	3	4	5	6
0	0	x	x	0	0	0	0
1	0	0	x	x	x	0	0
2	0	0	0	x	x	x	0
3	0	0	0	0	0	0	x
4	0	0	0	0	0	0	x
5	0	0	0	0	0	0	x
6	0	0	0	0	0	0	0

which the branches represented by (i, j) are listed in both the rows and columns in the increasing order of i. (The order j for each i can also be arranged in increasing order.) The coefficient of an element in the matrix is denoted by x if the branch in the row corresponding to this element is the predecessor of that in the column corresponding to the same element; and the coefficient is 0 if the branch in the row is not the predecessor of that in the column for the element. If the nodes in a network are denoted in such a way that the number of the initial node is smaller than the number of the terminal node in every branch, all coefficients below the main diagonal of the matrix are zero. As an example, the branch predecessor-successor matrix corresponding to the graph in Fig. 6-6(b) is shown in Table 6-5. Since branches (0, 1) and (0, 2) have no predecessor, the first two columns have no incidence; similarly, because branches (3, 6), (4, 6), and (5, 6) have no successor, the last three rows also have no incidence. These rows and columns can be omitted from the matrix.

Table 6-5

BRANCH PREDECESSOR-SUCCESSOR RELATIONSHIP

		Branch (i, j)										
		0,1	0,2	1,2	1,3	1,4	2,3	2,4	2,5	3,6	4,6	5,6
Branch (i, j)	0,1	0	0	x	x	x	0	0	0	0	0	0
	0,2	0	0	0	0	0	x	x	x	0	0	0
	1,2	0	0	0	0	0	x	x	x	0	0	0
	1,3	0	0	0	0	0	0	0	0	x	0	0
	1,4	0	0	0	0	0	0	0	0	0	x	0
	2,3	0	0	0	0	0	0	0	0	x	0	0
	2,4	0	0	0	0	0	0	0	0	0	x	0
	2,5	0	0	0	0	0	0	0	0	0	0	x
	3,6	0	0	0	0	0	0	0	0	0	0	0
	4,6	0	0	0	0	0	0	0	0	0	0	0
	5,6	0	0	0	0	0	0	0	0	0	0	0

6-7 FLOW IN A NETWORK WITH CAPACITY CONSTRAINTS

When a flow is imposed in a network, the node at which the flow enters the network is called a *source*, and the node from which the flow leaves the network is called a *sink*. The *unit* of the flow corresponds to the physical quantity passing through the network from its source to its sink. There may be more than one source or one sink in the network; however, problems with multiple sources and/or sinks can be reduced to the case of a single source and single sink. We shall consider here primarily steady state flows from a source to a sink in a directed network.

The *capacity* of a branch in a network is the maximum amount of flow per unit time permitted in the branch. The capacity of a directed branch is nonnegative; since a branch having zero capacity can be deleted from a network, a directed branch is regarded as having a *positive* capacity. If no capacity is imposed on a branch, it is assumed that the capacity of that branch is infinite.

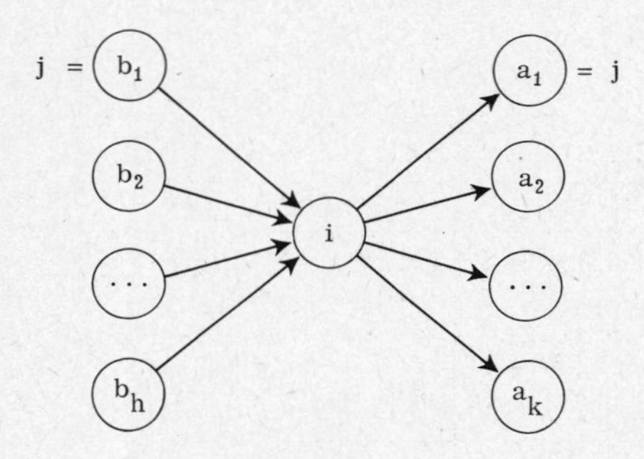

Figure 6-7

Let the positive capacity of a branch (i, j) be U_{ij}. Referring to Fig. 6-7, let x_{ij} be the flow in a branch leading from node i to a node j ($j = a_1, a_2, \ldots, a_k$ for all branches leading from i); let x_{ji} be the flow in a branch leading from a node j to the node i ($j = b_1, b_2, \ldots, b_h$ for all branches leading to i). Let F be the amount of net flow out of the source into the network, which must be the amount of net flow out of the network into the sink. Let node 0 be the source and node n be the sink in a network containing $n + 1$ nodes. Then, from Kirchhoff's node law, we have

$$\sum_{j=a_1}^{a_k} x_{ij} - \sum_{j=b_1}^{b_h} x_{ji} = \begin{cases} F, & i = 0, \\ 0, & i = 1, 2, \ldots, n-1, \\ -F, & i = n. \end{cases} \tag{6-1}$$

The capacity constraints are

$$0 \leq x_{ij} \leq U_{ij}, \qquad \text{for all } (i, j). \tag{6-2}$$

We may note that at node 0 or at node n the continuity condition is realized if we visualize that the flow F at the sink is returned to the source through an imaginary branch $(n, 0)$ with infinite capacity. Note also that one of the equations in (6-1) is redundant, since it can be obtained from the other n equations. Any flow in the network must satisfy (6-1) and (6-2).

One problem which arises naturally is the determination of the maximum flow in a network whose branches have fixed capacities. We shall attempt to maximize F in the equations for $i = 0$ and $i = n$ in (6-1), that is,

$$\sum_{j=a_1}^{a_k} x_{0j} - \sum_{j=b_1}^{b_h} x_{j0} = -\sum_{j=a_1}^{a_k} x_{nj} + \sum_{j=b_1}^{b_h} x_{jn} = \text{Max } F, \tag{6-3}$$

and the remaining equations $i = 1, 2, \ldots, n-1$ are constraint equations as follows:

$$\sum_{j=a_1}^{a_k} x_{ij} - \sum_{j=b_1}^{b_h} x_{ji} = 0. \tag{6-4}$$

Thus the problem of maximal flow in a directed network is completely defined by (6-2), (6-3) and (6-4).

Example 6-4. Formulate the maximum flow problem for the network in Example 6-1 (Fig. 6-1) in which node 0 is the source and node 6 is the sink.

$$
\begin{aligned}
i &= 0, & x_{01} + x_{02} &= F, \\
i &= 1, & -x_{01} + x_{12} + x_{13} + x_{14} &= 0, \\
i &= 2, & -x_{02} - x_{12} + x_{23} + x_{24} + x_{25} &= 0, \\
i &= 3, & -x_{13} - x_{23} + x_{36} &= 0, \\
i &= 4, & -x_{14} - x_{24} + x_{46} &= 0, \\
i &= 5, & -x_{25} + x_{56} &= 0, \\
i &= 6, & -x_{36} - x_{46} - x_{56} &= -F.
\end{aligned}
$$

The capacity constraints are given by

$$
\begin{aligned}
&0 \le x_{01} \le 8, && 0 \le x_{02} \le 15, && \\
&0 \le x_{12} \le 3, && 0 \le x_{13} \le 2, && 0 \le x_{14} \le 3, \\
&0 \le x_{23} \le 4, && 0 \le x_{24} \le 5, && 0 \le x_{25} \le 1, \\
&0 \le x_{36} \le 9, && 0 \le x_{46} \le 12, && 0 \le x_{56} \le 2.
\end{aligned}
$$

The problem is to determine a set of x_{ij} such that the continuity equations and capacity constraints are satisfied, and that the maximum flow F is

$$F^* = x_{01} + x_{02} = x_{36} + x_{46} + x_{56}.$$

Note that the coefficients of x_{ij} in the constraint equations can be expressed by the node incidence matrix in Table 6-6.

Table 6-6

NODE INCIDENCE MATRIX FOR EXAMPLE 6-4

		Branch (i, j)										
		0,1	0,2	1,2	1,3	1,4	2,3	2,4	2,5	3,6	4,6	5,6
Node i	0	1	1	0	0	0	0	0	0	0	0	0
	1	−1	0	1	1	1	0	0	0	0	0	0
	2	0	−1	−1	0	0	1	1	1	0	0	0
	3	0	0	0	−1	0	−1	0	0	1	0	0
	4	0	0	0	0	−1	0	−1	0	0	1	0
	5	0	0	0	0	0	0	0	−1	0	0	1
	6	0	0	0	0	0	0	0	0	−1	−1	−1

6-8 CHAIN FLOWS FROM SOURCE TO SINK

A *chain flow* in a network with directed branches of positive capacities is a constant flow, $x_{ij} = f$ for every branch (i, j) along a directed path and $x_{ij} = 0$ elsewhere. If the starting and terminal points of a directed path having branches of positive capacities are one and the same node, the path becomes a loop, and the chain flow becomes a *circular flow* in a loop. The flow in a path or loop is limited by the smallest positive capacity of all its branches. A branch having zero capacity is called a *flow blocking branch*, since no flow is possible in a path or a loop containing such a branch.

If a flow $F \geq 0$ exists in a directed network, the set of $x_{ij} \geq 0$ corresponding to F satisfies Kirchhoff's node law for all nodes incident to the directed branches. The flow F may be conceived intuitively as the superposition of all chain flows from the source to the sink. Although a *real* flow in a directed path must follow the positive capacities of its branches, usually leading to a chain flow from the source to the sink, a *fictitious* circular flow may exist if two or more chain flows have a common branch such that the real flow in that branch does not exceed its positive capacity. Let us define the flow in the direction of positive capacity as *positive flow*, and the flow in the opposite direction as *negative flow*, whether the flow is real or fictitious. If a positive flow x_{ij} is imposed on a branch (i, j) with a positive capacity U_{ij} which serves as a common branch for two or more flow chains, the excess capacity of the branch (i, j) after the flow x_{ij} takes place is defined by

$$g_{ij} = U_{ij} - x_{ij} \geq 0\,. \tag{6-5}$$

However, the deletion of the flow x_{ij} from the positive capacity of branch (i, j) also produces an excess capacity in the opposite direction, since a negative flow up to the magnitude of x_{ij} can be imposed on the branch without causing a real flow in the negative direction. Thus the excess capacity in the negative direction of branch (i, j) is defined by

$$g_{ji} = x_{ij} \geq 0\,. \tag{6-6}$$

It is this excess capacity in the negative direction that allows the formation of a circular flow in a loop instead of a chain flow. Since the negative flow in any branch of a loop must be eliminated by positive flows in the same branch on other paths upon superposition, this negative flow is termed a *fictitious flow*, and the circular flow containing this negative flow is called a *fictitious circular flow*.

In general, a flow path joining the source to the sink can be constructed by selecting a branch with positive capacity from node to node successively, starting from the source until the sink is reached. At each node, we select a succeeding branch which has the largest capacity. The maximum flow in this path is equal to the smallest capacity of all branches selected for the path; the branch which is used to full capacity is said to be *saturated*. Then, another path may be constructed in the same manner through other branches of the network which have positive capacities. If at least one branch of every possible path joining the source to the sink is saturated, no additional flow is possible. The maximum flow in the network is the sum of all positive chain flows from the source to the sink, provided that no circular flow has been formed in the process. Otherwise, the maximum flow is the *algebraic sum* of all positive chain flows and the circular flows resulting from the process. The decomposition of a set of x_{ij} corresponding to the maximum flow F in the network into chain flows and/or circular flows, may, in general, result in different sets of chain flows, depending on the order in specifying the chains. However, the superposion of each consistent set of chain flows should lead to the same maximum flow in the network.

Example 6–5. Determine the maximum flow in the network in Fig. 6–8(a) by the superposition of chain flows.

Starting with node 0, we select branches (0, 2), (2, 3), and (3, 6) for a path joining the source to the sink because at each node the succeeding branch has the largest or one of the largest capacities. This path has a chain flow of $f_1 = 2$, as shown in part (b) of the figure, since 2 is the

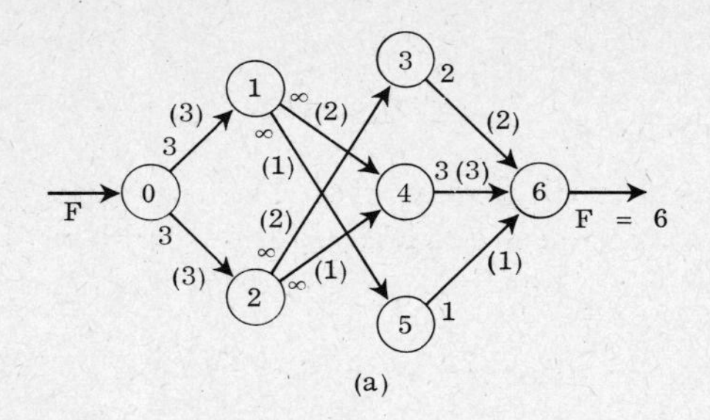

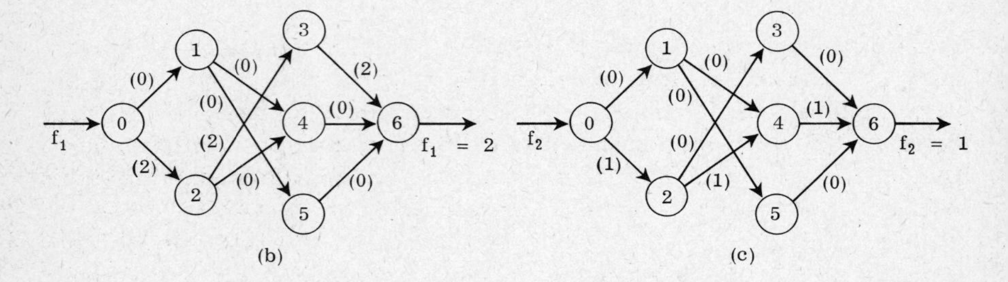

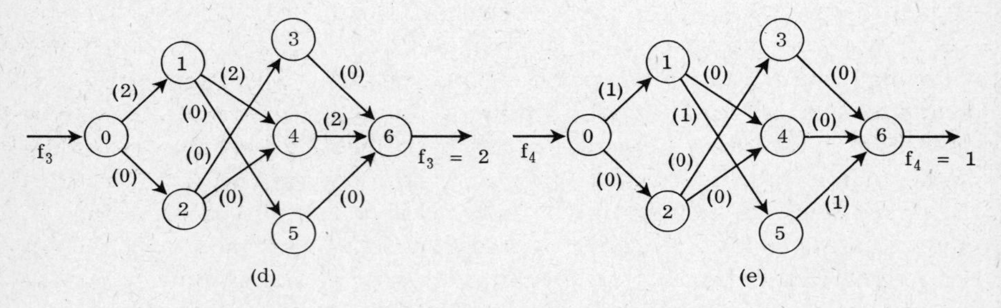

Figure 6–8

smallest capacity in all branches, and only branch (3, 6) is now saturated. Next, starting with node 0, we select branches (0, 2), (2, 4), and (4, 6) for the path which has a chain flow $f_2 = 1$, as shown in part (c), because branch (0, 2) has an unused capacity of only 1, which is the smallest capacity in all branches. Again, starting from node 0, we proceed to get $f_3 = 2$ in the path 0–1–4–6, and $f_4 = 1$ in chain 0–1–5–6. The chain flows corresponding to these paths are shown in parts (d) and (e), respectively. The maximum flow F is the sum of f_1, f_2, f_3, and f_4.

If we had started with a chain flow $f_1 = 3$ in the path 0–1–4–6, we would have come up next with a chain flow of $f_2 = 2$ in the chain 0–2–3–6, as shown in Fig. 6–9. It appears that at least one branch of every possible chain joining the source to the sink is saturated, but actually this is not true. If we try to impose a flow of $f_3 = 1$ in chain

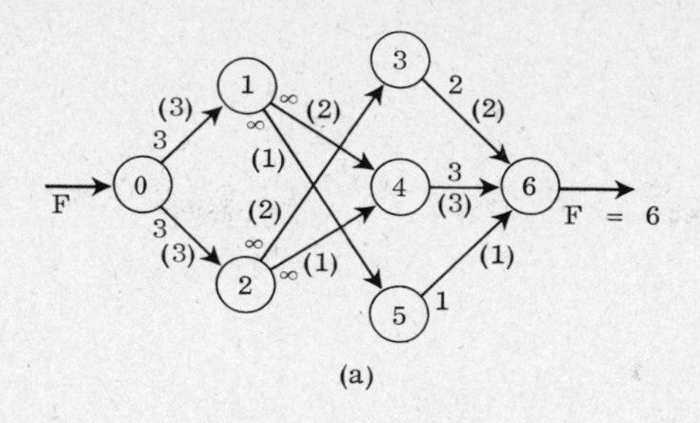

(a)

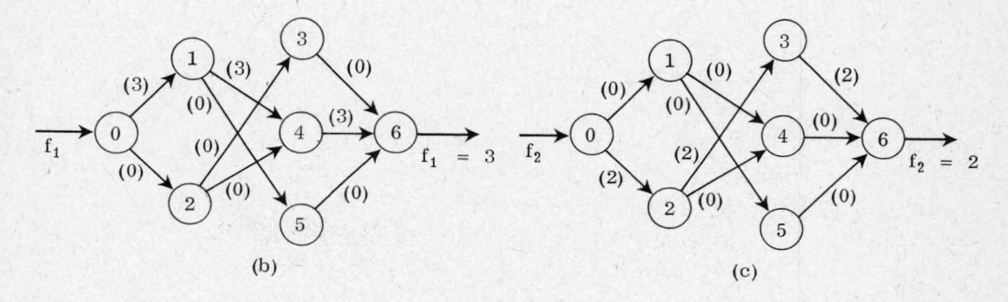

(b) (c)

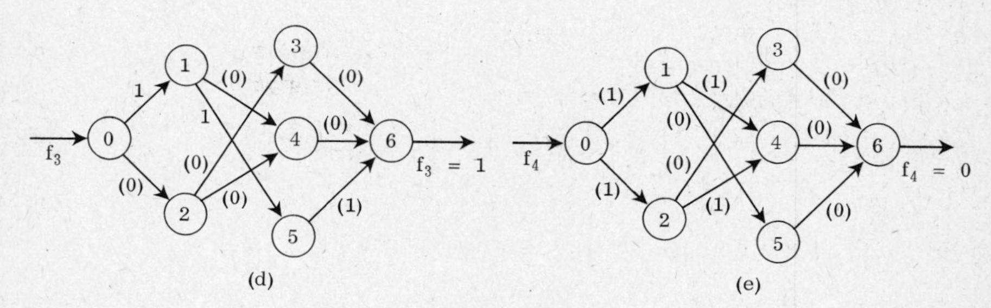

(d) (e)

Figure 6–9

0–1–5–6, the capacity of branch (0, 1) will be exceeded; however, if we also introduce a fictitious circular flow of $f_0 = 1$ in the direction of 0–2–4–1–0 with $f_4 = 0$, the actual flow will still follow the positive capacities of all branches. The maximum flow F then is the sum of f_1, f_2, f_3, and f_4; while f_0 affects only the flow pattern x_{ij} but not the maximum flow F.

6–9 MINIMUM CUT AND MAXIMUM FLOW

A *cut* separating the source and sink in a network is a set of branches which contains at least one branch from every path of positive capacity joining the source to the sink. The *value* of a cut is the sum of the capacities of all branches at the cut.

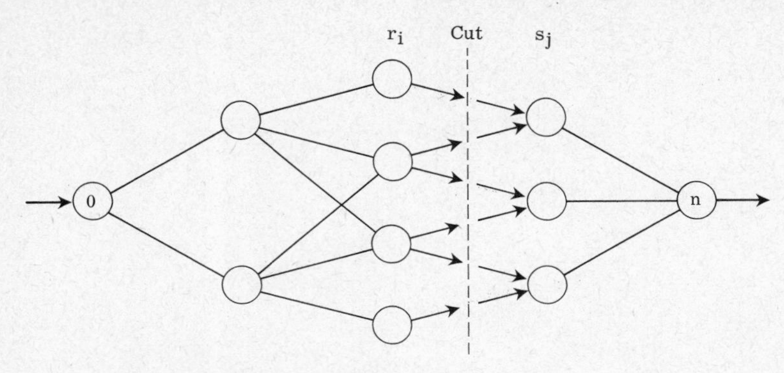

Figure 6-10

Let (r_i, s_j) be the set of branches at a cut which divides the network into two parts, one including the source 0 and all nodes connected to the source, and the other including the sink and all nodes connected to the sink. Thus, r_i represent nodes on the side of the source and s_j represent nodes on the side of the sink incident to the branches (r_i, s_j) at the cut, as shown in Fig. 6-10.

From the equations in (6-1), it can be seen that the total flow F entering the network from the source 0 is

$$F = \sum_{j=a_1}^{a_k} x_{0j} - \sum_{j=b_1}^{b_h} x_{j0} \, .$$

For each of the nodes connected to the source,

$$\sum_{j=a_1}^{a_k} x_{ij} - \sum_{j=b_1}^{b_h} x_{ji} = 0 \, , \qquad i = 1, 2, \ldots, n-1 \, .$$

Summing the equations for all nodes, including the source, on the side of the network which contains the source, we have

$$F = \sum_j x_{0j} - \sum_j x_{j0} + \sum_L \Big(\sum_j x_{ij} - \sum_j x_{ji} \Big) \, ,$$

in which L denotes the set of nodes, excluding the source, on the side of the source. Note that in the last equation the flows in any branch connecting two nodes on the side of the source will be considered once in a positive sense and once in a negative sense, and thus they will cancel each other, as shown by the opposite arrows in Fig. 6-11. The only uncanceled flows are represented by the flow F at the source, and the flows through the branches (r_i, s_j) at the cut. Summing the flows through

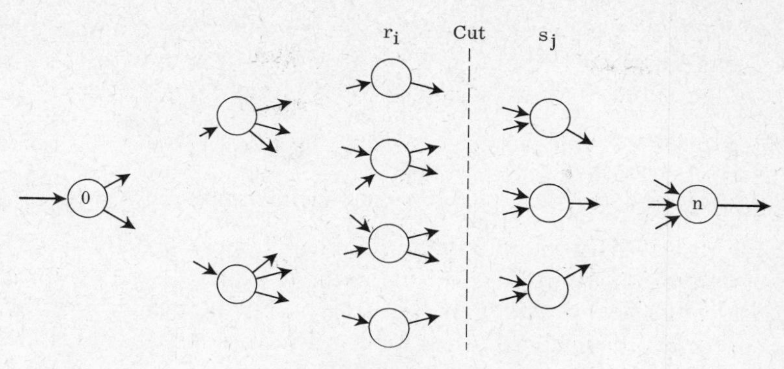

Figure 6-11

these branches, we have

$$F = \sum_{r_i} x_{r_i s_j} . \tag{6-7}$$

If we construct a positive flow path from the source to the sink and impose a chain flow on the path, then the excess capacity of branch (i, j), which has a positive capacity U_{ij} and an imposed positive flow x_{ij}, is

$$g_{ij} = U_{ij} - x_{ij} \geq 0 .$$

Hence

$$x_{ij} = U_{ij} - g_{ij} .$$

Substituting this expression into the flows through the branches at the cut in (6-7) we have

$$F = \sum_{r_i} (U_{r_i s_j} - g_{r_i s_j}) .$$

If we continue to delete chain flows for all possible paths in the network until there is no more excess capacity in any of the branches at the cut, then

$$g_{r_i s_j} = 0 , \tag{6-8}$$

and

$$F = \sum_{r_i} U_{r_i s_j} . \tag{6-9}$$

This last equation states that the flow F that can go through the network is limited by the value of the cut C (i.e., the sum of the capacities of all branches at the cut). If we can find all possible cut sets in the network and their values, we can conclude that the maximum flow allowed in the network is limited by the minimum cut value of the network. In other words, the maximum flow F^* equals the minimum cut C^*, or

$$\max F = F^* = C^* = \min C\,.$$

This is known as the *maximum-flow minimum-cut theorem.*

Example 6-6. Verify the maximum flow in Example 6-5 by the use of the maximum-flow minimum-cut theorem.

Consider the network for this problem as shown in Fig. 6-8(a). The set of branches (0, 1) and (0, 2) is a cut. The value of this cut is the sum of the capacities of branches (0, 1) and (0, 2), or $3 + 3 = 6$. The set of branches (0, 1), (2, 3), and (2, 4) is also a cut, and the value of this cut is infinite, since the capacities of both branches (2, 3) and (2, 4) are infinite. The set of branches (1, 4), (1, 5), (2, 3), and (2, 4) is another cut, the value of which is also infinite.

Of all the possible cuts, the cut through the set of branches (0, 1) and (0, 2) has a value of $3 + 3 = 6$, and the cut through the set of branches (3, 6), (4, 6), and (5, 6) has a value of $2 + 3 + 1 = 6$. Each of the remaining cuts has a value of infinity. Hence the minimum cut value is 6, which is also the value of the maximum flow.

6-10 LABELING TECHNIQUE FOR FINDING MAXIMUM FLOW

A systematic method of finding the maximum flow in a directed network has been developed on the basis of the relation between maximum flow and minimum cut. The computation procedure involves a sequence of operations, each of which either leads to a flow or higher value, or terminates when the existing flow reaches maximum.

We shall label each node in the network successively by two numbers (h, k), where h denotes the number of the node preceding the present node and k denotes the smallest positive capacity of all branches preceding the present node. Starting with the source (node 0), the labeling is $(-, \infty)$, which indicates that no node precedes node 0 and that the flow entering node 0 has no specified capacity or has infinite capacity. Next, consider the forward nodes that are connected directly to node 0 by branches, and label them one by one in the ascending order of node numbers until all nodes connected to node 0 by branches are labeled. In general, when we start considering a new node, we must examine backward to see if we have labeled all nodes with node numbers lower than that of the node under consideration. If not, label first any such backward nodes connected by branches to the node under consideration; otherwise, proceed to label forward nodes one by one in the ascending order of node numbers. No node can be labeled twice, i.e., from two or more nodes connected to it. For example, when we reach a new node j from node i through the connecting branch (i, j) in the labeling process, as shown in Fig. 6-12, it is possible that another node $i + p$ $(<j)$, also connected to node j through a branch, is either labeled or unlabeled previously. If node $i + p$ is already labeled, it cannot be labeled again from node j, nor can node j be labeled again from node $i + p$. If node $i + p$ has not been labeled previously, it should now be labeled, since it is connected to node j. After considering all backward nodes, we shall proceed to label the forward node $j + q$ $(>j)$ which has the lowest node number beyond j, and then the next node, etc., until all forward nodes connected directly to node j by branches are labeled. In labeling node j from node i $(i < j)$, we have (h, k) for node j as follows:

$$h_j = i\,,$$

$$k_j = \text{Min}\,(g_{ij}, k_i)\,.$$

In other words, h_j indicates that the flow to j comes from i, and k_j takes the smallest of the two values g_{ij} and k_i, representing respectively the excess capacity of branch (i, j) and the flow to node i from a previous node. This entire procedure is repeated until no additional nodes can be labeled. At that stage, if the sink is not labeled, the existing flow determined in previous operations is maximum; if the sink is labeled, the existing flow can be increased by the amount of the chain flow obtained in the current operation. The magnitude of this chain flow is given by the number k in the label (h, k) for the sink, and the path of this chain flow can be traced back to the source through the number h indicating the preceding node.

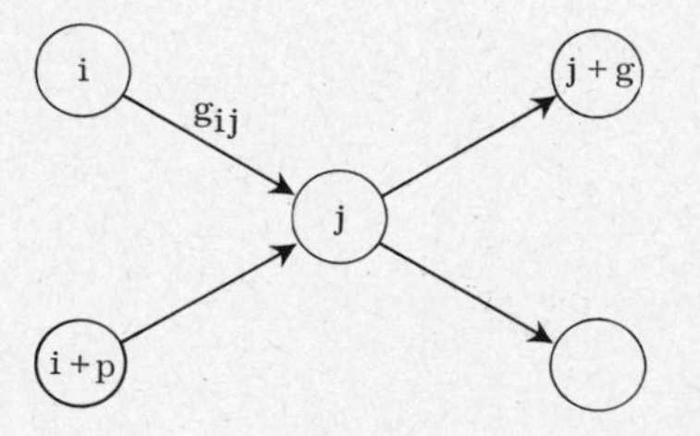

Figure 6-12

In the case that the existing flow is not optimal, the amount of flow in the last operation can be deducted from the positive capacity of every branch in the path, and at the same time this same value can be added to the negative capacity of each branch. The consideration of the negative capacities of directed branches may appear to be redundant, and often it is, but it becomes necessary whenever a circular flow is formed.

It may also be noted that an unlabeled backward node exists only when a circular flow occurs.

We can repeat the labeling process for the network with reduced capacities, each time a chain flow is deducted, until the sink cannot be labeled. Then the labeling process has the same effects of imposing chain flows in the network until the flows in all possible paths from the source to the sink have been blocked. Consequently, if the sink is not labeled when no additional nodes can be labeled, the minimum cut has been reached. *The maximum flow therefore is equal to the sum of all chain flows determined by the labeling technique.*

It can also be observed that since the positive and negative excess capacities for branch (i, j) in a directed network are given by

$$g_{ij} = U_{ij} - x_{ij} \geq 0 ,$$
$$g_{ji} = x_{ij} ,$$

then, at the completion of the labeling technique, the negative excess capacities for all branches leading away from the source 0 become

$$g_{j0} = x_{0j} , \qquad j = a_1, a_2, \ldots, a_k ,$$

and the positive excess capacities for all branches leading toward the sink n are

$$g_{nj} = x_{jn} , \qquad j = b_1, b_2, \ldots, b_h .$$

Hence the total flow F can be determined at the end of the labeling process from

$$F = \sum_{j=a_1}^{a_k} g_{j0} = \sum_{j=b_1}^{b_h} g_{nj} , \tag{6-10}$$

because at the source,

$$F = \sum_{j=a_1}^{a_k} x_{0j} ,$$

and at the sink,

$$F = \sum_{j=b_1}^{b_h} x_{jn} .$$

Example 6-7. Find the maximum flow in Example 6-1 by using the labeling technique.

We can first illustrate the labeling technique on the networks shown in Fig. 6-13. As shown in case (a), we label node 0 as $(-, \infty)$. Then, label node 1 as (0, 8), in which 0 refers to the preceding node 0, and 8 represents the smallest capacity of all branches preceding node 1; next label node 2 as (0, 15). When we label node 3 as (1, 2), we check to see

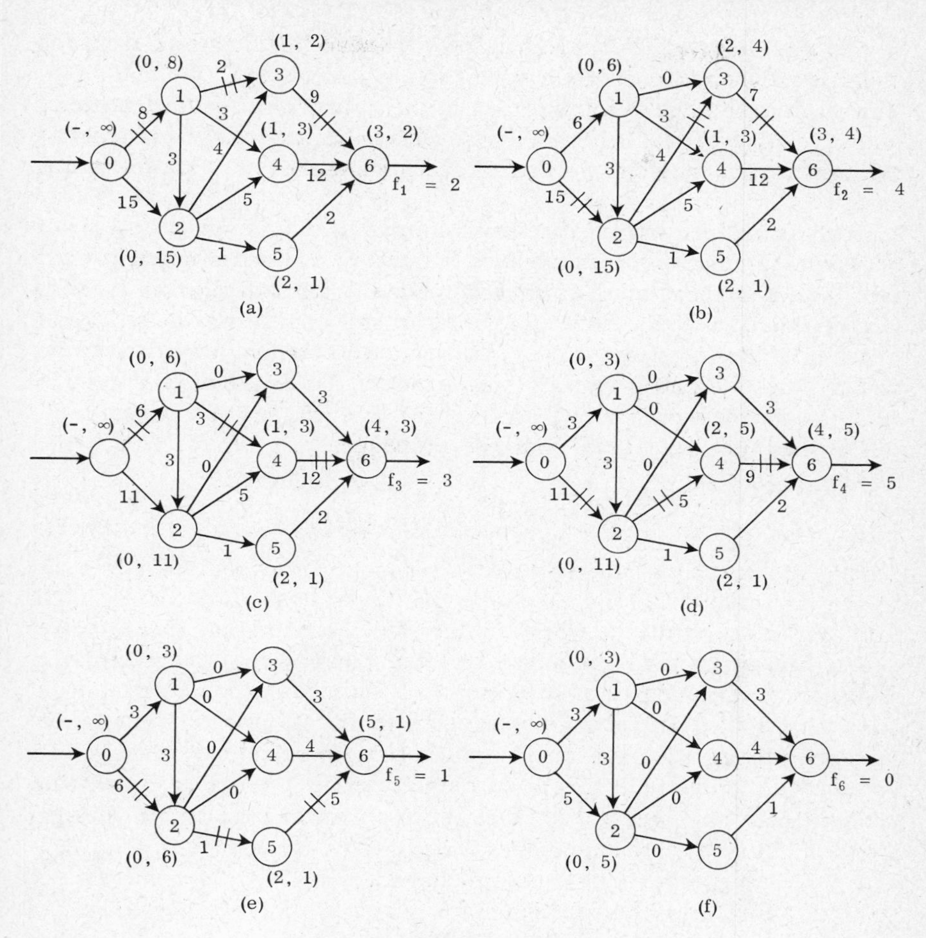

Figure 6–13

if node 2, as well as node 1, is labeled, as it indeed is. We then proceed to node 4 and label it as (1, 3). Note that for both nodes 3 and 4, a branch comes from node 2 as well as from node 1. However, we always proceed in a systematic manner and hence label nodes 3 and 4 as succeeding node 1 instead of node 2. We label node 5 as (2, 1), since it leads directly from node 2 only. Finally, we label node 6 as (3, 2) because 3 is the lowest index of all three nodes, 3, 4, and 5, leading to node 6.

Since the sink is labeled in case (a), the existing flow is not maximum. We find the existing flow $f_1 = 2$ from the second number in the label of node 6, and trace the chain flow back from node 6 to node 3, then node

1, and then node 0. Hence a chain flow of $f_1 = 2$ is deducted from the capacity of each branch in the path, i.e., branch (0, 1), (1, 3), and (3, 6). The result of this deduction is shown in the network of reduced capacity in case (b). Note that the negative excess capacities of the branches in the path are not shown in the figure because they are not pertinent in this case.

The labeling process for case (b) is similar to that for case (a). It is necessary to mention only that since branch (1, 3) now has zero capacity, the flow to node 3 must come from node 2 instead of from node 1. On the other hand, the flow to node 4 can still come from node 1, and hence node 4 is labeled as (1, 3) in order to use up first the capacity corresponding to the node with the lower index. This systematic approach leads to the networks shown in cases (c), (d), (e), and (f). In case (f), the sink cannot be labeled and the procedure terminates. Hence the maximum flow is

$$F^* = f_1 + f_2 + f_3 + f_4 + f_5 = 15 .$$

The labeling process may also be carried out in a tabulated form, as shown in Table 6-7. The steps of computation in the table are parallel to those shown on the networks in Fig. 6-13, except that the negative excess capacities in all branches are also shown. Note that each table shows the node predecessor-successor relationships in the network, with branch capacities between nodes as coefficients. In case (a), all branch capacities are positive, and therefore the corresponding capacities in the opposite direction are zero. A dash indicates that there is no branch incident between nodes. We start the problem in case (a) by entering row 0 ($i = 0$), which has branches incident with nodes 1 and 2 ($j = 1$ and $j = 2$), and note that $h = (-)$ and $k = \infty$. Next, we enter rows 1 and 2, which are the nodes indicated in row 0. In row 1 ($i = 1$), we let $h = 0$, which is the precedence of node 1, and $k = 8$, which is the smaller of the coefficients in columns 1 and k of row 0. Similarly, in row 2 ($i = 2$), we let $h = 0$, which is the precedence of node 2, and $k = 15$, which is the smaller of the coefficients in columns 2 and k of row 0. Then we look in row 1 for the nodes incident with branches from 1, and proceed to nodes 3 and 4, since node 0 and node 2 are already labeled. We therefore enter rows 3 and 4 and look for corresponding nodes on incident branches. Finally, we obtain the values of h and k in all rows; and $k = 2$ in the last row indicates that the chain flow is 2. We can then trace the path of chain flow for $h = 3$ in the last row ($j = 6$) to node 3 ($i = 3$), then for $h = 1$ in row 3 ($j = 3$) to node 1 ($i = 1$), and finally for $h = 0$ in row 1 ($j = 1$) to node 0 ($i = 0$). We also surround the capacity of each branch on the path of the chain flow by a square.

Table 6–7

SOLUTION OF EXAMPLE 6–7 BY LABELING TECHNIQUE

(a) $f_1 = 2$

i \ j	0	1	2	3	4	5	6	h	k
0	—	**8**	15	—	—	—	—	—	∞
1	0	—	3	**2**	3	—	—	0	8
2	0	0	—	4	5	1	—	0	15
3	—	0	0	—	—	—	**9**	1	2
4	—	0	0	—	—	—	12	1	3
5	—	—	0	—	—	—	2	2	1
6	—	—	—	0	0	0	—	3	2

(b) $f_2 = 4$

i \ j	0	1	2	3	4	5	6	h	k
0	—	6	**15**	—	—	—	—	—	∞
1	2	—	3	0	3	—	—	0	6
2	0	0	—	**4**	5	1	—	0	15
3	—	2	0	—	—	—	**7**	2	4
4	—	0	0	—	—	—	12	1	3
5	—	—	0	—	—	—	2	2	1
6	—	—	—	2	0	0	—	3	4

(Cont.)

Table 6–7

(Continued)

(c) $f_3 = 3$

i \ j	0	1	2	3	4	5	6	h	k
0	—	6	11	—	—	—	—	—	∞
1	2	—	3	0	3	—	—	0	6
2	4	0	—	0	5	1	—	0	11
3	—	2	4	—	—	—	3	—	—
4	—	0	0	—	—	—	12	1	3
5	—	—	0	—	—	—	2	2	1
6	—	—	—	6	0	0	—	4	3

(d) $f_4 = 5$

i \ j	0	1	2	3	4	5	6	h	k
0	—	3	11	—	—	—	—	—	∞
1	5	—	3	0	0	—	—	0	3
2	4	0	—	0	5	1	—	0	11
3	—	2	4	—	—	—	3	—	—
4	—	3	0	—	—	—	9	2	5
5	—	—	0	—	—	—	2	2	1
6	—	—	—	6	3	0	—	4	5

Table 6–7

(Continued)

(e)

$f_5 = 1$

i \ j	0	1	2	3	4	5	6	h	k
0	—	3	$\boxed{6}$	—	—	—	—	—	∞
1	5	—	3	0	0	—	—	0	3
2	9	0	—	0	0	$\boxed{1}$	—	0	6
3	—	2	4	—	—	—	3	—	—
4	—	3	5	—	—	—	4	—	—
5	—	—	0	—	—	—	$\boxed{2}$	2	1
6	—	—	—	6	8	0	—	5	1

(f)

$f_6 = 0$

i \ j	0	1	2	3	4	5	6	h	k
0	—	3	5	—	—	—	—	—	∞
1	5	—	3	0	0	—	—	0	3
2	10	0	—	0	0	0	—	0	5
3	—	2	4	—	—	—	3	—	—
4	—	3	5	—	—	—	4	—	—
5	—	—	1	—	—	—	1	—	—
6	—	—	—	6	8	1	—		

When the chain flow in case (a) is deducted from the positive capacity of each branch in the path and added to the negative capacity of that branch, the table shown in (b) results. For example, $U_{01} = 8 - 2 = 6$, but $U_{10} = 0 + 2 = 2$, etc. The same labeling technique may then be applied to (b) until we obtain a chain flow of $k = 4$ in the last row.

Passing over the tables in (c), (d), and (e), we finally reach case (f) in which the sink cannot be labeled. Hence we can sum up the chain flows obtained from cases (a) through (e) inclusive:

$$F^* = 2 + 4 + 3 + 5 + 1 = 15 .$$

In view of the expression in (6–10), however, the total flow in the network can also be determined from the sum of all numbers in column 0, or the sum of all numbers in the last row of the table in (f). Hence

$$F^* = 5 + 10 = 15 ,$$

or

$$F^* = 6 + 8 + 1 = 15 .$$

Example 6–8. Determine the maximum flow in the network of Fig. 6–9(a) for Example 6–5 by the use of the labeling technique.

The solution this problem is given in Table 6–8, which is similar to the previous example except for the negative flow in (c). As indicated in (c), the chain flow is imposed on the path 0–2–4–1–5–6. Since the branch 1–4 has a positive capacity from node 1 to node 4, the flow from node 4 to node 1 is a negative flow. This condition is taken care of automatically if we follow the label technique mechanically. Thus in (c) we check row 0 and find that the branch 0–1 has zero capacity, but branch 0–2 has a capacity of 1. We therefore bypass row 1 and enter row 2. Since node 0 is already labeled, we proceed to nodes 3 and 4. However, in row 3 we can only go back to node 2, which is already labeled, but in row 4 we can go back to node 1, which is not yet labeled. Since the flow from node 4 to node 1 is negative flow, we indicate that $h = 4$ and $k = 1(-)$ in row 1. The steps from node 1 to node 5 and then to node 6 are straightforward.

In comparing the intuitive solution in Fig. 6–9 and the solution by labeling technique in Table 6–8, we note that the flow pattern in Fig. 6–9(b) corresponds to the chain flow in case (b) in the table, and that the flow pattern in Fig. 6–9(c) corresponds to the chain flow in case (a) in the table. The superposition of the flow patterns in Fig. 6–9(d) and (e) leads to the chain flow in case (c) in the table.

The flows in all branches of the network resulting from the maximum flow F^* at the source can be obtained by superposition of all chain flows.

Table 6-8

SOLUTION OF EXAMPLE 6-8 BY LABELING TECHNIQUE

(a) $f_1 = 2$

$i \backslash j$	0	1	2	3	4	5	6	h	k
0	—	3	$\boxed{3}$	—	—	—	—	—	∞
1	0	—	—	—	∞	∞	—	0	3
2	0	0	—	$\boxed{\infty}$	∞	—	—	0	3
3	—	0	0	—	—	—	$\boxed{2}$	2	3
4	—	0	0	—	—	—	3	1	3
5	—	—	0	—	—	—	1	1	3
6	—	—	—	0	0	0	—	3	2

(b) $f_2 = 3$

$i \backslash j$	0	1	2	3	4	5	6	h	k
0	—	$\boxed{3}$	1	—	—	—	—	—	∞
1	0	—	—	—	$\boxed{\infty}$	∞	—	0	3
2	2	0	—	∞	∞	—	—	0	1
3	—	0	2	—	—	—	0	2	1
4	—	0	0	—	—	—	$\boxed{3}$	1	3
5	—	—	0	—	—	—	1	1	3
6	—	—	—	2	0	0	—	4	3

(Cont.)

Table 6–8

(Continued)

(c) $f_3 = 1$

i \ j	0	1	2	3	4	5	6	h	k
0	—	0	[1]	—	—	—	—	—	∞
1	3	—	—	—	∞	[∞]	—	4	1(—)
2	2	0	—	∞	[∞]	—	—	0	0
3	—	—	2	—	—	—	0	2	1
4	—	[3]	0	—	—	—	0	2	1
5	—	—	0	—	—	—	[1]	1	1
6	—	—	—	2	3	0	—	5	1

(d) $f_4 = 0$

i \ j	0	1	2	3	4	5	6	h	k
0	—	0	0	—	—	—	—	—	∞
1	3	—	—	—	∞	∞	—	—	—
2	3	0	—	∞	∞	—	—	—	—
3	—	—	2	—	—	—	0	—	—
4	—	2	1	—	—	—	0	—	—
5	—	1	—	—	—	—	0	—	—
6	—	—	—	2	3	1	—	—	—

Table 6-9

BRANCH FLOWS FOR EXAMPLE 6-8

		Branch (i, j)									
		0,1	0,2	1,4	1,5	2,3	2,4	3,6	4,6	5,6	6,0
Loop	(a)	0	2	0	0	2	0	2	0	0	2
	(b)	3	0	3	0	0	0	0	3	0	3
	(c)	0	1	−1	1	0	1	0	0	1	1
Sum		3	3	2	1	2	1	2	3	1	6

This can be accomplished conveniently in the form of a loop-branch incidence matrix if we imagine a branch $(n, 0)$ linking the sink n to the source 0. Thus in Table 6-9 the columns indicate the branches in the network for this example, including the fictitious branch (6, 0), and the rows include the loops corresponding to the chain flows (a), (b), and (c) of Table 6-8. The amount of flow in each branch incident to a loop is given as a coefficient in the matrix. The sum of the flows in each branch is given in the last row, and the total flow F is given by the sum in branch (6, 0), which is the flow out of the network through the sink.

6-11 NETWORKS WITH UNDIRECTED BRANCHES

In networks with undirected branches, each of the undirected branches may be replaced by two direct branches such that the method of analysis for a directed network may be applicable. Two types of network problems involving undirected branches are usually encountered: (1) The capacities of the undirected branches can accommodate flows in one direction or the other but not both, and (2) the capacities of the undirected branches are to be divided to accommodate flows in both directions. Examples of these two types of problems have been given in Examples 6-2 and 6-3, respectively.

When a flow is allowed in either direction i–j or j–i of an undirected branch (i, j), with branch capacities U_{ij} and U_{ji} for directions i–j and j–i respectively, the undirected branches may be replaced by two fictitious directed branches, provided that the flow in one of the branches is zero while a flow is imposed on the other. Let us first assume that a flow x_{ij} satisfying the capacity constraint $U_{ij} \geq 0$ is imposed on the directed branch (i, j), and that a flow x_{ji} satisfying the capacity constraint $U_{ji} \geq 0$ is imposed on the directed branch (j, i). Note that U_{ij}

and U_{ji} need not be equal. For example, in a pipeline network, the capacity for up-grade transport is different from that for down-grade transport. If we substitute the set of flows $x_{ij} \geq 0$ and $x_{ji} \geq 0$ by another set of flows defined by

$$x'_{ij} = x_{ij} - \text{Min}\,(x_{ij}, x_{ji}) , \tag{6-11}$$

$$x'_{ji} = x_{ji} - \text{Min}\,(x_{ij}, x_{ji}) , \tag{6-12}$$

then for $x_{ij} \geq x_{ji}$,

$$x'_{ij} = x_{ij} - x_{ji} \geq 0 , \qquad x'_{ji} = 0 ,$$

and for $x_{ij} \leq x_{ji}$,

$$x'_{ij} = 0 , \qquad x'_{ji} = x_{ji} - x_{ij} \geq 0 .$$

This change of variables does not in any way alter the flow outside of branches (i, j) and (j, i), since we have essentially subtracted a loop flow from i to j and j to i. However, the results satisfy the requirement that the real flow can be in one direction only. In other words, if we replace each of the undirected branches in a network by a pair of fictitious directed branches and solve the resulting directed network, the only required modification on the imposed flows in the network is the substitution of the flows in each pair of the fictitious branches by the relations in (6-11) and (6-12).

In the case that the capacity of an undirected branch is divided for two-way flows, we can simply replace this branch by two directed branches, each of which may have different capacity as assigned. If the branch capacity of an undirected branch (i, j) is either U_{ij} or U_{ji}, corresponding to the directions i–j and j–i, respectively, we assume that $U_{ij} = U_{ji}$ so that the capacity can be divided between the two directed branches, (i, j) and (j, i). Let the capacity of the directed branch (i, j) be $\bar{U}_{ij}$ and the capacity of the directed branch (j, i) be $\bar{U}_{ji}$. Then

$$\bar{U}_{ij} + \bar{U}_{ji} = U_{ij} = U_{ji} .$$

The conditions of the imposed flows on the directed branches (i, j) and (j, i) are

$$0 \leq x_{ij} \leq \bar{U}_{ij} \qquad \text{and} \qquad 0 \leq x_{ji} \leq \bar{U}_{ji} .$$

We can therefore solve the resulting directed network without further modification.

Example 6-9. Determine the maximum flow for the network in Example 6-2. Since branch 5-6 in this network is undirected, we can replace it by two fictitious directed branches 5-6 and 6-5, as shown by the double lines in Fig. 6-14. We can determine the maximum flow in the resulting

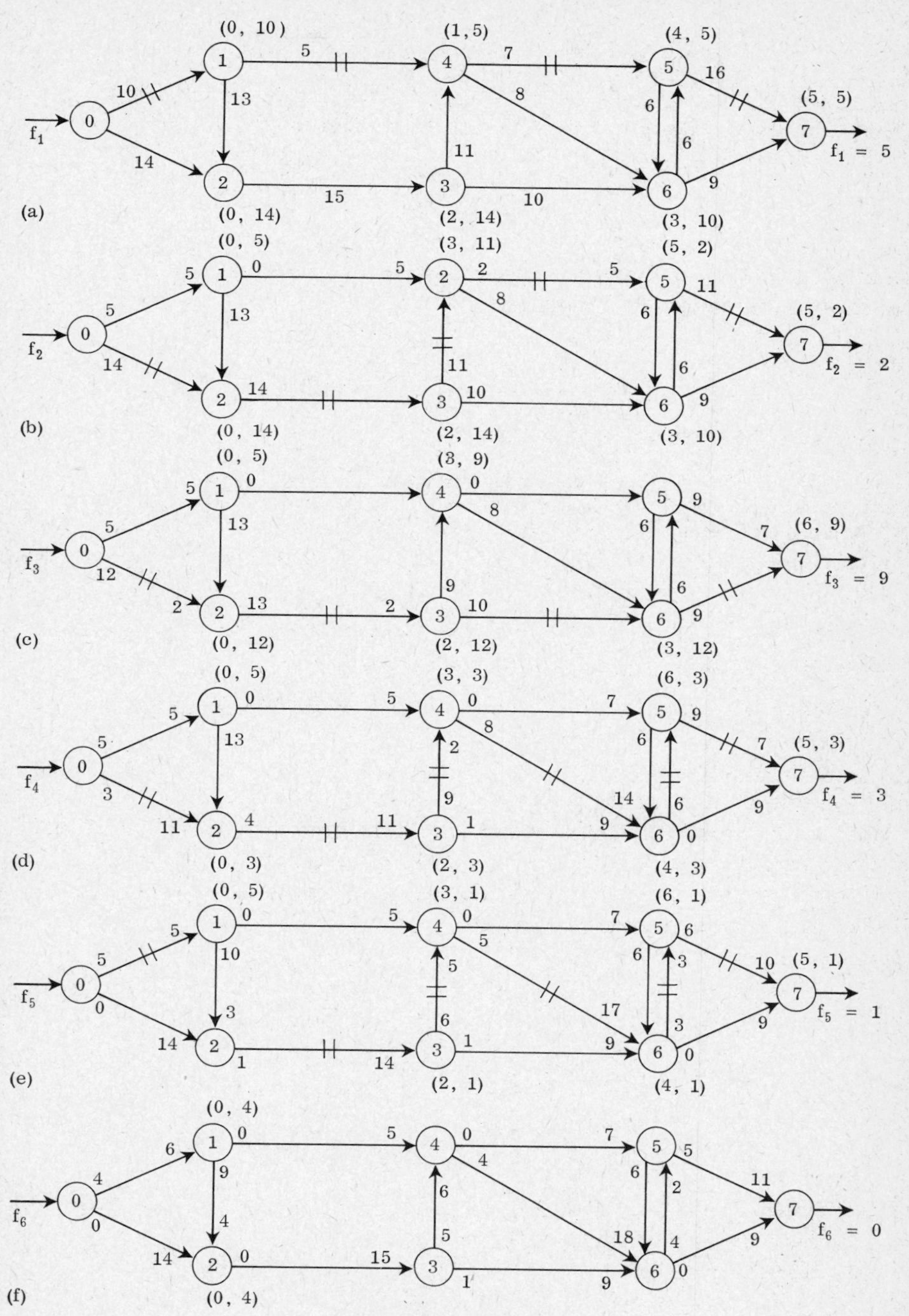

Figure 6-14

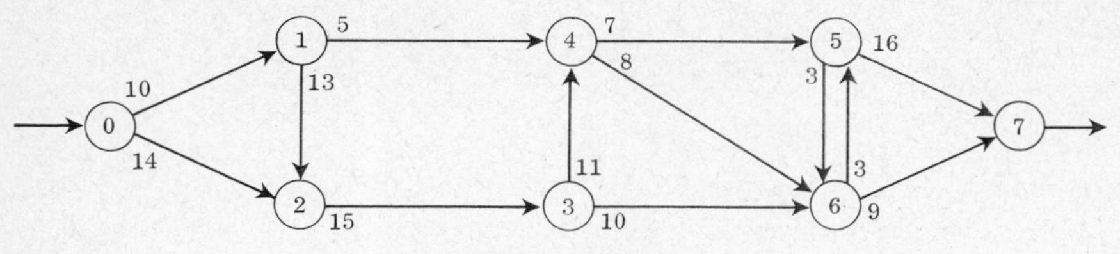

Figure 6–15

network by the labeling technique, as shown by various stages of operations in the figure. Finally, we have

$$F^* = f_1 + f_2 + f_3 + f_4 + f_5$$
$$= 5 + 2 + 9 + 3 + 1 = 20\ .$$

The flows in the fictitious branches are

$$x_{56} = 0\ , \qquad x_{65} = 6\ .$$

Since x_{56} is already zero, it is not necessary to substitute this set of flows into the equations in (6–11) and (6–12). If we do, however, we will obtain

$$x'_{56} = 0 - 0 = 0\ , \qquad x'_{65} = 6 - 0 = 6\ ,$$

which are identical to x_{56} and x_{65}, respectively.

Example 6–10. Determine the maximum flow for the network in Example 6–3.

The undirected branch 5–6 in the network for this problem is replaced by two directed branches 5–6 and 6–5, as shown in Fig. 6–15. The solution to this problem using labeling technique is exactly the same as that shown in Fig. 6–14 for the previous example, except that the chain flow f_5 should be excluded, since the capacity of branch 6–5 is exhausted after f_4 is deducted from it. Hence

$$F^* = f_1 + f_2 + f_3 + f_4$$
$$= 5 + 2 + 9 + 3 = 19\ .$$

REFERENCES

6–1. Ore, O., *Theory of Graphs*, The Mathematical Society, Providence, R. I., 1962.

6–2. Seshu, S., and M. B. Reed, *Linear Graphs and Electrical Networks*, Addison-Wesley, Reading, Mass., 1961.

6–3. Fenves, S. J., and F. H. Branin, Jr., "Network-Topological Formulation of Structural Analysis," *Journal of the Structural Division, ASCE*, **89,** No. ST4 (1963), 483–514.

6-4. FORD, L. R., JR., and D. R. FULKERSON, *Flow in Networks*, Princeton University Press, Princeton, N. J., 1962.

6-5. BERGE, C., and A. GHOUILA-HOURI, *Programming, Games and Transportation Networks*, Wiley, New York, 1965.

6-6. PINNELL, C., and G. T. SATTERLY, JR., "Analytical Methods in Transportation: Systems Analysis for Arterial Street Operation," *Journal of Engineering Mechanics Division, ASCE*, **89,** No. EM6 (1963), 67–95.

PROBLEMS

P6–1 and P6–2. Construct the node incidence matrix for the network shown in the figures for each problem. Also, construct the loop incidence matrix containing the basic loops.

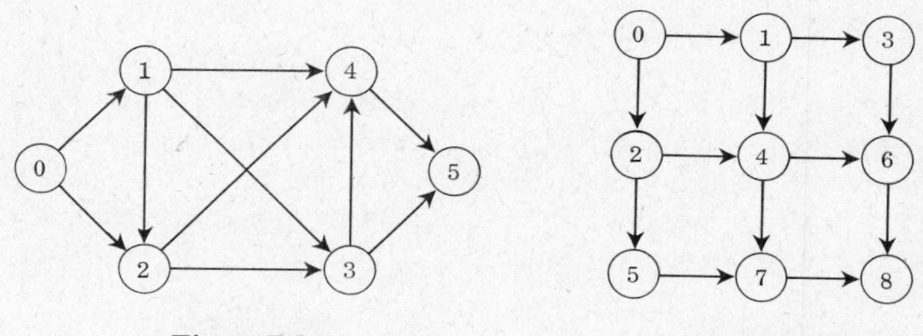

Figure P6–1

Figure P6–2

P6–3 and P6–4. For network shown in the figures for each problem, rearrange the nodes in such a way that the number of the initial node is smaller than the number of terminal node in every branch.

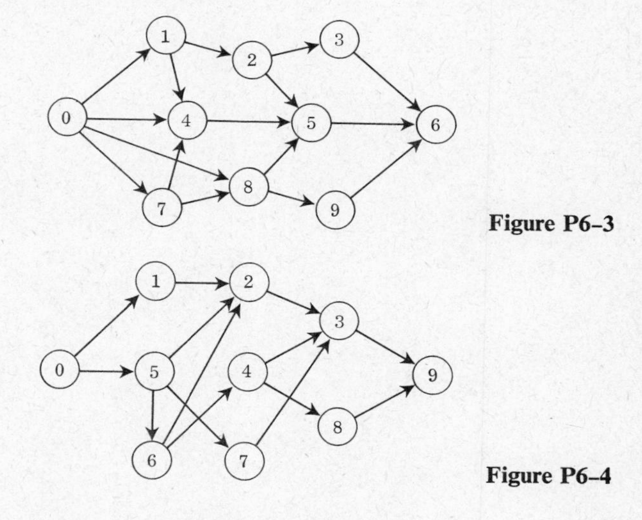

Figure P6–3

Figure P6–4

P6–5 and P6–6. Determine the maximum flow in the network with branch capacities shown in the figures for each problem by constructing flow chains from the source to the sink.

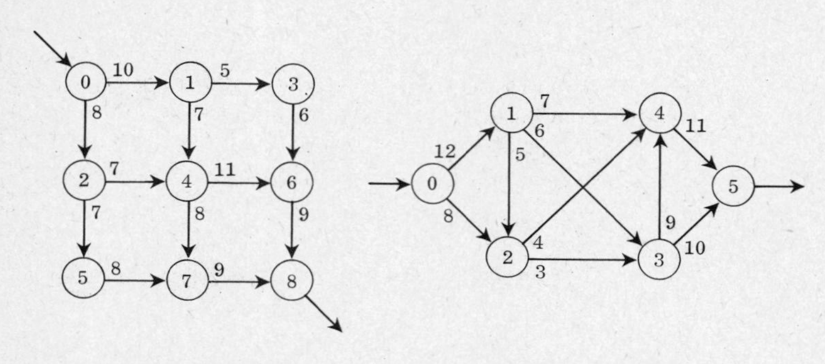

Figure P6–5

Figure P6–6

P6–7. Repeat Problem P6–5 by using the labeling technique on the graph.

P6–8. Repeat Problem P6–6 by using the labeling technique in matrix form.

P6–9 and P6–10. Determine the maximum flow in the network with branch capacities shown in the figures for each problem by using the labeling technique on the graph and in the matrix form.

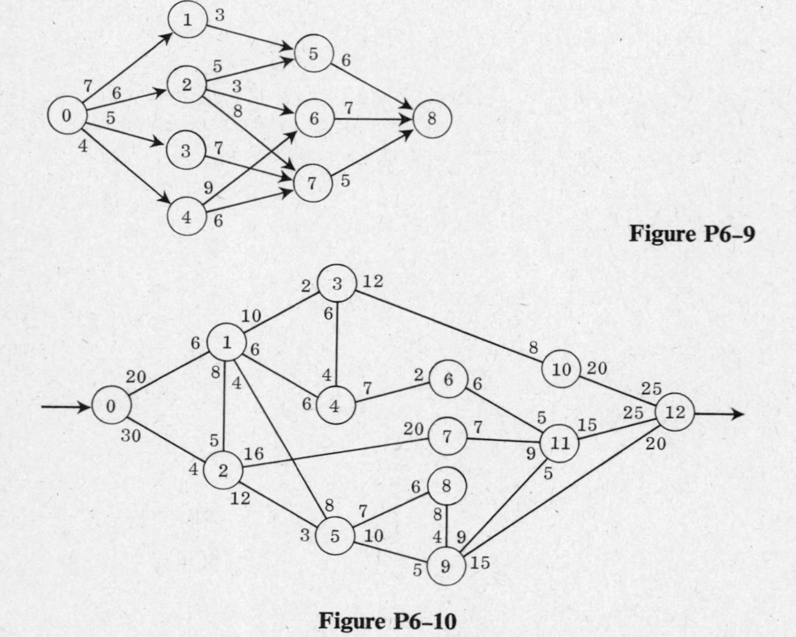

Figure P6–9

Figure P6–10

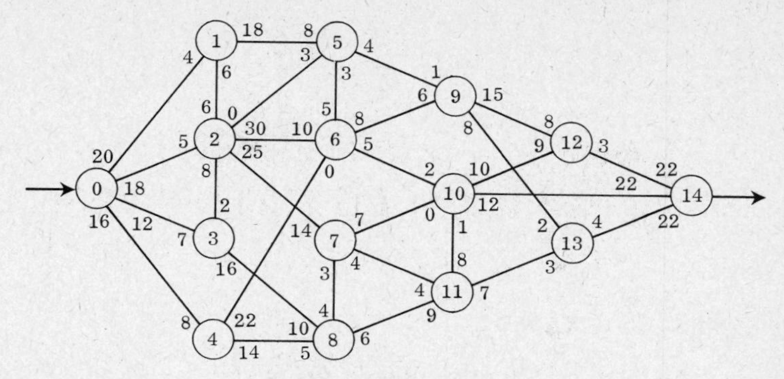

Figure P6–11

P6–11. Determine the maximum flow in the network with the branch capacities shown in Fig. P6–11 by using the labeling technique on the graph.

CHAPTER 7

MINIMUM COST FLOW PROBLEMS

7-1 ELEMENTARY EXAMPLES

In this chapter, we shall consider linear programming problems of finding minimum cost solutions, problems which can also be treated as flow problems in networks. The term *cost* is used in a broad sense here, since it may be expressed in terms of distance, time, weight, etc. as well as monetary value. The concepts of network flows are introduced to facilitate the formulation and solution of such problems.

Example 7-1. A contractor has six pieces of building equipment of the same kind available in cities 1 and 2 (three pieces each), and he has jobs in cities 3, 4, and 5 which require 2, 3, and 1 piece of equipment, respectively. These cities are located as shown in Fig. 7-1, and the unit costs of shipping between cities are indicated at the branches of the network. When there is no direct link between two cities, shipment may be made through an intermediate city. How should the shipment be made in order to minimize the total cost?

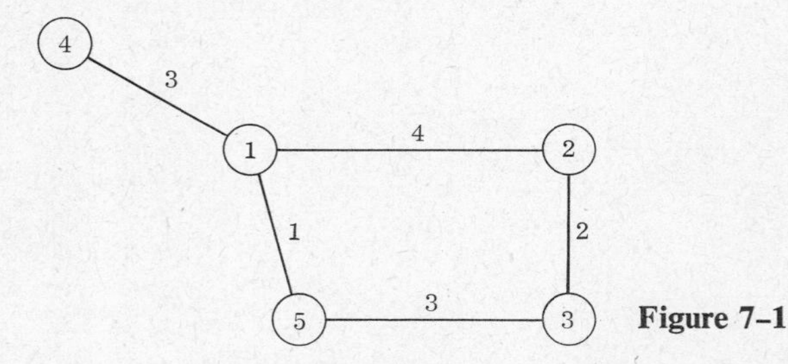

Figure 7-1

This problem involves the transshipment of equipment through cities which lie between the origins and destinations. For example, the shipment from city 1 to city 3 must go through either city 2 or 5, which may be regarded as the temporary destination for shipment from city 1 and at the same time the origin of the transshipment to city 3. Thus in transshipment problems every city is a potential origin and simultaneously a potential destination.

As we may note from Fig. 7-1, the most logical shipping pattern would be to send the three pieces of equipment in city 1 to city 4, two of the

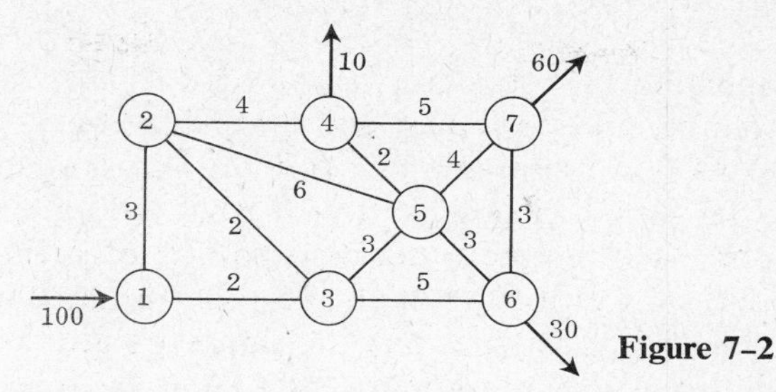

Figure 7-2

three pieces of equipment in city 2 to city 3, and the remaining one piece in city 2 to city 5, passing through either city 1 or city 3.

Example 7-2. In the highway network in Fig. 7-2 the normal traveling time in minutes is shown on the branches. If 100 cars are expected to enter from junction 1, with 10 cars departing at junction 4, 30 cars at junction 6 and 60 cars at junction 7, how should the cars be distributed in the network so that the total traveling time (in car-minutes) will be minimum?

This is a transshipment problem which can be solved by using the network in Fig. 7-2. For example, 10 cars may go from junction 1 to 2 and then to 4; 90 cars may go from junction 1 to 3; hence 30 cars will go to junction 6, and 60 cars will go to junction 5 and then to 7. If the network is more complicated, we cannot identify the routes readily by inspection, but can search systematically for routes involving minimum total traveling time.

Example 7-3. In Fig. 7-3, the distances in miles between nodes, wherever they are linked, are given on the branches of the network. Determine the shortest path from node 0 to node 10.

We can try to find the shortest path by exhaustive enumeration; however, the procedure can be simplified by systematic elimination. It has been found that two routes, one through nodes 0-1-4-8-10 and the other through nodes 0-1-6-9-10, have the same minimum distance of 17 mi. These routes are indicated by heavy lines in Fig. 7-3.

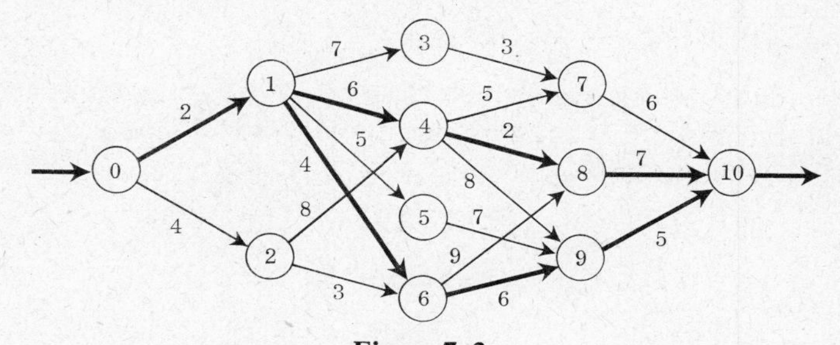

Figure 7-3

7-2 TRANSSHIPMENT PROBLEM

In the transshipment problem, we consider a group of n cities, each of which produces and consumes certain quantities of a product. If production exceeds consumption in a city, the excess quantity will be available as the net supply to other cities; if consumption exceeds production, the difference will create a net demand for shipment from other cities. For a group of n cities in which the total production equals the total consumption, we can determine the shipping pattern such that the total cost of shipment can be minimized. For unbalanced systems, we can introduce a fictitious city to which all surpluses may be shipped, but we cannot create a fictitious city from which all deficits are generated. We shall confine our discussion to systems with balanced total production and total consumption.

Unlike the direct transportation problem in which the cities are classified as origins and destinations, each city may be both an origin and a destination in the transshipment problem. We shall consider the balance of gross supply and demand at each city i ($i = 1, 2 \ldots, n$) in the formulation of the problem. Let

a'_i = local production in city i,
b'_i = local consumption in city i,
a_i = net supply in city i, if any,
b_i = net demand in city i, if any,
x'_{ii} = gross supply or gross demand in city i in a balanced system,
x_{ii} = quantity transshipped through city i,
x_{ij} = quantity shipped from city i to city j,
x_{ji} = quantity shipped from city j to city i,
c_{ij} = unit cost of shipment from city i to city j,
c_{ji} = unit cost of shipment from city j to city i.

The gross supply in city i is the sum of all quantities shipped from all other cities plus the local production, or

$$x'_{ii} = \sum_{j \neq i} x_{ji} + a'_i\,, \qquad i = 1, 2, \ldots, n\,. \tag{7-1}$$

The gross demand in city i, which must equal the gross supply in the city in a balanced system, is the sum of all quantities shipped to other cities plus the local consumption, or

$$x'_{ii} = \sum_{i \neq j} x_{ij} + b'_i\,, \qquad i = 1, 2, \ldots, n\,. \tag{7-2}$$

Then for all i and j,

$$x_{ij} \geq 0\,. \tag{7-3}$$

The objective function may be written as

$$\sum_{i=1}^{n}\sum_{j=1}^{n} c_{ij}x_{ij} = \text{Min } z\,, \tag{7–4}$$

since the summations of both i and j in the range from 0 to 1 cover all possible shipments between cities. Hence it is not necessary to include both x_{ij} and x_{ji}, or c_{ij} and c_{ji} in (7–4).

The transshipment problem can therefore be treated as a linear programming problem in which the constraint equations are represented by (7–1) and (7–2). These two sets of equations can be reduced to n equations as follows:

$$\sum_{j\neq i} x_{ji} + a'_i = \sum_{i\neq j} x_{ij} + b'_i\,, \qquad i = 1, 2, \ldots, n\,,$$

or

$$\sum_{i\neq j} x_{ij} - \sum_{j\neq i} x_{ji} = a'_i - b'_i\,, \qquad i = 1, 2, \ldots, n\,. \tag{7–5}$$

This set of n constraint equations can further be simplified by noting the relations:

(1) If local production exceeds local consumption, i.e., for $a'_i \geq b'_i$, we have

$$a_i = a'_i - b'_i\,, \qquad b_i = 0\,.$$

(2) If local production lags behind local consumption, i.e., for $a'_i \leq b'_i$, we get

$$a_i = 0\,, \qquad b_i = b'_i - a'_i = -(a'_i - b'_i)\,.$$

Using the abbreviated notation Min (a'_i, b'_i), which is defined as the smallest of the quantities in the parenthesis, that is, a'_i and b'_i, we can write

$$a_i = a'_i - \text{Min }(a'_i, b'_i)\,,$$
$$b_i = b'_i - \text{Min }(a'_i, b'_i)\,.$$

Hence the equations in (7–5) can also be established in terms of the net supplies and demands, instead of local productions and consumptions in the cities. That is, for $i = 1, 2, \ldots, n$,

$$\sum_{i\neq j} x_{ij} - \sum_{j\neq i} x_{ji} = \begin{cases} a_i\,, & \text{if } a'_i \geq b'_i\,, \\ -b_i\,, & \text{if } a'_i \leq b'_i\,. \end{cases} \tag{7–6}$$

Example 7–4. Formulate the transshipment problem between the five cities in Example 7–1 as a linear programming problem given that $a'_1 > b'_1$, $a'_2 > b'_2$, $a'_3 < b'_3$, $a'_4 < b'_4$, and $a'_5 < b'_5$, and that the net supplies are denoted by $a_1 = 3$, $a_2 = 3$, $a_3 = 0$, $a_4 = 0$, and $a_5 = 0$, while the net demands are denoted by $b_1 = 0$, $b_2 = 0$, $b_3 = 2$, $b_4 = 3$, and $b_5 = 1$.

There are $n = 5$ constraint equations, one for each city, which are obtained from (7-6) as follows:

$$\begin{aligned}
i &= 1\,, & x_{12} + x_{13} + x_{14} + x_{15} - x_{21} - x_{31} - x_{41} - x_{51} &= a_1\,,\\
i &= 2\,, & x_{21} + x_{23} + x_{24} + x_{25} - x_{12} - x_{32} - x_{42} - x_{52} &= a_2\,,\\
i &= 3\,, & x_{31} + x_{32} + x_{34} + x_{35} - x_{13} - x_{23} - x_{43} - x_{53} &= -b_3\,,\\
i &= 4\,, & x_{41} + x_{42} + x_{43} + x_{45} - x_{14} - x_{24} - x_{34} - x_{54} &= -b_4\,,\\
i &= 5\,, & x_{51} + x_{52} + x_{53} + x_{54} - x_{15} - x_{25} - x_{35} - x_{45} &= -b_5\,.
\end{aligned}$$

The nonnegative constraints are

$$x_{ij} \geq 0\,, \qquad i = 1, 2, 3, 4, 5\,; \qquad j = 1, 2, 3, 4, 5\,.$$

The objective function is given by

$$\sum_{i=1}^{5} \sum_{j=1}^{5} c_{ij} x_{ij} = \text{Min } z\,.$$

It is seen that x_{ij} include the quantities of transshipment x_{11}, x_{22}, x_{33}, x_{44}, and x_{55}, as well as all variables in the constraint equations. It should also be noted that if no direct route exists between some i and j (i.e., when $c_{ij} = \infty$), we must specify that $x_{ij} = 0$, corresponding to those $c_{ij} = \infty$, before attempting to solve this linear programming problem.

If this problem involves only direct transportation without transshipment, then all $x_{ij} = 0$ except x_{13}, x_{14}, x_{15}, x_{23}, x_{24}, and x_{25}. The set of constraint equations becomes

$$\begin{aligned}
x_{13} + x_{14} + x_{15} &= a_1\,,\\
x_{23} + x_{24} + x_{25} &= a_2\,,\\
-x_{13} - x_{23} \qquad &= -b_3\,,\\
-x_{14} - x_{24} \qquad &= -b_4\,,\\
-x_{15} - x_{25} \qquad &= -b_5\,,
\end{aligned}$$

which are identical to those obtained for the classical transportation problem.

7-3 TRANSSHIPMENT ARRAY

The constraint equations in (7-1) and (7-2) in the transshipment problem may also be simplified individually by considering the following relations:

(1) If local production exceeds local consumption, that is, $a_i' \geq b_i'$, then the gross supply can be determined by

$$x_{ii}' = x_{ii} + a_i'\,.$$

By substituting this relation into (7-1) and (7-2), we obtain, respectively, for city i,

$$-x_{ii} + \sum_{j \neq i} x_{ji} = 0$$

and

$$-x_{ii} + \sum_{i \neq j} x_{ij} = a'_i - b'_i .$$

The first expression is the restatement of the fact that when local production exceeds local consumption, all quantities shipped into city i will also be shipped out; hence the sum of all quantities shipped into city i is the transshipment through the city. The second expression states that the sum of all quantities shipped out of city i equals the transshipment through city i plus the surplus $a'_i - b'_i = a_i$.

(2) If local production lags behind local consumption, that is, $a'_i \leq b'_i$, then the gross demand can be determined by

$$x'_{ii} = x_{ii} + b'_i .$$

By substituting this relation into (7-1) and (7-2), we obtain, respectively, for city i,

$$-x_{ii} + \sum_{j \neq i} x_{ji} = b'_i - a'_i$$

and

$$-x_{ii} + \sum_{i \neq j} x_{ij} = 0 .$$

The first expression states that when local production lags behind local consumption, an amount of $b'_i - a'_i = b_i$ will be retained in city i from the sum of all quantities shipped into the city before transshipment. The second expression is a restatement that the sum of all quantities to be shipped out of city i equals the amount available for transshipment.

Using the same notation of a'_i and b'_i defined in the previous section, we can therefore summarize the expressions for transshipment as follows:

$$-x_{ii} + \sum_{i \neq j} x_{ij} = a_i , \qquad i = 1, 2, \ldots, n ; \tag{7-7}$$

$$-x_{ii} + \sum_{j \neq i} x_{ji} = b_i , \qquad i = 1, 2, \ldots, n . \tag{7-8}$$

We can interchange i and j in the set of equations in (7-8) such that j will become the designation for a city and i will be a dummy index for summation as follows:

$$-x_{jj} + \sum_{i \neq j} x_{ij} = b_j , \qquad j = 1, 2. \ldots, n . \tag{7-9}$$

This interchange has not altered any physical meaning of the problem. However, we may now regard the equations in (7-7) as row equations and those in (7-9) as column equations in a square array, as shown in

Table 7-1

STANDARD TRANSSHIPMENT ARRAY

		Destination city j				Net supply
		1	2	...	n	
Origin city i	1	$-x_{11}$ 0	x_{12} c_{12}	... ...	x_{1n} c_{1n}	a_1 π_1
	2	x_{21} c_{21}	$-x_{22}$ 0	... ...	x_{2n} c_{2n}	a_2 π_2
	...	... ...	... ...	... ...	... ...	... ...
	n	x_{n1} c_{n1}	x_{n2} c_{n2}	... ...	$-x_{nn}$ 0	a_n π_n
Net demand		b_1 $-\pi_1$	b_2 $-\pi_2$	... ...	b_n $-\pi_n$	

Table 7-1, which is called a *standard transshipment array*. In the array, each city is identified as an origin city i when the row equation relates the transshipment to the quantities to be shipped out from the city, and each city is identified as a destination city j when the column equation relates the transshipment to the quantities shipped into it. Each cost coefficient c_{ij} representing the unit cost of shipment from city i to city j is given in the problem. The cost coefficients c_{ij} and c_{ji} may or may not be equal, depending on the conditions of shipping. The unit cost of shipment within a city is assumed to be zero, or $c_{ii} = 0$, since each city is regarded as a point or node in the network. The unit cost of shipment between cities with no direct access is assumed to be infinite. The terms π_i and $-\pi_i$ ($i = 1, 2, \ldots, n$) are simplex multipliers, which will be explained later.

Example 7-5. Formulate Example 7-1 in a standard transshipment array.

The row equations and column equations formulated for this problem are shown in the transshipment array in Table 7-2, in which the cost coefficients and the net demands and supplies in cities are also given. A solution of this problem is seen intuitively to be

$$x_{11} = 1\,, \quad x_{14} = 3\,, \quad x_{15} = 1\,, \quad x_{21} = 1\,, \quad x_{23} = 2\,;$$
$$x_{ij} = 0 \quad \text{for all other combinations of } i \text{ and } j.$$

Table 7-2

TRANSSHIPMENT ARRAY FOR EXAMPLE 7-5

		Destination city					Row sum
		1	2	3	4	5	
Origin city	1	$-x_{11}$ 0	x_{12} 4	x_{13} ∞	x_{14} 3	x_{15} 1	3
	2	x_{21} 4	$-x_{22}$ 0	x_{23} 2	x_{24} ∞	x_{25} ∞	3
	3	x_{31} ∞	x_{32} 2	$-x_{33}$ 0	x_{34} ∞	x_{35} 3	0
	4	x_{41} 3	x_{42} ∞	x_{43} ∞	$-x_{44}$ 0	x_{45} ∞	0
	5	x_{51} 1	x_{52} ∞	x_{53} 3	x_{54} ∞	$-x_{55}$ 0	0
Column sum		0	0	2	3	1	

Thus each of the row equations with nonzero elements is balanced, and each of the column equations with nonzero elements is also balanced. Equations with all elements equal to zero are automatically satisfied.

7-4 EQUIVALENT TRANSPORTATION PROBLEM

The transshipment array is remarkably similar to the transportation array except that all variables x_{ii} ($i = 1, 2, \ldots, n$) along the main diagonal are preceded by a negative sign. Hence the method of solution for the transportation problem will be applicable to the transshipment problem if $-x_{ii}$ can be replaced by another set of nonnegative variables

$$\bar{x}_{ii} = t - x_{ii}\,, \qquad i = 1, 2, \ldots, n\,, \tag{7-10}$$

in which t is any positive constant which is sufficiently large so that all $\bar{x}_{ii}$ will always be positive. When unit shipping costs are nonnegative, the transshipment through any city cannot exceed the total net demands or supplies in all cities. Hence we may let

$$t = \sum_{i=1}^{n} a_i = \sum_{j=1}^{n} b_j\,.$$

Table 7–3

EQUIVALENT TRANSPORTATION ARRAY

		Destination city j				
		1	2	...	n	
Origin city	1	$\bar{x}_{11}$	x_{12}	...	x_{1n}	$t+a_1$
		0	c_{12}	...	c_{1n}	
	2	x_{21}	$\bar{x}_{22}$	...	x_{2n}	$t+a_2$
		c_{21}	0	...	c_{2n}	
	...	...	...	...	...	...
		...	...	...	...	...
	n	x_{n1}	x_{n2}	...	$\bar{x}_{nn}$	$t+a_n$
		c_{n1}	c_{n2}	...	0	
		$t+b_1$	$t+b_2$	...	$t+b_n$	

Thus $\bar{x}_{ii}$ in (7–10) may be regarded as slack variables such that

$$x_{ii} = t - \bar{x}_{ii}\,, \qquad i = 1, 2, \ldots, n\,; \tag{7–11}$$

$$x_{jj} = t - \bar{x}_{jj}\,, \qquad j = 1, 2, \ldots, n\,. \tag{7–12}$$

Substituting these expressions into (7–7) and (7–9), respectively, we have

$$\bar{x}_{ii} + \sum_{i \neq j} x_{ij} = t + a_i\,, \qquad i = 1, 2, \ldots, n\,; \tag{7–13}$$

$$\bar{x}_{jj} + \sum_{i \neq j} x_{ij} = t + b_j\,, \qquad j = 1, 2, \ldots, n\,. \tag{7–14}$$

If these two sets of equations are treated as the row and column equations, respectively, of an equivalent transportation problem, the corresponding transportation array may be as given in Table 7–3. Since the objective function in (7–4) remains unchanged, the equivalent problem can be solved as a standard transportation problem. After all $\bar{x}_{ii}$, as well as other x_{ij}, are obtained, the transshipment x_{ii} may be determined from (7–11).

Example 7–6. Find an optimal solution by considering the transshipment problem in Example 7–1 as an equivalent transportation problem.

In this problem with $n = 5$, we have five row equations and five column equations:

$$t = \sum a_i = \sum b_j = 6\,,$$

$$t + a_1 = 6 + 3 = 9\,, \qquad t + b_1 = 6 + 0 = 6\,,$$

$$t + a_2 = 6 + 3 = 9\,, \qquad t + b_2 = 6 + 0 = 6\,,$$

$$t + a_3 = 6 + 0 = 6\,, \qquad t + b_3 = 6 + 1 = 7\,,$$

$$t + a_4 = 6 + 0 = 6\,, \qquad t + b_4 = 6 + 3 = 9\,,$$

$$t + a_5 = 6 + 0 = 6\,, \qquad t + b_5 = 6 + 2 = 8\,.$$

With the given cost coefficients, an initial solution may be obtained, as shown in Table 7–4.

In solving this problem as an equivalent transportation problem, we try to obtain an initial solution according to the minimum cost rule. Since the main diagonal of the matrix has $c_{ii} = 0$ ($i = 1, 2, 3, 4, 5$), it is logical to fill these positions first. However, because of the locations of $c_{ij} = \infty$, we cannot always fill up all positions along the main diagonal with a number greater than or equal to $t = 6$. In case (a) of Table 7–4, for example, $x_{33} = 5$, in spite of the fact that $a_3 = 6$ and $b_3 = 8$. Nevertheless, all five positions in the main diagonal are filled and only four other basic variables are added outside of the main diagonal, since $2n - 1 = 9$. By introducing the simplex multipliers u_i and v_j, we note that if $u_i = \pi_i$ ($i = 1, 2, 3, 4, 5$), then $v_i = -\pi_i$, since all diagonal elements are basic variables. Upon the test of optimality, we find that all $\bar{c}_{ij} \geq 0$ for all nonbasic variables. Hence the initial solution is optimal.

We have thus obtained $\bar{x}_{ii}$ as well as x_{ij} for $i \neq j$ from the transportation array. To obtain x_{ii}, we have from (7–11),

$$x_{11} = t - \bar{x}_{11} = 0\,,$$

$$x_{22} = t - \bar{x}_{22} = 0\,,$$

$$x_{33} = t - \bar{x}_{33} = 1\,,$$

$$x_{44} = t - \bar{x}_{44} = 0\,,$$

$$x_{55} = t - \bar{x}_{55} = 0\,.$$

Thus the shipment goes only through city 3. The minimum cost based on this shipping pattern is given by

$$z^* = (3)(3) + (3)(2) + (1)(3) = 18\,.$$

An alternative solution is also given as case (b) in Table 7–4, in which the shipment goes only through city 1. The minimum cost based on the alternative shipping pattern is

$$z^* = (3)(3) + (1)(1) + (1)(4) + (2)(2) = 18\,.$$

Table 7-4

SOLUTION OF EXAMPLE 7-6

(a) An optimal solution

6 0	4	∞	3 3	0 1	9 0
4	6 0	3 2	∞	∞	9 4
∞	2	5 0	∞	1 3	6 2
3	∞	∞	6 0	∞	6 −3
1	∞	3	∞	6 0	6 −1
6 0	6 −4	8 −2	9 3	7 1	

(b) An alternative optimal solution

5 0	4	∞	3 3	1 1	9 0
1 4	6 0	2 2	∞	∞	9 4
∞	2	6 0	∞	3	6 2
3	∞	∞	6 0	∞	6 −3
1	∞	3	∞	6 0	6 −1
6 0	6 −4	8 −2	9 3	7 1	

7–5 NETWORK SOLUTION OF TRANSSHIPMENT ARRAYS

In the last section, we used an indirect procedure for solving the transshipment problem, i.e., converting it first to an equivalent transportation problem and then solving the latter problem. It is possible to solve this problem directly on the basis of the standard transshipment array in Table 7–1, provided that a systematic procedure can be established to find an initial feasible solution, to test its optimality, and to change the basis if the solution is not optimal.

It is noteworthy that every basis of a transshipment problem is triangular and that the variables along the main diagonal can be made a part of every basis that is feasible. The property that any basis of a transshipment problem is triangular can be verified in a way similar to that for a transportation problem, because such a basis does not depend on the sign of the variables. Furthermore, the transshipment variables along the main diagonal will be retained in the basis once they have entered it, because they are preceded by a negative sign, which makes them ineligible as departing variables. Thus in an $n \times n$ transshipment array, which nominally has $(2n - 1)$ basic variables, the n transshipment variables always lie on the main diagonal. Only the remaining $(n - 1)$ basic variables outside of the main diagonal can be assigned to different positions in seeking an optimal feasible solution.

An initial feasible solution of the $(n - 1)$ basic variables can be obtained by the construction of a subgraph of $(n - 1)$ branches connecting the n nodes in the network. In order that the solution be feasible, the subgraph must be connected so that a path exists between nodes. This connected subgraph, having n nodes and $(n - 1)$ branches, is a tree and contains no loop. Thus, *if a network of n nodes representing the transshipment problem has a feasible solution, there exists a tree corresponding to the $(n - 1)$ basic variables other than the n transshipment variables.* To find an initial feasible solution, we first connect the nodes to form a tree having $(n - 1)$ directed branches and evaluate these $(n - 1)$ basic variables corresponding to the directed branches of the tree. Once these variables are determined, the n transshipment variables may be evaluated by balancing the rows and columns in the transshipment array.

After all basic variables, including the transshipment variables, have been assigned values for an initial feasible solution, the optimality test for the transportation problem may be applied, i.e., the feasible solution is optimal if, for all nonbasic variables,

$$\bar{c}_{ij} \geq 0 .$$

If the solution is not optimal, we can change the basis by introducing a new basic variable to replace the departing variable, as in the transpor-

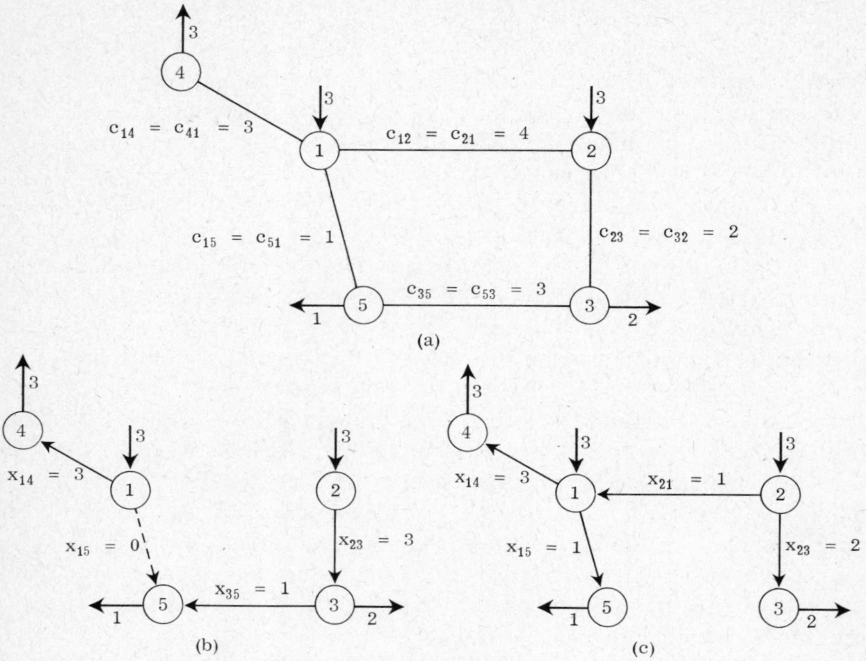

Figure 7-4

tation problem. However, we shall exclude all basic variables along the main diagonal in the selection of the departing variable, since they are always preceded by negative signs. Hence it is seen that if $u_i = \pi_i$, then $v_j = -\pi_i$, and conversely. The remaining steps in the solution are essentially the same as those in the transportation problem.

Example 7-7. Solve Example 7-1 by the use of the standard transshipment array.

This problem, shown in Fig. 7-4(a), has two solutions represented by (A) and (B) in Table 7-5. In case (A), the initial solution is represented by the tree in part (b) of the figure. Since $x_{15} = 0$, there is no shipment between cities 1 and 5. However, this link (or another link, such as that between cities 1 and 2) is needed to connect all cities to form a connected subgraph. The variable $x_{15} = 0$ (or another degenerate basic variable, such as $x_{12} = 0$) is also needed in the transshipment array so that the total number of basic variables will be $2n - 1$. This initial solution is optimal since all $\bar{c}_{ij} \geq 0$. In case (B), an alternative initial solution is represented by the tree routes in Fig. 7-4(c). This initial solution is also found to be optimal.

Table 7-5

SOLUTION OF EXAMPLE 7-7

(A) Transshipment through city 3

<table>
<tr><td>0
0</td><td>
4</td><td>
∞</td><td>3
3</td><td>0
1</td><td>3
0</td></tr>
<tr><td>
4</td><td>0
0</td><td>3
2</td><td>
∞</td><td>
∞</td><td>3
4</td></tr>
<tr><td>
∞</td><td>
2</td><td>−1
0</td><td>
∞</td><td>1
3</td><td>0
2</td></tr>
<tr><td>
3</td><td>
∞</td><td>
∞</td><td>0
0</td><td>
∞</td><td>0
−3</td></tr>
<tr><td>
1</td><td>
∞</td><td>
3</td><td>
∞</td><td>0
0</td><td>0
−1</td></tr>
<tr><td>0
0</td><td>0
−4</td><td>2
−2</td><td>3
3</td><td>1
1</td><td></td></tr>
</table>

(B) Transshipment through city 1

<table>
<tr><td>−1
0</td><td>
4</td><td>
∞</td><td>3
3</td><td>1
1</td><td>3
0</td></tr>
<tr><td>1
4</td><td>0
0</td><td>2
2</td><td>
∞</td><td>
∞</td><td>3
4</td></tr>
<tr><td>
∞</td><td>
2</td><td>0
0</td><td>
∞</td><td>
3</td><td>0
2</td></tr>
<tr><td>
3</td><td>
∞</td><td>
∞</td><td>0
0</td><td>
∞</td><td>0
−3</td></tr>
<tr><td>
1</td><td>
∞</td><td>
3</td><td>
∞</td><td>0
0</td><td>0
−1</td></tr>
<tr><td>0
0</td><td>0
−4</td><td>2
−2</td><td>3
3</td><td>1
1</td><td></td></tr>
</table>

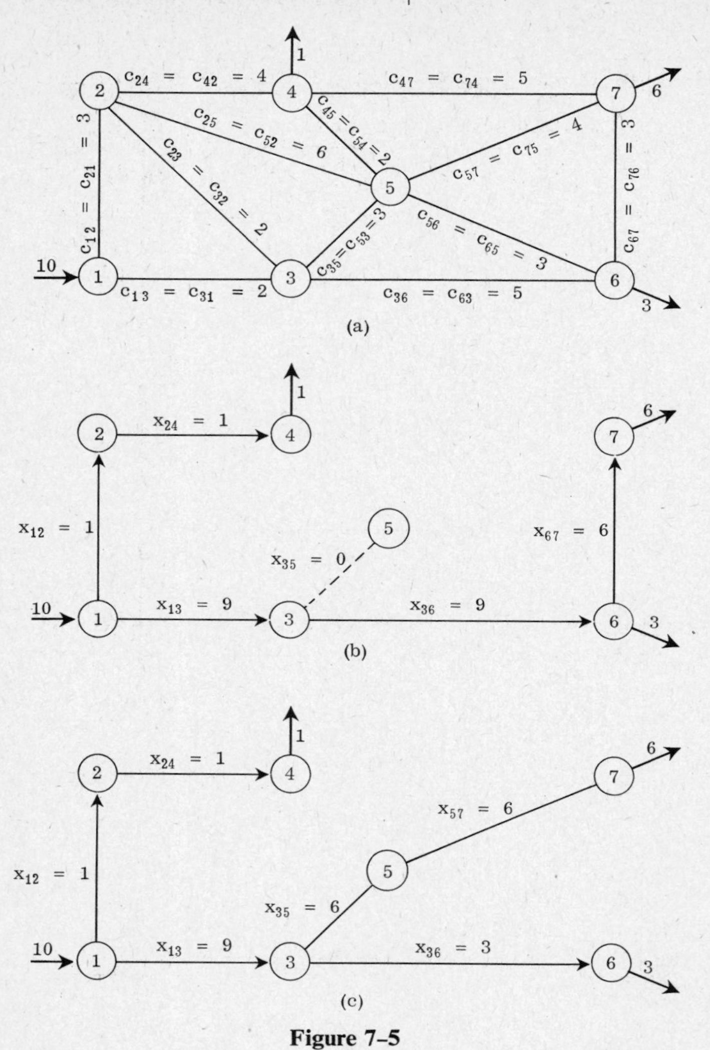

Figure 7-5

Example 7-8. Figure 7-5(a) shows a highway network linking seven cities with entering and departing cars. The unit traveling time for each route is shown on the route. The number of cars entering the network from a city is indicated by the quantity near an incoming arrow and the number of cars leaving the network to a city is indicated by the quantity near an outgoing arrow. Determine the minimum total traveling time of all cars in the network.

The initial solution is represented by the tree in Fig. 7-5(b), showing the values of the variables in Cycle 0 of Table 7-6 (p. 252). In the test of optimality, it is found that $\bar{c}_{57}$ is negative, and the solution is not optimal.

Let $x_{57} = \theta$ be the entering variable, and the rows and columns in the array are rebalanced as shown. In selecting the departing variable, however, it must be remembered that the transshipment variables along the main diagonal are preceded by a negative sign. Hence the value of θ is governed by $x_{67} = 6 - \theta = 0$ outside the diagonal. With x_{67} as a departing variable and $\theta = 6$, a new basis is obtained, as shown in Cycle 1. This latter solution is found to be optimal, and is represented by the routes shown in Fig. 7-5(c).

If we follow the transshipment array row by row, we can interpret the movement of cars from city to city. For example, by referring to Cycle 1 of Table 7-6 and Fig. 7-5(c), we see that one car goes from city 1 to city 2, and nine cars go from city 1 to city 3. The one car going into city 2, in turn, moves on to city 4 as its destination; the nine cars going into city 3, in turn, break up into two groups, with six cars going to city 5 and three cars going to city 6. While the three cars at city 6 have reached their destination, the six cars at city 5 move on to city 7 before they finally stop. Thus a positive number in the transshipment array indicates the quantity entering a city, and a negative number indicates the quantity leaving a city.

7-6 CATERER PROBLEM

Another extension of the application of transportation problem is the so-called *caterer problem*, which is in reality the paraphrase of a problem concerning the number of spare engines required to assure given operational levels of a fleet of airplanes. It can be seen that the general approach is also applicable to the maintenance of operating equipment for a construction project to be completed in a specified period of time. We shall restate the caterer problem as follows:

In maintaining the operational levels of a fleet of haulers over a period of n days, the number of spare engines required on the jth day is denoted by b_j ($j = 1, 2, \ldots, n$). Engines disabled on any day are immediately sent to the repair shop which provides services at two rates—a lower cost e per engine for slower service in p days and a higher cost f per engine for faster service in q days ($e < f$ and $p > q$). Let the number of disabled engines on the ith day be denoted by a_i. Any engine disabled on the ith day cannot be repaired in time for operation on the jth day (where $j > i$) unless $j - i \geq q$. If the level b_j cannot be replenished in time by the repaired engines, new spare engines must be purchased at a cost of d each ($d > f > e$). At the beginning of the n-day period, no spare engine is in the stock and the earlier demands must be met through new purchases. However, repaired engines may be used later for spare if they are returned from the repair shop in time to meet the demand. The disabled engines will not be repaired near the end of the n-day period if they cannot be

Table 7-6

SOLUTION OF EXAMPLE 7-8

Cycle 0 $\theta = 6$

0 0	1 3	9 2	 ∞	 ∞	 ∞	 ∞	10 0
 3	-1 0	 2	1 4	 6	 ∞	 ∞	0 -3
 2	 2	-9 0	 ∞	$0+\theta$ 3	$9-\theta$ 5	 ∞	0 -2
 ∞	 4	 ∞	0 0	 2	 ∞	 5	0 -7
 ∞	 6	 3	 2	$0-\theta$ 0	 3	θ 4	0 -5
 ∞	 ∞	 5	 ∞	 3	$-6+\theta$ 0	$6-\theta$ 3	0 -7
 ∞	 ∞	 ∞	 5	 4	 3	0 2	0 -10
0 0	0 3	0 2	1 7	0 5	3 7	6 10	

returned from the repair shop for reuse on or before the last day of operation. Disabled engines thus retired are assumed to have no salvage value. Determine the layout for purchasing and repairing engines which will minimize the total cost.

If b_j, the required number of spare engines on the jth day, is regarded as the demand at the jth destination, then a_i, the number of disabled engines on the ith day, may be considered as the supply at the ith origin, because each disabled engine on the ith day will be repaired at a cost and become useful again on the jth day ($j > i$). If repair cannot be done in time for meeting the demands, the shortage will be alleviated by new purchases. Toward the end of the n-day period, the disabled engines are retired as surplus. Thus the problem has the basic structure of a transportation problem with surplus and shortage, in which some routes are inaccessible.

Table 7–6

(Continued)

Cycle 1 Optimal

0 0	1 3	9 2	 ∞	 ∞	 ∞	 ∞	10 0
 3	−1 0	 2	1 4	 6	 ∞	 ∞	0 −3
 2	 2	−9 0	 ∞	6 3	3 5	 ∞	0 −2
 ∞	 4	 ∞	0 0	 2	 ∞	 5	0 −7
 ∞	 6	 3	 2	−6 0	 3	6 4	0 −5
 ∞	 ∞	 5	 ∞	 3	0 0	 3	0 −7
 ∞	 ∞	 ∞	 5	 4	 3	0 2	0 −9
0 0	0 3	0 2	1 7	0 5	3 7	6 9	

Let x_{ij} be the number of engines sent to the repair shop on the ith day and returned as spare on the jth day. Then

$$\sum_{j=1}^{n} x_{ij} \leq a_i\,, \qquad i = 1, 2, \ldots, n\,,$$

and

$$\sum_{i=1}^{n} x_{ij} \leq b_j\,, \qquad j = 1, 2, \ldots, n\,.$$

Introducing the slack variables x_{i0} and x_{0j}, we have

$$\sum_{j=1}^{n} x_{ij} + x_{i0} = a_i\,, \qquad i = 1, 2, \ldots, n\,; \tag{7–15}$$

$$\sum_{i=1}^{n} x_{ij} + x_{0j} = b_j\,, \qquad j = 1, 2, \ldots, n\,. \tag{7–16}$$

Summing all n equations in (7-15) over i and all n equations in (7-16) over j, and denoting

$$\sum_{i=1}^{n}\sum_{j=1}^{n} x_{ij} = x_{00} ,$$

we get

$$x_{00} + \sum_{i=1}^{n} x_{i0} = \sum_{i=1}^{n} a_i , \tag{7-17}$$

$$x_{00} + \sum_{j=1}^{n} x_{0j} = \sum_{j=1}^{n} b_j . \tag{7-18}$$

The cost coefficients for the problem are given by the costs of replenishing the spare engines as follows.

1) For retiring the surplus disabled engines;

$$c_{i0} = 0 , \qquad i = 1, 2, \ldots, n .$$

2) For new purchases in case of shortage:

$$c_{0j} = d , \qquad j = 1, 2, \ldots, n .$$

3) For the general case ($i = 1, 2, \ldots, n ;\quad j = 1, 2, \ldots, n$):

$$\begin{aligned} c_{ij} &= e , \qquad j - i \geq p ; \\ c_{ij} &= f , \qquad q \leq j - i < p ; \\ c_{ij} &= \infty , \qquad j - i < q . \end{aligned}$$

Note also that $c_{00} = 0$, since x_{00} is not an independent variable. Thus for given values of p and q, all c_{ij} for the problem can be determined. Therefore for

$$x_{ij} \geq 0 , \qquad i = 1, 2, \ldots, n ; \quad j = 1, 2, \ldots, n , \tag{7-19}$$

the objective function is seen to be the minimization of the total cost, i.e.,

$$\text{Min } z = \sum_{i=0}^{n}\sum_{j=0}^{n} c_{ij}x_{ij} . \tag{7-20}$$

Hence the problem is completely defined.

Note that every day in the n-day period in this problem belongs to the set $i = 1, 2, \ldots, n$ (the days when the engines are disabled) and at the same time belongs to the set $j = 1, 2, \ldots, n$ (the days when the spare engines are required). This is analogous to the situation in the transshipment problem in which every city is regarded as an origin and a destination at the same time. Thus the transportation array for the caterer's problem will have an $(n + 1) \times (n + 1)$ cost matrix, including a surplus column and a shortage row.

Table 7–7

SOLUTION FOR EXAMPLE 7–9

Cycle 0 $\qquad \theta = 1$

$14-\theta$ / 0	2 / 8	3 / 8	$1+\theta$ / 8	8	8	20 / 0
0	∞	5	2 / 2	2	2	2 / −6
0	∞	∞	$1-\theta$ / 5	$2+\theta$ / 2	2	3 / −3
0	∞	∞	∞	$4-\theta$ / 5	θ / 2	4 / 0
$1+\theta$ / 0	∞	∞	∞	∞	$5-\theta$ / 5	6 / 0
5 / 0	∞	∞	∞	∞	∞	5 / 0
20 / 0	2 / 8	3 / 8	4 / 8	6 / 5	5 / 5	

Cycle 1

13 / 0	2 / 8	3 / 8	2 / 8	8	8	20 / 0
0	∞	5	2 / 2	2	2	2 / −6
0	∞	∞	5	3 / 2	2	3 / −6
0	∞	∞	∞	3 / 5	1 / 2	4 / −3
2 / 0	∞	∞	∞	∞	4 / 5	6 / 0
5 / 0	∞	∞	∞	∞	∞	5 / 0
20 / 0	2 / 8	3 / 8	4 / 8	6 / 5	5 / 0	

Example 7-9. Consider a numerical example of the caterer's problem in which $n = 5$, $p = 2$, and $q = 1$. A cost matrix of 6×6 ($i, j = 0, 1, 2, 3, 4$, and 5) can be established, as shown in Table 7-7, in which $e = 2$, $f = 5$, and $d = 8$. Also, $a_i = 2, 3, 4, 6$, and 5 for $i = 1, 2, 3, 4$, and 5, respectively, and $b_j = 2, 3, 4, 6$, and 5 for $j = 1, 2, 3, 4$, and 5, respectively. That is, the operational level is to be kept constant daily so that if a_i is the number of disabled engines on any day, the number of the spare engines required on the same day ($i = j$) is $b_j = a_i$. Hence $\sum a_i = \sum b_j = 20$. Determine the minimum cost solution.

Since an initial solution can be obtained by inspection, Phase I is not used. At the end of Cycle 0, the solution is not optimal, since

$$\bar{c}_{35} = 2 - 5 = -3 .$$

Let $x_{35} = \theta$ be the entering variable, and upon the setting of $\theta = 1$, we see that $x_{23} = 1 - \theta = 0$ becomes the departing variable. The new basis in Cycle 1 is found to be optimal. Hence the minimum cost is

$$\begin{aligned} z^* &= (2 + 3 + 2)(8) + (3 + 4)(5) + (2 + 3 + 1)(2) \\ &= 56 + 35 + 12 = 103 . \end{aligned}$$

We may also note from the breakdown that the cost of buying new engines is 56, the cost of using fast-repair service is 35, and the cost of using slow-repair service is 12.

Needless to say, it is possible to have more than one set of basic variables which may produce the same minimum cost. However, the total number of basic variables including x_{00} must be $2n - 1 = 11$ in each optimal set.

7-7 THE SHORTEST-ROUTE PROBLEM

Another application of network analysis is the determination of the shortest route between two points in the network in which the distances between all nodes are known. The same formulation can be used to determine the minimum-cost route between two points in the network if the shipping costs between all nodes are given.

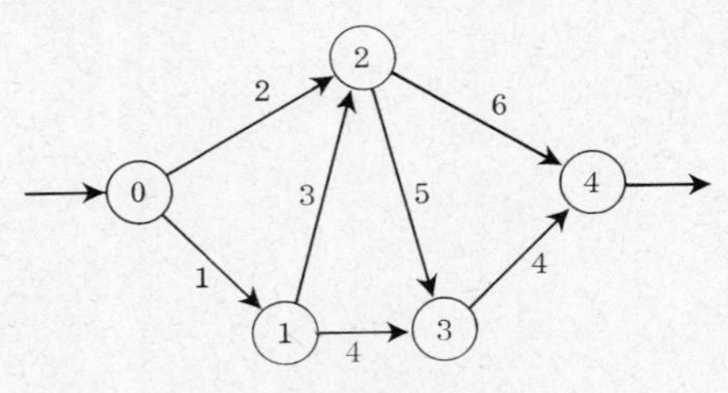

Figure 7-6

For example, consider, the network in Fig. 7-6, in which the distance d_{ij} between nodes i and j (for all i and j) are given. In order to determine the shortest route between nodes 1 and 4, we can assume that one unit of flow enters the network from the source at node 0 and leaves the network through the sink at node 4. Let x_{ij} be the flow in a branch (i, j). If the flow passes through that branch, $x_{ij} = 1$; otherwise $x_{ij} = 0$. In general, we can express Kirchhoff's node equations in the form

$$\sum_j x_{ij} - \sum_j x_{ji} = \begin{cases} 1, & i = 0, \\ 0, & i = 1, 2, \ldots, n-1, \\ -1, & i = n, \end{cases} \tag{7-21}$$

for all $x_{ij} \geq 0$. The objective function is given by

$$\text{Min } z = \sum_i \sum_j d_{ij} x_{ij} . \tag{7-22}$$

For this particular example, the node equations for nodes 0, 1, 2, 3, and 4 are as follows:

$$\begin{aligned}
x_{01} + x_{02} \qquad\qquad\qquad\qquad\qquad\qquad &= 1, \\
-x_{01} \qquad + x_{12} + x_{13} \qquad\qquad\qquad\qquad &= 0, \\
-x_{02} - x_{12} \qquad + x_{23} + x_{24} \qquad &= 0, \\
-x_{13} - x_{23} \qquad + x_{34} &= 0, \\
-x_{24} - x_{34} &= -1;
\end{aligned}$$

and the objective function is

$$\text{Min } z = x_{01} + 2x_{02} + 3x_{12} + 4x_{13} + 5x_{23} + 6x_{24} + 4x_{34} .$$

In general, the solution of such a problem is lengthy, since the number of variables and number of constraint equations can be large. For this particular example, however, it can easily be seen that the optimal solution is given by

$$x_{02} = x_{24} = 1 \qquad \text{and} \qquad x_{01} = x_{12} = x_{13} = x_{23} = x_{34} = 0 ,$$

for which the optimal value of z is $z^* = 8$.

The solution of the shortest-route problem may be simplified if the flow problem defined by (7-21) and (7-22) is regarded as the primal problem, and a solution is sought for its dual. Let y_i be the dual variables corresponding to the number of equations for the nodes in the network. These variables are unrestricted in sign, since all constraints in the primal problem are equations, and each of them represents the distance y_i from a common reference point to node i. Then, for the given example in which

$i = 0, 1, 2, 3, 4$, the dual problem can be constructed according to the primal-dual relationship in Table 4–11 (Chapter 4) as follows:

$$
\begin{aligned}
y_0 - y_1 \quad\quad\quad\quad\quad\quad &\leq 1\,, \\
y_0 \quad\quad - y_2 \quad\quad\quad\quad &\leq 2\,, \\
y_1 - y_2 \quad\quad\quad\quad &\leq 3\,, \\
y_1 \quad\quad - y_3 \quad\quad &\leq 4\,, \\
y_2 - y_3 \quad\quad &\leq 5\,, \\
y_2 \quad\quad - y_4 &\leq 6\,, \\
y_3 - y_4 &\leq 4\,;
\end{aligned}
$$

and

$$\text{Max } A = y_0 - y_4\,.$$

In examining the objective function and the constraints of the dual, it is seen that all values of y_i will be negative if the common point of reference is so chosen that $y_0 < y_1 < \cdots < y_n$. The physical interpretation of the problem can be made more obvious if we let $\bar{y}_i = -y_i$ for $i = 0, 1, \ldots, 4$. Then the resulting dual problem with $\bar{y}_i \geq 0$ becomes

$$\text{Max } A = \bar{y}_4 - \bar{y}_0$$

subject to

$$
\begin{aligned}
\bar{y}_1 - \bar{y}_0 &\leq 1 \\
\bar{y}_2 - \bar{y}_0 &\leq 2 \\
\bar{y}_2 - \bar{y}_1 &\leq 3\,, \\
\bar{y}_3 - \bar{y}_1 &\leq 4\,, \\
\bar{y}_3 - \bar{y}_2 &\leq 5\,, \\
\bar{y}_4 - \bar{y}_2 &\leq 6\,, \\
\bar{y}_4 - \bar{y}_3 &\leq 4\,.
\end{aligned}
$$

Hence max $A = A^*$ represents the maximum possible value of $(\bar{y}_4 - \bar{y}_0)$ if all branches of the network are scanned for the shortest route between nodes 0 and 4.

For the general problem defined by (7–21) and (7–22), a dual with $y_i \geq 0$ for $i = 0, 1, \ldots, n$ may be expressed in the form:

$$\text{Max } A = y_n - y_0\,, \tag{7–23}$$

subject to

$$y_j - y_i \leq d_{ij}\,, \qquad \text{for all branches } (i, j)\,, \tag{7–24}$$

in which

$$y_j > y_i \qquad \text{for} \qquad j > i\,, \qquad i, j = 1, 2, \ldots, n\,.$$

The solution for the dual problem may be obtained by searching systematically the paths leading from the source to the sink, assuming that $y_0 = 0$.

7-8 DUAL SOLUTION OF THE SHORTEST ROUTE

In seeking for a solution of the shortest-route problem, as defined by its dual in (7-23) and (7-24), we are interested in the set of branches constituting the shortest route, as well as the total distance for this route. The former is governed by the inequalities in (7-24), while the latter is automatically obtained by (7-23). Therefore, we shall examine further the relations in (7-24). Let $y_{ij} \geq 0$ be the slack variable in branch (i, j), which is incident to i and j at its ends. If the nodes in the directed network are so arranged that the node at the tail is always smaller than the node at the arrowhead of every directed branch, we have $i < j$. Then the inequalities in (7-24) may be replaced by equations as follows:

$$y_j - y_i + y_{ij} = d_{ij}\,, \qquad \text{for all branches } (i, j)\,. \tag{7-25}$$

If a branch (i, j) lies in the set of branches constituting the shortest route between the source and the sink, it must satisfy the condition $y_{ij} = 0$, or

$$y_j - y_i = d_{ij}\,. \tag{7-26}$$

Let us start to construct a path from node 0 forward to every other node in the increasing order of node i $(i = 1, 2, \ldots, n)$. At any node i where one or more merging branches meet, the value of y_i corresponding to the shortest of all possible paths leading from the source up to that node (i.e., the minimum value of y_i) is denoted by y'_i. (See Fig. 7-7.) Beginning with node 0, let $y_0 = y'_0$. Then, at any node i, the shortest possible distance to *each* of the nodes j on the other ends of all bursting branches incident to node i is given by

$$y_j = y'_i + d_{ij}\,. \tag{7-27}$$

This value of y_j for any branch (i, j) incident on j will be examined later together with those for other branches incident on j in determining $\min y_j = y'_j$ for the node j. The procedure can be continued from node

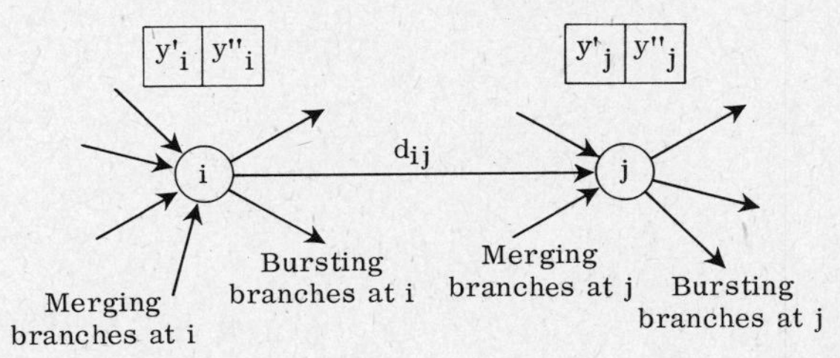

Figure 7-7

to node in the increasing order of i until $i = n$. Then the resulting $y'_n = \min y_n$. Since this process moves in the forward direction from the source to every other node, it is called the *forward pass*, and the set of y_i thus obtained is denoted by y'_i $(i = 0, 1, \ldots, n)$, ending with $y'_n = \min y_n$.

If we use the same procedure in the backward direction, constructing a path from node n to every other node in the decreasing order of node i $(i = 1, 2, \ldots, n)$, then the value of y_n for node n is taken to be $y''_n = \min y_n$. At any node j where one or more bursting branches meet, the value of y_j corresponding to the shortest of all possible paths leading back from the sink to that node (i.e., the maximum value of y_j) is denoted by y''_j. (See Fig. 7-7.) Going backward from node j, the shortest possible distance to *each* of the nodes i on the other ends of all merging branches incident to node j is given by

$$y_i = y''_j - d_{ij}\,. \tag{7-28}$$

This value of y_i for any branch (i, j) incident on j will be examined later together with those for other branches incident on i in determining $\max y_i = y''_i$ for node i. The procedure can be continued from node to node in the decreasing order of i until $i = 0$. Then the resulting y''_0 should become zero. Since this process moves in the backward direction from the sink to every other node, it is called the *backward pass*, and the set of y_i thus obtained is denoted by y''_i $(i = 0, 1, \ldots, n)$ starting with $y''_n = \min y_n$.

In examining the results of forward pass and backward pass, it is seen that a branch (i, j) lies in the set of branches constituting the shortest route $A^* = y'_n$ if it satisfies all of the following conditions:

$$y_i = y'_i = y''_i\,, \tag{7-29a}$$

$$y_j = y'_j = y''_j\,, \tag{7-29b}$$

$$y'_j - y'_i = y''_j - y''_i = d_{ij}\,. \tag{7-29c}$$

The computation procedure based on these equations is shown in the following numerical examples.

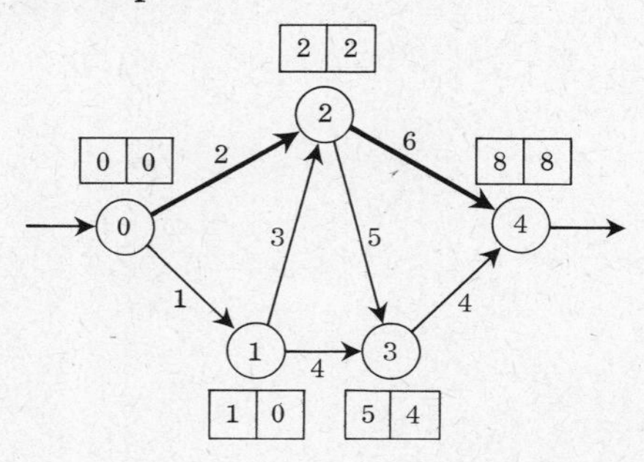

Figure 7-8

Example 7–10. Determine the shortest route from node 0 to node 4 in the network in Fig. 7–6.

The network shown in Fig. 7–6 is reproduced in Fig. 7–8, in which two numbers corresponding to y'_i and y''_i are placed near each node i. (Each left-hand box is for y'_i and each right-hand box for y''_i.)

In the forward pass, we start with $y'_0 = 0$ for node 0. For other nodes, we have from (7–27),

$$
\begin{aligned}
i &= 1\,, & y_1 &= y'_0 + d_{01} = 0 + 1 = 1 & &(\min y_1 = 1 = y'_1)\,;\\
i &= 2\,, & y_2 &= y'_0 + d_{02} = 0 + 2 = 2\,, & &\\
& & y_2 &= y'_1 + d_{12} = 1 + 3 = 4 & &(\min y_2 = 2 = y'_2)\,;\\
i &= 3\,, & y_3 &= y'_1 + d_{13} = 1 + 4 = 5\,, & &\\
& & y_3 &= y'_2 + d_{23} = 2 + 5 = 7 & &(\min y_3 = 5 = y'_3)\,;\\
i &= 4\,, & y_4 &= y'_2 + d_{24} = 2 + 6 = 8\,, & &\\
& & y_4 &= y'_3 + d_{34} = 5 + 4 = 9 & &(\min y_4 = 8 = y'_4)\,.
\end{aligned}
$$

In the backward pass, we start with $y''_4 = \min y_4 = 8$. For other nodes, we get from (7–28)

$$
\begin{aligned}
i &= 3\,, & y_3 &= y''_4 - d_{34} = 8 - 4 = 4 & &(\max y_3 = 4 = y''_3)\,;\\
i &= 2\,, & y_2 &= y''_4 - d_{24} = 8 - 6 = 2\,, & &\\
& & y_2 &= y''_3 - d_{23} = 4 - 5 = -1 & &(\max y_2 = 2 = y''_2)\,;\\
i &= 1\,, & y_1 &= y''_3 - d_{13} = 4 - 4 = 0\,, & &\\
& & y_1 &= y''_2 - d_{12} = 2 - 3 = -1 & &(\max y_1 = 0 = y''_1)\,;\\
i &= 0\,, & y_0 &= y''_2 - d_{02} = 2 - 2 = 0\,, & &\\
& & y_0 &= y''_1 - d_{01} = 0 - 1 = -1 & &(\max y_0 = 0 = y''_0)\,.
\end{aligned}
$$

In checking the relations in (7–29), it is seen that for branch 0–1,

$$
\begin{gathered}
y'_0 = y''_0 = 0\,, \qquad y'_2 = y''_2 = 2\,,\\
y'_2 - y'_0 = y''_2 - y''_0 = 2 - 0 = 2\,,
\end{gathered}
$$

and for branch 0–2,

$$
\begin{gathered}
y'_2 = y''_2 = 2\,, \qquad y'_4 = y''_4 = 8\,,\\
y'_4 - y'_2 = y''_4 - y''_2 = 6\,.
\end{gathered}
$$

Hence the shortest route is 0–2–4, as indicated by a darker line in Fig. 7–8, and the total distance is $A^* = y'_4 = 8$.

Example 7–11. Determine the shortest route from node 0 to node 10 in the network shown in Fig. 7–9.

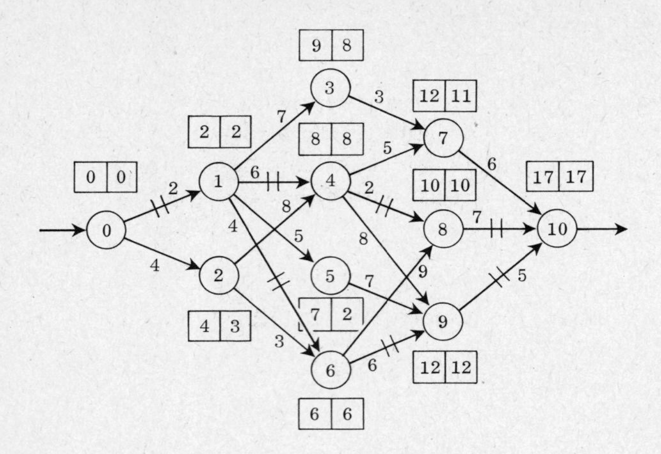

Figure 7-9

The method of solution for this problem is identical to that of the previous example. Hence only the results are shown on the figure. Note that for this network, two routes are tied for the shortest distance, and that each branch on these routes is marked by a pair of short parallel lines. It is seen that these routes are 0–1–4–8–10 and 0–1–6–9–10.

7-9 TRAVELING SALESMAN PROBLEM

The so-called traveling salesman problem involves the determination of the shortest path through a number of points, reaching each point only once, and ending at the starting point. This problem finds application not only in determining the least distance or traveling time, but also in finding the shortest hook-up pipelines and in sequencing operations of machines.

In general, the cost coefficients for a traveling salesman problem with n points can be given in an $n \times n$ matrix, as shown in Table 7-8. Since a path cannot be constructed from any point to itself, the cost coefficients c_{ii} ($i = 1, 2, \ldots, n$) along the main diagonal are taken to be infinite. For $i \neq j$, coefficient c_{ij} may or may not be equal to c_{ji}. The problem is to find a set of n coefficients, one from each row and one in each column, such that the sequence of coefficients form a continuous path that is the shortest. That is,

$$x_{ij} = \begin{cases} 1 & \text{if } (i, j) \text{ is linked}, \\ 0 & \text{if not}. \end{cases} \tag{7-30}$$

Furthermore, n branches in the path must be linked to form a sequence (p, q), (q, r), (r, s), $\ldots$, (u, v), (v, p) such that

$$x_{pq} = x_{qr} = x_{rs} = \cdots = x_{uv} = x_{vp} = 1, \tag{7-31}$$

Table 7–8

COST COEFFICIENTS FOR THE TRAVELING SALESMAN PROBLEM

j \ i	1	2	3	...	n
1	∞	c_{12}	c_{13}	...	c_{1n}
2	c_{21}	∞	c_{23}	...	c_{2n}
3	c_{31}	c_{32}	∞	...	c_{3n}
...	...	...	...	...	...
n	c_{n1}	c_{n2}	c_{n3}	...	∞

where

$$
\begin{aligned}
p &= 1, 2, \ldots, n\,; \\
q &= 1, 2, \ldots, n\,, \qquad \text{except} \qquad q \neq p\,; \\
r &= 1, 2, \ldots, n\,, \qquad \text{except} \qquad r \neq p, q\,; \\
s &= 1, 2, \ldots, n\,, \qquad \text{except} \qquad s \neq p, q, r\,; \\
&\vdots
\end{aligned}
$$

Thus

$$\sum_{j=1}^{n} x_{ij} = 1\,, \qquad i = 1, 2, \ldots, n\,,$$

and

$$\sum_{i=1}^{n} x_{ij} = 1\,, \qquad j = 1, 2, \ldots, n\,.$$

The objective function is

$$\text{Min } z = \sum_{i=1}^{n} \sum_{j=1}^{n} c_{ij} x_{ij}\,.$$

This problem is remarkably similar to the assignment problem except that the enumeration of possible paths for comparison is necessary in order to obtain the shortest path. At a quick glance, it may appear simple to search through all paths and select the shortest one. However, the number of possible paths multiplies rapidly with the increase of the number of points. Even if the cost (with respect to distance, time, etc.) of traveling from i to j equals that from j to i, the total number of possible paths is $\frac{1}{2}(n - 1)!$. Hence, for any such problem with a large number of points, a heuristic approach is often used to obtain an approximate solution. The discussion of short cut methods for special cases or approximate solutions for the general case is beyond the scope of this book.

REFERENCES

7-1. ORDEN, A., "The Transshipment Problem," *Management Science*, **2**, No. 3 (1956), 276–285.

7-2. JACOBS, W., "The Caterer Problem," *Naval Research Logistics Quarterly*, **1**, No. 2 (1954), 159–165.

7-3. PRAGER, W., "On the Caterer Problem," *Management Science*, **3**, No. 1 (1956), 15–23.

7-4. WATTLEWORTH, J. A., and SHULDINER, P. W., "Analytical Methods in Transportation: Left-turn Penalties in Traffic Assignment Models," *Journal of Engineering Mechanics Division, ASCE*, **89**, No. EM6 (1963), 97–126.

7-5. DANTZIG, G. B., D. R. FULKERSON, and S. JOHNSON, "Solution of a Large-Scale Traveling Salesman Problem," *Journal of Operations Research Society of America*, **2** (1954).

PROBLEMS

P7–1 to P7–5. The unit cost coefficients on the branches in network shown for the problem are given in the respective figures. The incoming and outgoing flows are also shown. Determine the shipping pattern which minimizes the total cost by considering each problem as an equivalent transportation problem.

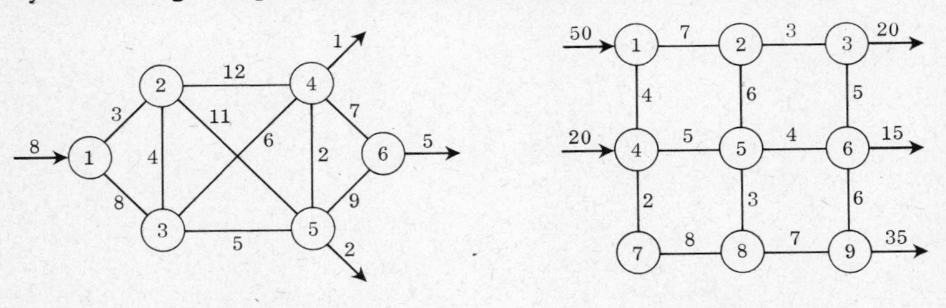

Figure P7–1 **Figure P7–2**

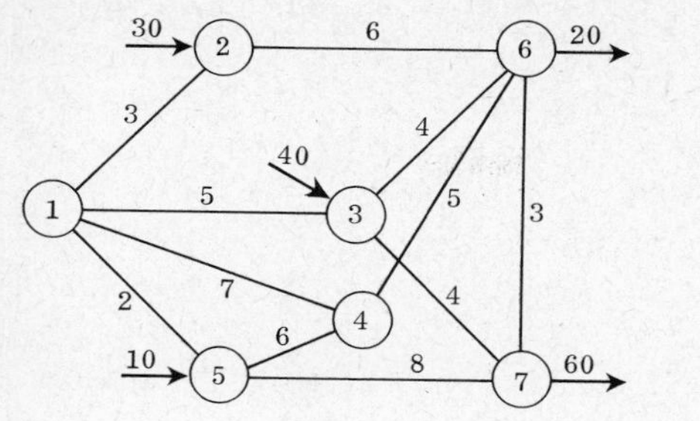

Figure P7-3

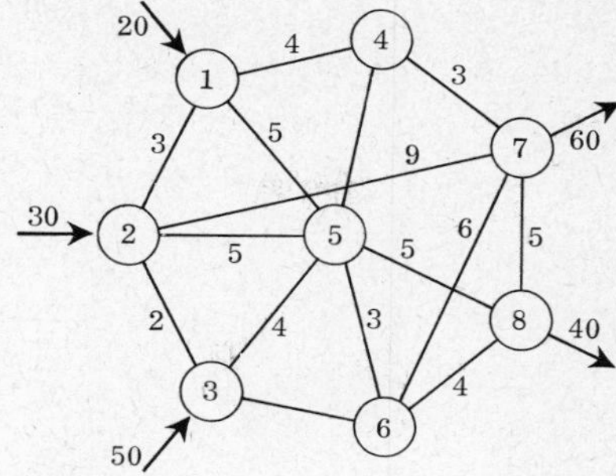

Figure P7-4

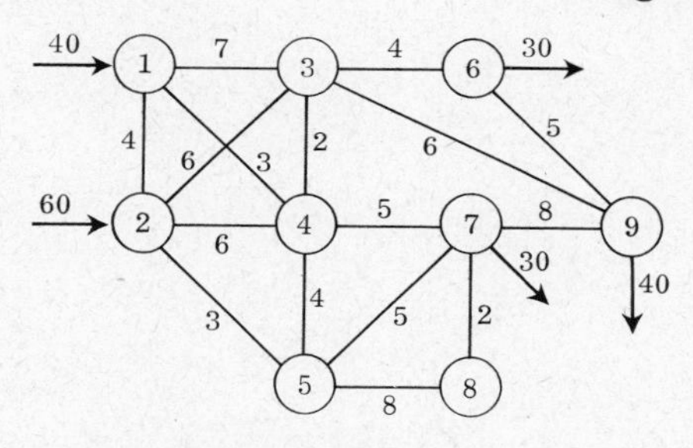

Figure P7-5

P7-6 to P7-10. For Problems P7-1 to P7-5, respectively, solve each problem by using the network solution.

P7-11. Solve the caterer problem given that $n = 5$, $p = 2$, $q = 1$, $d = 6$, $e = 2$, and $f = 4$. Also given are

$$a_1 = b_1 = 50\,, \qquad a_2 = b_2 = 40\,, \qquad a_3 = b_3 = 30\,,$$
$$a_4 = b_4 = 60\,, \qquad a_5 = b_5 = 70\,.$$

P7-11. In the caterer problem, let us assume that there are three different services in the repair shop (instead of two) as follows:

Slow	p days	cost e
Medium	q days	cost f
Fast	s days	cost g

where $p > q > s$ and $e < f < g$. How should the formulation of the problem be modified? Using the same notation, set up the mathematical expressions for the problem. With the numerical values given below, express the problem in the standard transportation array, and find the optimal solution, given that

$$n = 7\,, \qquad p = 3\,, \qquad q = 2\,, \qquad s = 1\,,$$
$$d = 8\,, \qquad e = 2\,, \qquad f = 4\,, \qquad g = 6\,.$$

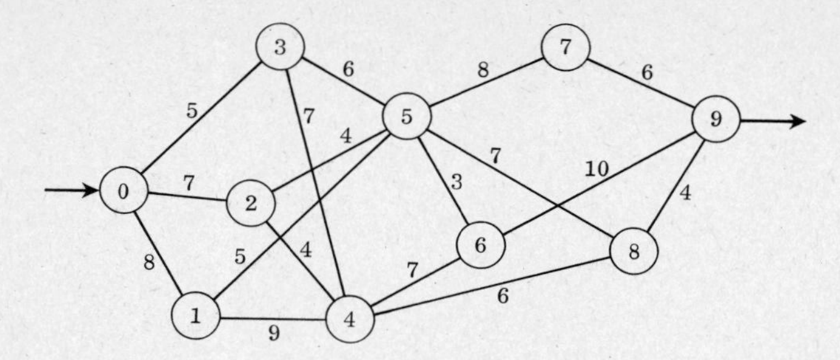

Figure P7–13

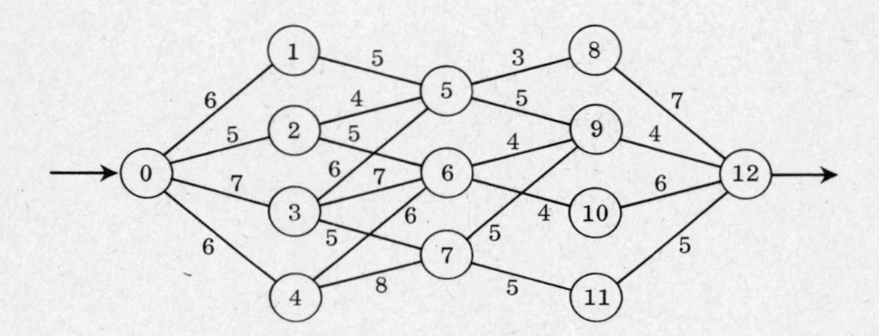

Figure P7–14

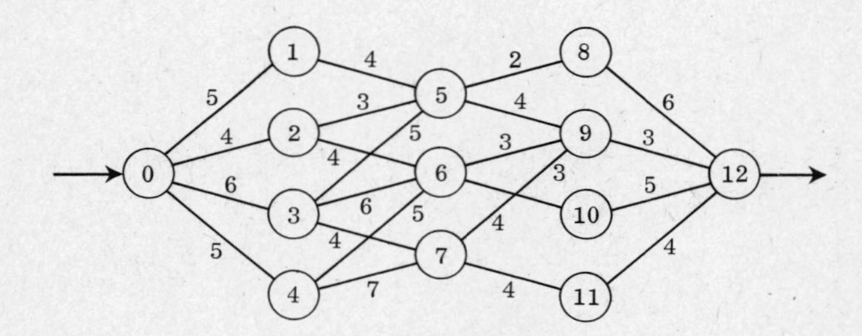

Figure P7–15

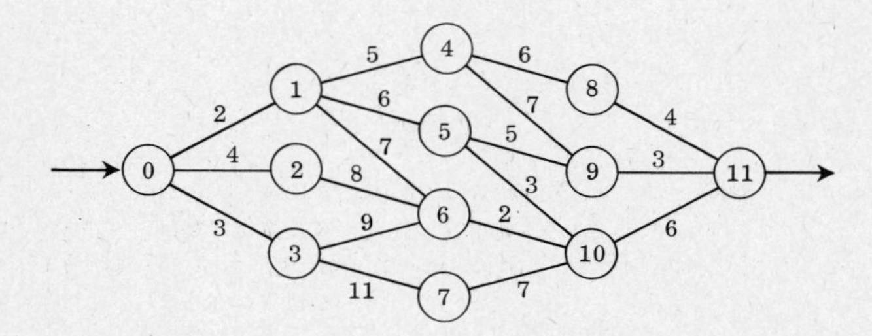

Figure P7–16

Also, you are given that $i = j$, $a_j = b_j$, and

$$a_1 = 4\,, \quad a_2 = 2\,, \quad a_3 = 5\,, \quad a_4 = 3\,,$$
$$a_5 = 8\,, \quad a_6 = 7\,, \quad a_7 = 6\,.$$

P7–13 to P7–16. Determine the shortest route from the source to the sink in the network shown in the respective figures for the problems.

CHAPTER 8

CRITICAL-PATH SCHEDULING

8–1 ELEMENTARY EXAMPLES

The problems of planning and scheduling engineering projects can be represented by networks indicating various activities in the proper order of execution of such projects. A network showing the dependency or precedence relationships of various activities in a project can be used effectively for the control of performance time and cost of the project, especially when the project is of enormous size and consists of widely diversified activities. The network provides a master plan in time scale for achieving the objectives of expediting the completion, allocating available resources and/or controlling the cost of the project. The critical-path method is used to identify the most critical activities or elements in the plan.

An *activity* is any subdivision of a project whose execution requires time and resources, including manpower and equipment. The time required to perform an activity is called the *duration* of the activity. The beginning and the end of activities are signposts or milestones, indicating the progress of the project. The instantaneous time denoting any such beginning or end is called an *event*. The time at which a specified event occurs is called the *event time*.

If an activity is denoted by a directed branch between two nodes in a network, the arrow indicates the direction of time flow from one event to another, the events being denoted by the nodes. Usually, either the branches or the nodes in a network are identified, depending upon whether the emphasis is placed on the role of the activities or on the significance of the events. On the other hand, it is possible to place activities on the nodes if such an arrangement is preferred. The precedence relationships of activities in different forms of networks can be illustrated by examples.

Example 8–1. In conducting a traffic survey for the planning of an urban mass transit system, the entire project consists of the following major activities:

A. Define the scope of the study.
B. Establish the procedure of the survey.

C. Design questionnaires for the survey.
D. Hire and organize a staff.
E. Train the staff to be familiar with the questionnaires.
F. Select sampling stations along the proposed routes.
G. Assign the staff to sampling stations.
H. Conduct the field survey.

Determine the precedence relationships of these activities and represent them in a network.

In the construction of a network for planning, we must establish the activities involved in a project, and the logical sequence and dependency of these activities. If the activities are clearly defined and their interrelationships are precisely described, there will be one and only one set of precedence relationships in a network describing such a project, although the network may be represented in different forms. In reality, however, the division of the project into various activities is not clear-cut. In the traffic survey, for example, the activity "Hire and organize a staff" may be broken down to several smaller activities such as:

a. Advertise in local newspapers for qualified applicants.
b. Contact promising applicants after the deadline for response.
c. Interview applicants and make selections.

If such breakdowns are introduced, not only are more activities involved but the interrelationships of various activities may also be affected. Hence it is a matter of judgment to decide how detailed the activities should be. Obviously, if subdivided activities are too refined, the size of the network becomes unwieldly; on the other hand, if the specified activities are all-inclusive, the network does not serve the intended purpose, simple though it may be. In general, the division of activities for a network depends also to some extent on the level of management which makes use of it. For the top management, a simple network representing major activities of the project may be sufficient, but a project engineer who is directly

Table 8-1

PRECEDENCE RELATIONSHIPS OF ACTIVITIES FOR EXAMPLE 8-1

Activity	Description	Predecessors	Successors
A	Define study	—	B, D
B	Establish procedure	A	C
C	Design questionnaires	B	E, F
D	Organize a staff	A	E, F
E	Train staff	C, D	H
F	Select sampling stations	C, D	G
G	Assign staff to stations	F	H
H	Conduct field survey	E, G	—

Table 8-2

SEQUENCE OF EVENTS FOR EXAMPLE 8-1

Event no.	Description
0	Begin activity *A*, the first activity in the project
1	Finish activity *A*
2	Finish activity *B*
3	Finish activities *C* and *D*
4	Finish activity *F*
5	Finish activities *E* and *G*
6	Finish activity *H*, the last activity in the project

responsible for its performance must know more intimately the details of activities. Fortunately, we can always start with a simple network consisting of fewer activities; and if we are not satisfied with the results scheduling, the activities and their interrelationships should be reexamined and refined so that a better network can be constructed.

In this example, we shall consider the given classification of major activities for the project as adequate, and agree on the precedence relationships of activities as shown in Table 8-1, or as given by the sequence of events in Table 8-2. The logical sequence of activities in the project is represented by a network in Fig. 8-1(a), in which the network is said to be *activity-on-branch*, and the identifiers are *activity-oriented*. The network in Fig. 8-1(b) is also activity-on-branch, but the identifiers are *event-oriented*. The network in Fig. 8-1(c) is said to be *activity-on-node*, and the branches are used to denote the precedence relationships of activities only.

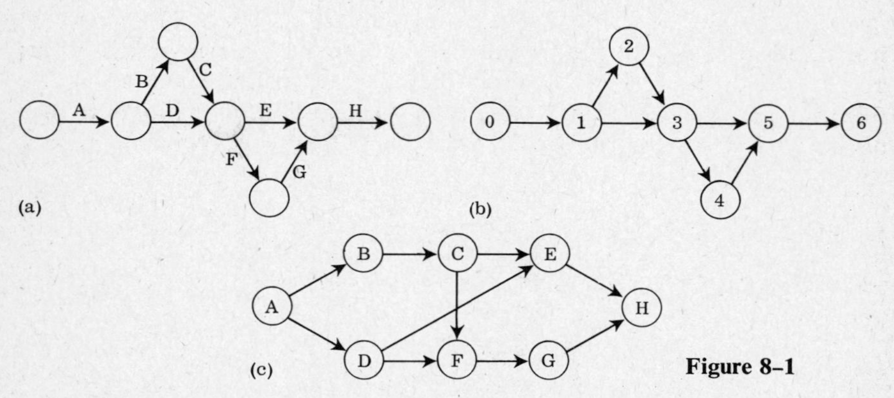

Figure 8-1

Example 8-2. A construction project consists of nine different activities, the sequence of which is given in Table 8-3. Represent the precedence relationships of activities in the project in an activity-on-branch network, and number the events in an order such that the number for the node at

Table 8-3

PRECEDENCE RELATIONSHIPS OF ACTIVITIES FOR EXAMPLE 8-2

Activity	Description	Predecessors	Duration
A	Site clearing	—	4
B	Removal of trees	—	3
C	General excavation	*A*	8
D	Grading general area	*A*	7
E	Excavation for trenches	*B*, *C*	9
F	Placing formwork and reinforcement for concrete	*B*, *C*	12
G	Installing sewer lines	*D*, *E*	2
H	Open ditch excavation	*D*, *E*	5
I	Pouring concrete	*F*, *G*	6

the tail end of every branch is smaller than that at the arrowhead of the same branch. Also, express the precedence relationships of all activities in the matrix form.

The activity-on-branch network representing the precedence relationships of activities in the project in Table 8-3 is shown in Fig. 8-2, in which the durations of activities are also indicated in parentheses on the branches. In numbering the events, we start with the initial event 0, denoting the beginning of activities *A* and *B*, which have no predecessors, and proceed to other events in sequence. Note that event 2 follows event 1 logically, since the former cannot be numbered until the tails of both *B* and *C* are numbered so that the latter must be numbered first. Next, event 3 can be numbered because the tails of both *D* and *E* are numbered, but event 4 must wait until the tail of *G*, as well as that of *F*, is numbered. Finally, event 5 is numbered as the terminal event.

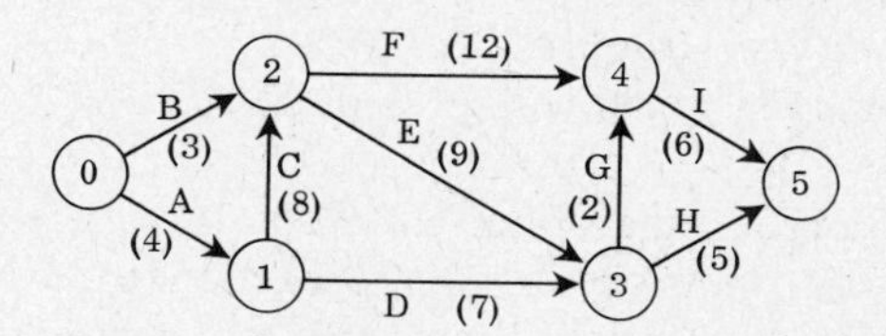

Figure 8-2

The predecessor-successor relationships of various activities on branches can be given by the matrix in Table 8-4 which is sometimes referred to as the *implicit precedence matrix*. Note that the activities on branches (i, j) are arranged in the ascending order of i first. The predecessors of an activity in a column heading are located by the rows in which the x's are marked in that column. For example, column *C* has an x in row *A*, since activity *A* is the immediate predecessor of *C*; column *D* also has an x in row *A*, since activity *A* is also the immediate predecessor of *D*;

Table 8-4

IMPLICIT PRECEDENCE MATRIX FOR EXAMPLE 8-2

Predecessor \ Successor	*A* (0, 1)	*B* (0, 2)	*C* (1, 2)	*D* (1, 3)	*E* (2, 3)	*F* (2, 4)	*G* (3, 4)	*H* (3, 5)	*I* (4, 5)
A (0, 1)			x	x					
B (0, 2)					x	x			
C (1, 2)					x	x			
D (1, 3)							x	x	
E (2, 3)							x	x	
F (2, 4)									x
G (3, 4)									x
H (3, 5)									
I (4, 5)									

column E has an x in both rows B and C, because activities B and C are immediate predecessors of E; etc. Conversely, the successors of an activity in a row heading are given by the columns in which the x's are found in that row. For example, row A has an x in columns C and D, since activities C and D are the immediate successors of A; and row B has an x in columns E and F, because activities E and F are the immediate successors of B; etc. Thus all predecessors and all successors of any activity (i, j) can be obtained by following through immediate predecessors and successors step by step. For example, all successors of D can be obtained by entering the table from row D to find columns G and H marked by x, and then form rows F and G to find column I marked by x in both rows.

A systematic approach may be used to obtain all predecessors and successors of any activity from the implicit precedence matrix. The resulting matrix, which contains all precedence relationships of all activities, will provide useful information on the dependency of any activity on the other; such a matrix is referred to as the *explicit precedence matrix*. Referring to Table 8-5, which is a duplication of Table 8-4 before X's are added, we shall construct the complete sequence of successors for each activity. We start with the first activity, A, by entering row A to find the C and D successors of A. In order to trace quickly the successors of C and D, we enter rows C and D, respectively. Each of these operations can be facilitated by following the column heading down until it hits the

Table 8-5

PRECEDENCE RELATIONSHIPS OF ALL ACTIVITIES IN EXAMPLE 8-2

Successor / Predecessor	*A* (0, 1)	*B* (0, 2)	*C* (1, 2)	*D* (1, 3)	*E* (2, 3)	*F* (2, 4)	*G* (3, 4)	*H* (3, 5)	*I* (4, 5)
A (0, 1)			*x*	*x*	*X*	*X*	*X*	*X*	*X*
B (0, 2)					*x*	*x*	*X*	*X*	*X*
C (1, 2)					*x*	*x*	*X*	*X*	*X*
D (1, 3)							*x*	*x*	*X*
E (2, 3)							*x*	*x*	*X*
F (2, 4)									*x*
G (3, 4)									*x*
H (3, 5)									
I (4, 5)									

dashed diagonal line in the table and turn right. Hence we first move down column *C* to row *C* and observe that *E* and *F* are successors of *C*; then *X*'s are added to row *A* under columns *E* and *F*. Similarly, we move down column *D* to row *D*, and observe that *G* and *H* are successors of *D*; then *X*'s are added to row *A* under columns *G* and *H*. If we now move down column *E*, which is a successor of *A*, we find in row *E* that *G* and *H* are successors of *E*. Since *X*'s were already placed in row *A* under columns *G* and *H* previously, it is not necessary to repeat it. When we move down column *F* to row *F*, we observe that *I* is the successor of *F*, and that an *X* should be added in row *A* under column *F*. Thus all successors of *A* are now found. A similar procedure is used to obtain all successors of other activities.

Table 8-5 gives all predecessors and all successors of all activities in the project. If we enter from the column heading of an activity and move downward to search for *x*'s and *X*'s, the activities corresponding to the rows containing these *x*'s and *X*'s constitute all predecessors of this activity. Similarly, if we enter from the row heading of an activity and move horizontally to the right to search for *x*'s and *X*'s, the activities corresponding to the columns containing these *x*'s and *X*'s constitute all successors of that activity. Note that *x*'s represent immediate predecessors or successors, while *X*'s denote indirect predecessors or successors in the explicit precedence matrix.

8-2 NETWORKS FOR PROJECT PLANNING AND CONTROL

A network for project planning and control may be regarded as a flow diagram in time units. During a brief period in its development, there appeared to be two diverging approaches to the formulation and solution of networks for project planning and control, even though both approaches are based on activity-on-branch networks. On the one hand, the so-called *critical path method* (CPM) was developed primarily for the evaluation of performance time and the total cost of projects consisting of relatively well-defined activities, and has found applications in construction planning and in other project management problems. Thus an activity-oriented network is preferred in following through each activity. The durations of the activities can be estimated fairly accurately and may be considered as deterministic quantities. On the other hand, *a program evaluation and review technique* (PERT) was developed for the management of very large or long range projects for which the nature and duration of many activities involve high degrees of uncertainty. Hence an event-oriented network clearly marks the occasions when reports of progress should be made to the management. The duration of each activity can be established on the basis of several estimates, together with the probability of their occurrences. Let t_a be the optimistic time estimate, t_b be the most likely time estimate, and t_c be the pessimistic time estimate of an activity. Then the duration of the activity is taken to be the weighted mean value t as follows:

$$t = \tfrac{1}{6}(t_a + 4t_b + t_c)\,.$$

As both methods were later revised for improvement, the attractive features of one were soon incorporated in the other. We shall not differentiate the two approaches in our discussion inasmuch as both of them can be represented by an activity-on-branch network. More recently, the activity-on-node network has been found by some to be more convenient for applications. However, this approach has not yet gained wide acceptance in practice and shall not be treated in detail here.

The fundamental idea of critical-path scheduling for an activity-on-branch network is to find an optimal time path from the initial event to the terminal event in the network such that the completion of the project will not be delayed unnecessarily. Basically, each activity has its own precedence. It cannot begin until all activities preceding it have been completed, and it must be completed before all activies following it can begin. Each event represents the instantaneous occurrence when all activities merging to that event have been completed. A path in the network is a sequence of activities between two specified events; there may be more than one path linking one event to another, provided these events are not the beginning and end events of a single activity. A network describing a project has only one initial event which has no predecessors, and only one terminal event which has no successors. The minimum time required for the completion of all activities in a project is governed by the longest time path linking the initial event and the terminal event of the project. This time path includes all activities which must be completed on time if a minimum of time to complete the project is expected, and such a path called a *critical time path*, or *critical path*.

The concept of critical-path scheduling can easily be refined and extended. For example, the optimal schedule may be adjusted for working days in the calendar instead of consecutive days. The information obtained from critical-path scheduling may be used as the basis for leveling manpower and resources required for the project. Furthermore, if a crash program is desired, it is possible to examine the time path in the network to determine how this can be done most effectively. We shall stress the basic concepts of the critical-path method, but must omit most details pertaining to specific applications.

8–3 ORGANIZATION OF ACTIVITY-ON-BRANCH NETWOKS

After the activities and their interrelationships in a project are clearly defined, a network representing the project can be uniquely established. In an activity-on-branch network, activities are represented by branches, and events are denoted by nodes of the network. The procedure for the organization and construction of such a network will be explained further.

An activity is graphically represented by a solid line with an arrow pointing to the direction of the time flow. An event toward which some activities may be merging and from which others may be bursting is graphically denoted by a circle. Each activity may be identified by a letter along the branch, and each node may be identified by a number inside the circle. The initial event in the order of execution is the beginning of the project, and the terminal event represents the end of the project. A network thus constructed represents graphically a project plan. For the convenience of sequencing, events are numbered in such a way that the head of each arrow always has a larger number than its tail. Hence no event will be numbered until the tail of each arrow merging into that event has been numbered first. After all events are numbered, each activity (i, j) may be identified by its beginning and end events i and j, provided that no two activities have the same designations. Thus, in Fig. 8–3(a), which is the combination of Fig. 8–1(a) and (b) in Example 8–1, activities $A, B, C, \ldots, H$ may be denoted by (0, 1), (0, 2), (2, 3), . . . , (5, 6), respectively.

Dummy activities are sometimes introduced in a network for the purposes of providing unique activity designations and maintaining the cor-

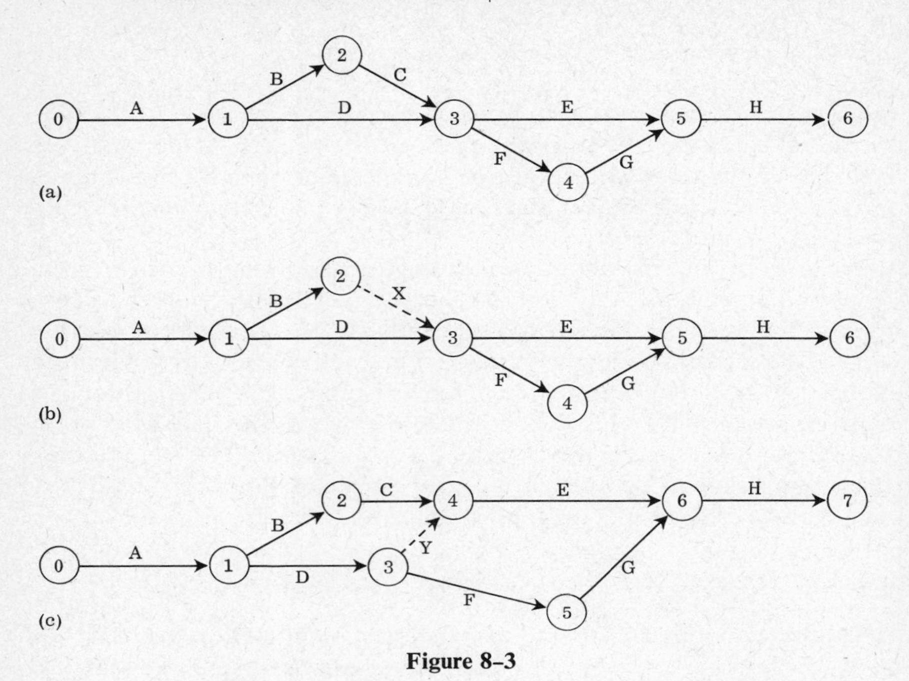

Figure 8-3

rect order in the sequence. A *dummy activity* is assumed to have no time duration and is graphically represented by a dashed line with an arrow. Referring to the network in Fig. 8-3(a), for example, if activity C were eliminated, activities B and D could not be uniquely identified by the beginning and ending events, since both would have been the same i and j. However, if a dummy activity X is introduced, as shown in part (b) of the figure, the unique designations for B and D will be preserved. Furthermore, if the problem in part (a) is changed so that activity E cannot start until both C and D are completed but that F can start after D alone is completed, the order in the new sequence can be indicated by the addition of a dummy activity Y, as shown in part (c). In general, dummy activities can be introduced without disturbing the network sequence except at locations where their presence is necessary.

Some exogenous factors which are not activities within the project may nevertheless restrain the progress or completion of the project. Such restraints include deliveries of material and equipment, availability of capital, weather conditions, etc. Each of these restraints can be represented graphically by an arrow at the appropriate location of the network. In Fig. 8-3(a), for example, if the project cannot proceed after event 3 without appropriation of funds to the traffic study group, an arrow for capital restraint can be added at that event.

Table 8–6

PRECEDENCE RELATIONSHIPS OF ACTIVITIES IN EXAMPLE 8–3

Activity	Description	Predecessors	Duration
A	Preliminary design	—	6
B	Evaluation of design	*A*	1
C	Contract negotiation	—	8
D	Preparation of fabrication plant	*C*	5
E	Final design	*B*, *C*	9
F	Fabrication of product	*D*, *E*	12
G	Shipment of product to owner	*F*	3

The durations of activities for a project are estimated on the basis of normal conditions of available labor and material unless emergencies call for a crash program. All durations thus determined are understood to be *normal durations* if they are not specified to be *crash durations*. Also in our discussion only one time estimate will be used for each activity duration. The unit of activity duration may be in any consistent time unit, such as days or hours.

It is clear that the network model describing a project must be correctly formulated, since the results of scheduling depend on the model selected. Additional examples of network construction are therefore given.

Example 8–3. Construct a network for the project of supplying a special fabricated steel product to a customer as outlined by the sequence of activities in Table 8–6. Also, the fabrication of the product cannot begin until the material is delivered to the plant and the shipment cannot begin until trucks are made available.

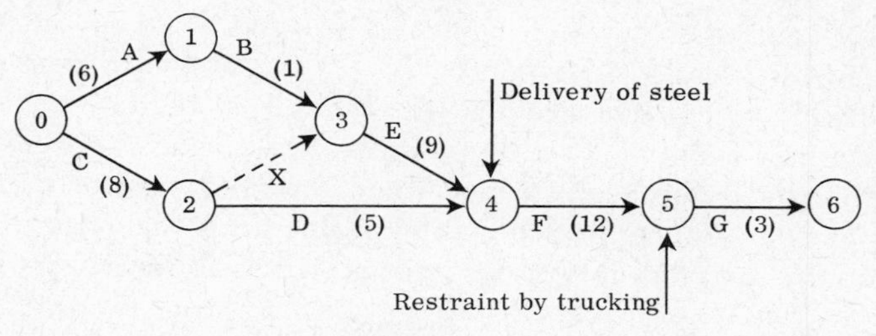

Figure 8–4

The network for the planning of this project is shown in Fig. 8–4. A dummy is introduced between nodes 2 and 3 to indicate the precedence relationship because the predecessor of *D* is *C* only, but those of *E* are *B* and *C*. A simple rule to observe is that if an activity has more than one immediate predecessor, say *B* and *C*, and another activity has either *B* or

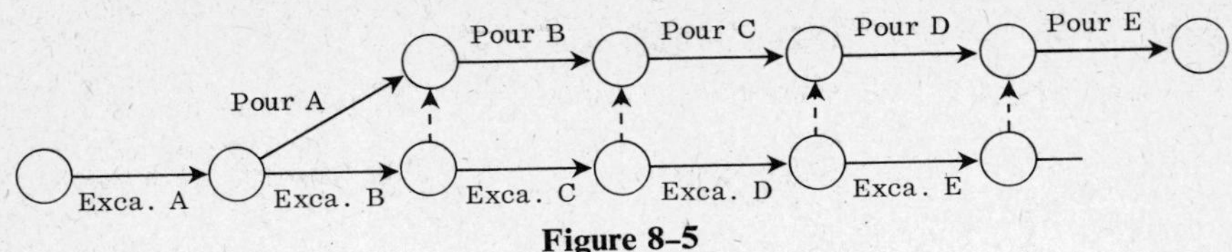

Figure 8-5

C but not both) as a predecessor, regardless of how many other predecessors the latter may have, the situation calls for a dummy activity to maintain the precedence relationships. Note also that arrows have been added for restraints at nodes 4 and 5 in the figure.

Example 8-4. Construct a network for the excavation and paving of a long sidewalk which must be done in sections. In other words, first excavate section A and then pour concrete for section A. However, excavation for a subsequent section B is permitted while concrete is being poured at A. This procedure is repeated for sections B, C, D, etc.

This is another example for the use of dummy activities, as shown in Fig. 8-5. The network is self-explanatory.

8-4 FORMULATION OF THE SCHEDULING PROBLEM

The scheduling problem for a network plan is to determine an optimal schedule which is consistent with the given logical sequence and durations of various activities in the network, and one which requires a minimum time for the completion of the project. Consider a network with $n + 1$ nodes, the initial event being 0 and the terminal event being n. Let the event times at these nodes be $x_0, x_1, x_2, \ldots, x_n$, respectively, which have increasing value in time units. These event times are constrained by the durations of activities between events. The duration of any activity (i, j) between events i and j is denoted by D_{ij} (≥ 0). If the events in the directed network are so arranged that the event x_i at the tail end is always smaller than the event x_j at the arrowhead of every activity (i, j), we have $i < j$. Then the difference of event times x_j and x_i must be greater than or equal to D_{ij}. Hence the scheduling problem for the determination of the minimum completion time z can be stated as follows:

$$\text{Min} \qquad z = x_n - x_0\,, \tag{8-1}$$

subject to

$$x_j - x_i - D_{ij} \geq 0\,, \qquad \text{all} \quad (i, j)\,; \tag{8-2}$$

$$x_j > x_i \quad \text{for} \quad j > i\,, \qquad i, j = 0, 1, 2, \ldots, n\,. \tag{8-3}$$

Note that x_i or x_j is unrestricted in sign, since the reference point of the time scale has not been specified. Without losing generality, it is convenient to assume that $x_0 = 0$. Then all other x_i must be nonnegative. By introducing a nonnegative slack variable x_{ij} for each activity (i, j), the inequality constraints may be replaced by equation constraints. Thus the problem can be restated as follows:

$$\text{Min} \quad z = x_n , \tag{8–4}$$

subject to

$$x_j - x_i - D_{ij} - x_{ij} = 0 , \quad \text{all} \quad (i, j) , \tag{8–5}$$

$$x_{ij} \geq 0 , \quad \text{all} \quad (i, j) ; \tag{8–6}$$

$$x_j > x_i \geq 0 , \quad i, j = 0, 1, 2, \ldots, n . \tag{8–7}$$

This set of relations may also be obtained in a manner similar to the formulation of the shortest route problem in Section 7–7. In the case of critical-path scheduling, however, we regard the performance of each activity (i, j) as the occurrence of the branch (i, j) in the critical time path. Of all the possible time paths leading from the initial event to the terminal event, the longest time path is the critical path for the completion of the project. We shall illustrate this latter approach with a specific example.

Example 8–5. In the activity-on-branch network in Fig. 8–6, the duration of each activity is shown on the branch. (a) Formulate the critical-path scheduling problem as the determination of the longest time path from the initial to the terminal event, and (b) construct the dual problem which requires the determination of the minimum time for the completion of the project.

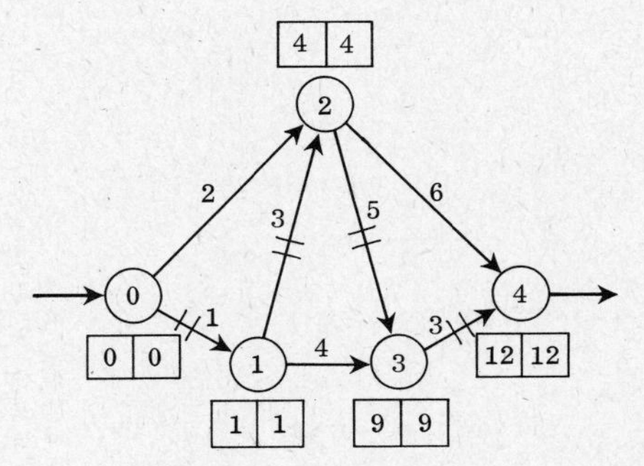

Figure 8–6

We regard the selection of a time path from the initial to the terminal event as passing a unit time flow along the path from the source to the sink in the network. Let y_{ij} be the flow in branch (i, j) such that $y_{ij} = 1$ if the branch is in the critical path, or $y_{ij} = 0$ if the branch is not in it. Hence the longest time path A can be expressed in terms of the flow variables y_{ij}, which are subjected to the constraints of Kirchhoff's flow law. Thus

$$\text{Max} \quad A = y_{01} + 2y_{02} + 3y_{12} + 4y_{13} + 5y_{23} + 6y_{24} + 3y_{34},$$

subject to

$$\begin{array}{lllllllll} i = 0, & y_{01} + y_{02} & & & & & & & = 1, \\ i = 1, & -y_{01} & + y_{12} + y_{13} & & & & & & = 0, \\ i = 2, & & -y_{02} - y_{12} & & + y_{23} + y_{24} & & & & = 0, \\ i = 3, & & & -y_{13} - y_{23} & & + y_{34} & & & = 0, \\ i = 4, & & & & & -y_{24} - y_{34} & & & = -1; \end{array}$$

and

$$y_{ij} \geq 0.$$

This problem may be expressed in the form of a primal problem for the primal-dual relationship in Table 4–11 (Chapter 4) if the objective function is restated as follows:

$$\text{Min } (-A) = -y_{01} - 2y_{02} - 3y_{12} - 4y_{13} - 5y_{23} - 6y_{24} - 3y_{34}.$$

The dual problem may be constructed by first letting x_i $(i = 0, 1, 2, 3, 4)$ be the set of dual variables. Since all constraints in the primal problem are equations, the dual variables x_i are unrestricted in sign. Then, from the primal-dual relationship, we get

$$\text{Max } (-z) = x_0 - x_4,$$

subject to

$$\begin{aligned} x_0 - x_1 \quad\quad\quad\quad\quad\quad & \leq -1, \\ x_0 \quad\quad - x_2 \quad\quad\quad\quad & \leq -2, \\ x_1 - x_2 \quad\quad\quad\quad & \leq -3, \\ x_1 \quad\quad - x_3 \quad\quad & \leq -4, \\ x_2 - x_3 \quad\quad & \leq -5, \\ x_2 \quad\quad - x_4 & \leq -6, \\ x_3 - x_4 & \leq -3, \end{aligned}$$

and

$$-\infty < x_i < +\infty, \qquad i = 0, 1, 2, 3, 4.$$

Note that a minus sign is introduced for $(-z)$ in the dual problem in order to be consistent with $(-A)$ in the primal problem. However, the minus sign aids the physical interpretation later and is not a mathematical necessity since the maximization refers to $(x_0 - x_4)$. Thus, the dual problem may also be expressed in the form of the minimization of the completion time z as follows:

$$\text{Min} \qquad z = x_4 - x_0$$

subject to

$$\begin{aligned} x_1 - x_0 &\geq 1\,, \\ x_2 - x_0 &\geq 2\,, \\ x_2 - x_1 &\geq 3\,, \\ x_2 - x_1 &\geq 4\,, \\ x_3 - x_2 &\geq 5\,, \\ x_4 - x_2 &\geq 6\,, \\ x_4 - x_3 &\geq 3\,. \end{aligned}$$

This set of relations could have been obtained directly from (8-1), (8-2), and (8-3).

Since the reference point for the time scale has not been specified, we can assume that $x_0 = 0$ for convenience. If we proceed to solve this problem (according to the procedure to be explained in the next section), then, for $x_0 = 0$, we get $x_1 = 1$, $x_2 = 4$, $x_3 = 9$, $x_4 = 12$, and min $z = 12$.

8-5 CRITICAL TIME PATH

Since the minimum time required for the completion of all activities in a project is governed by the longest time path linking the initial event and the terminal event of the project, the solution of the critical scheduling problem is equivalent to the determination of the longest time path in a network having a unit time flow from the source to the sink. If a branch (i, j) lies in the set of branches constituting the longest time path between the source and the sink, it must satisfy the condition $x_{ij} = 0$, that is, it must be completed on time if a minimum of time to complete the project is expected. Hence from (8-5) we have

$$x_j - x_i = D_{ij}\,. \tag{8-8}$$

Let us start to construct a path from node 0 (initial event) to every other node in the increasing order of node i ($i = 0, 1, 2, \ldots, n$). At any node where one or more merging branches meet (Fig. 8-7), the value of x_i corresponding to the longest of all possible paths leading from node 0 to node i (i.e., the maximum value of x_i) is denoted by x_i'. (At node 0, let $x_0 = x_0'$.) Starting from node i, the time path to each of the nodes j on the other ends of all bursting branches incident to node i is given by

$$x_j = x_i' + D_{ij}\,. \tag{8-9}$$

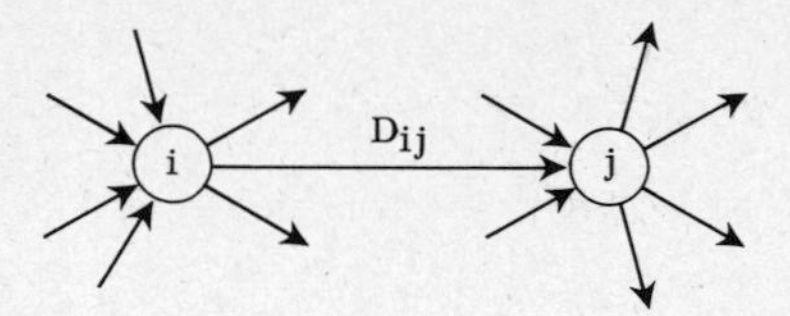

Figure 8-7

This value of x_j for any branch (i, j) incident on i will be examined later, together with those for other branches incident on j, in determining $\max x_j = x'_j$ for node j. The procedure can be continued from node to node in the increasing order of i until $i = n$. Then the resulting $x'_n = \max x_n$. Since this process moves in the forward direction from node 0 to every other node, it is called the *forward pass*, and the set of x_i thus obtained is denoted by x'_i $(i = 0, 1, \ldots, n)$, ending with $x'_n = \max x_n$. Note that each x'_i represents the earliest time that event i may take place, and it is called the *earliest event time* for event i.

If we use the same procedure in the backward direction, starting with node n to every other node in the decreasing order of node i $(i = 1, 2, \ldots, n)$, then the value of x_n for node n is taken to be $x''_n = \max x_n$. At any node x_j where one or more bursting branches meet (Fig. 8-7), the value of x_j corresponding to the longest of all possible time paths leading back from the sink n to node j (i.e., the minimum value of x_j) is denoted by x''_j. Going backward from node j, the time path to each node i on the other ends of all merging branches incident to node j is given by

$$x_i = x''_j - D_{ij} . \tag{8-10}$$

This value of x_i for any branch (i, j) incident on j will be examined later with those for other branches incident on i in determining $\min x_i = x''_i$ for node i. The procedure can be continued from node to node in the decreasing order of i until $i = 0$. Then the resulting x''_0 should become zero. Since this process moves in the backward direction from node n to every other node, it is called the *backward pass*, and the set of x_i thus obtained is denoted by x''_i $(i = 0, 1, 2, \ldots, n)$ starting with $x''_n = \max x_n$. Note that each x''_i represents the latest time that event i may take place, and it is called the *latest event time* for event i.

In examining the results of forward pass and backward pass, it is seen that a branch (i, j) lies in the set of branches constituting the longest time path $\max x_n$ if it satisfies all of the following conditions:

$$x_i = x'_i = x''_i , \tag{8-11a}$$

$$x_j = x'_j = x''_j , \tag{8-11b}$$

$$x'_j - x'_i = x''_j - x''_i = D_{ij} . \tag{8-11c}$$

The computational procedure on these equations is explained in the numerical examples.

Example 8-6. Determine the critical time path for the network in Example 8-5.

The earliest event times x'_i $(i = 0, 1, 2, 3, 4)$ and the latest event time x''_i are computed by the forward pass and the backward pass, respectively. The numbers x'_i and x''_i corresponding to each node i are recorded in a pair of boxes near node i in Fig. 8-6. (The left-hand box is for x'_i and the right-hand box for x''_i.)

In the forward pass, we start with $x'_0 = 0$ for node 0. For other nodes, we obtain from (8-9)

$$
\begin{array}{lll}
i = 1\,, & x_1 = x'_1 + D_{01} = 0 + 1 = 1 & (\max x_1 = 1 = x'_1)\,; \\
i = 2\,, & x_2 = x'_0 + D_{02} = 0 + 2 = 2\,, & \\
 & x_2 = x'_1 + D_{12} = 1 + 3 = 4 & (\max x_2 = 4 = x'_2)\,; \\
i = 3\,, & x_3 = x'_1 + D_{13} = 1 + 4 = 5\,, & \\
 & x_3 = x'_2 + D_{23} = 4 + 5 = 9 & (\max x_3 = 9 = x'_3)\,; \\
i = 4\,, & x_4 = x'_2 + D_{24} = 4 + 6 = 10\,, & \\
 & x_4 = x'_3 + D_{34} = 9 + 3 = 12 & (\max x_4 = 12 = x'_4)\,.
\end{array}
$$

In the backward pass, we start with $x''_4 = \max x_4 = 12$. For other nodes, we get from (8-10),

$$
\begin{array}{lll}
i = 3\,, & x_3 = x''_4 - D_{34} = 12 - 3 = 9 & (\min x_3 = 9 = x''_3)\,; \\
i = 2\,, & x_2 = x''_4 - D_{24} = 12 - 6 = 6\,, & \\
 & x_2 = x''_3 - D_{23} = 9 - 5 = 4 & (\min x_2 = 4 = x''_2)\,; \\
i = 1\,, & x_1 = x''_3 - D_{13} = 9 - 4 = 5\,, & \\
 & x_1 = x''_2 - D_{12} = 4 - 3 = 1 & (\min x_1 = 1 = x''_1)\,; \\
i = 0\,, & x_0 = x''_2 - D_{02} = 4 - 2 = 2\,, & \\
 & x_0 = x''_1 - D_{01} = 1 - 1 = 0 & (\min x_0 = 0 = x''_0)\,.
\end{array}
$$

To determine the critical activities, we use the relations in (8-11). Thus we check each of the branches (i, j) in which $x'_i = x''_i$ and $x'_j = x''_j$. In this particular example, these relations hold for all nodes. Hence we have to check all branches (i, j) for the condition $x'_j - x'_i = x''_j - x''_i = D_{ij}$:

$$
\begin{array}{lll}
\text{for } (0, 1)\,, & x'_1 - x'_0 = 1 - 0 = 1 & \text{(critical)}\,; \\
\text{for } (0, 2)\,, & x'_2 - x'_0 = 4 - 0 = 4 & \text{(not critical)}\,; \\
\text{for } (1, 2)\,, & x'_2 - x'_1 = 4 - 1 = 3 & \text{(critical)}\,; \\
\text{for } (1, 3)\,, & x'_3 - x'_1 = 9 - 1 = 8 & \text{(not critical)}\,; \\
\text{for } (2, 3)\,, & x'_3 - x'_2 = 9 - 4 = 5 & \text{(critical)}\,; \\
\text{for } (2, 4)\,, & x'_4 - x'_2 = 12 - 4 = 8 & \text{(not critical)}\,; \\
\text{for } (3, 4)\,, & x'_4 - x'_3 = 12 - 9 = 3 & \text{(critical)}\,.
\end{array}
$$

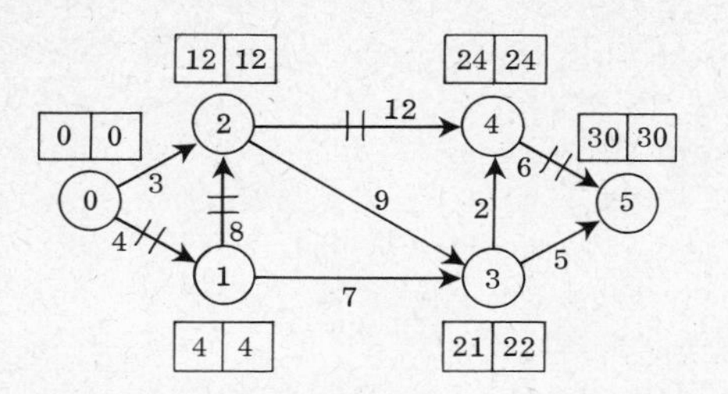

Figure 8-8

Hence the critical path is seen to be 0-1-2-3-4, and the completion time corresponding to this path is 12.

Example 8-7. Determine the critical time path for the network in Example 8-2.

The determination of the critical path for the network in Example 8-2 has been carried out out in Fig. 8-8 according to the procedure explained in the previous example. However, it can also be conveniently given in a matrix indicating node predecessor-successor relationships. As shown in Table 8-7, the activity duration of each branch (i, j) is shown in the ith row and the jth column. The earliest event times x'_i $(i = 0, 1, 2, \ldots, 5)$ are shown in the vertical stub, while the latest event times x''_j $(j = 0, 1, 2, \ldots, 5)$ are given in the horizontal stub. Each critical activity is indicated by a cell which is marked at both the upper left-hand and lower right-hand corners. The procedure of computation will be explained in detail.

The earliest event times are computed by first setting $x'_0 = 0$ in row $i = 0$ under column heading x'_i. To move forward to x'_1, examine all activities merging into node 1 under column $j = 1$. Since there is only

Table 8-7

MATRIX REPRESENTATION OF CPM

i \ j	0	1	2	3	4	5	x'_i
0		4	3				0
1			8	7			4
2				9	12		12
3					2	5	21
4						6	24
5							30
x''_j	0	4	12	22	24	30	

one activity (0, 1) in that column, x'_1 is obtained from (8–9)

$$x_1 = D_{01} + x'_0 = 4 + 0 = 4 \qquad (\max x_1 = x''_1) .$$

This computation can be carried out in row $i = 0$ of the table with the result recorded in row $i = 0$ under the column heading x'_i. We also mark the upper left-hand corner of the cell representing activity (0, 1) to denote that this activity has the maximum finish time leading to node 1. Similarly, to go from x'_1 to x'_2, we look for all activities merging into node 2 under column $j = 2$. In this case, we observed from rows $i = 0$ and $i = 1$ in the table, respectively,

$$x_2 = D_{02} + x'_0 = 3 + 0 = 3 ,$$
$$x_2 = D_{12} + x'_1 = 8 + 4 = 12 \qquad (\max x_2 = x'_2) .$$

Thus x'_2 is taken to be 12 and recorded in row $i = 2$. We then mark the upper left-hand corner of the cell representing activity (1, 2) which leads to $\max x_2 = x'_2$. In general, to move forward to the next node, examine all activities merging into that node and select the $\max x_i = x'_i$. The procedure is repeated until all earliest event times are computed; then $\max x_5 = x'_5$.

In computing the latest event times, the nodes are taken in descending order starting from $x''_5 = x'_5 = \max x_5$, as recorded in the row x''_j under column $j = 5$. Then x''_4 is obtained from (8–10)

$$x_4 = x''_5 - D_{45} = 30 - 6 = 24 \qquad (\min x_4 = x''_4) ,$$

since (4, 5) is the only activity bursting from node 4 and it is found in row $i = 4$. Hence the cell representing this activity in the table is marked at the lower right-hand corner. Similarly, x''_3 is obtained by considering the activities bursting from node 3 that are found in row $i = 3$, or

$$x_3 = x''_5 - D_{35} = 30 - 5 = 25 ,$$
$$x_3 = x''_4 - D_{34} = 24 - 2 = 22 \qquad (\min x_3 = x''_3) .$$

These computations can be carried out in columns $i = 4$ and $i = 5$, respectively, and the cell representing activity (3, 4) is marked at the lower right-hand corner. The procedure is repeated until all latest event times are computed.

The cells in the table that have been marked twice, once during the computation of x'_i and once during the computation of x''_j, represent activities for which $x'_i = x''_i$ and $x'_j = x''_j$. We can also compute the slack times for these activities to determine whether they are critical. That is, if all relations in (8–11) are satisfied, these activities are critical. Finally, a continuous path is constructed by linking the critical activities in the proper sequence [here, (0, 1), (1, 2), (2, 4), and (4, 5)].

8-6 SLACKS IN NONCRITICAL ACTIVITIES

So far, we have been primarily concerned with the determination of critical activities in the time path of a project network. We shall now turn our attention to the noncritical activities in which the slack variables are not zero. We are particularly interested in the scheduling of the noncritical activities in order to reduce project cost.

Let t_{ij} be the start time of an activity (i, j), and T_{ij} be the finish time of the same activity (i, j). The earliest time that an activity (i, j) can start without interfering with completion times of its predecessors is denoted by t'_{ij}, and the latest time that this activity (i, j) can start without interfering with the completion times of its successors is denoted by t''_{ij}. The earliest finish time T'_{ij} and the latest finish time T''_{ij} are similarly defined. These activity start times and finish times can be related to the earliest event times, x'_i or x'_j and to the latest event times, x''_i or x''_j as follows:

$$t'_{ij} = x'_i \,, \tag{8-12a}$$

$$T'_{ij} = x'_i + D_{ij} \,, \tag{8-12b}$$

$$T''_{ij} = x''_j \,, \tag{8-12c}$$

$$t''_{ij} = x''_j - D_{ij} \,. \tag{8-12d}$$

The *total slack*, or *total float*, in an activity (i, j) is defined by

$$x^t_{ij} = T''_{ij} - T'_{ij} = t''_{ij} - t'_{ij} \,. \tag{8-13}$$

For a critical activity, $x^t_{ij} = 0$.

Unless specific reasons are given explicitly to the contrary, it is advisable to follow a schedule of the earliest event times, i.e., to start and finish all activities as early as possible. Such a schedule is called the *early time schedule*. Once the slack in an activity is used, however, it may interfere with the slacks of its successors. The part of the total slack in the activity (i, j) which is free from interference by its successors is called the *free slack*, and the part which is the difference between total and free slack is called the *dependent slack*. If a portion of the total slack inthe activity (i, j) is entirely independent of successors or predecessors, that portion is called the *independent slack*. According to these definitions, the total slack x^t_{ij}, the free slack x^f_{ij}, the dependent slack x^d_{ij}, and the independent slack x^i_{ij} can be related to the event times as follows:

$$x^t_{ij} = x''_j - x'_i - D_{ij} \,, \tag{8-14a}$$

$$x^f_{ij} = x'_j - x'_i - D_{ij} \,, \tag{8-14b}$$

$$x^d_{ij} = x^t_{ij} - x^f_{ij} = x''_j - x'_j \,, \tag{8-14c}$$

$$x^i_{ij} = x'_j - x''_i - D_{ij} \,, \qquad x^i_{ij} \geq 0 \,. \tag{8-14d}$$

Table 8-8

ACTIVITY TIMES AND ACTIVITY SLACKS

Activity	Duration	Earliest activity time		Latest activity time		Total slack
		Start	Finish	Start	Finish	
(i, j)	D_{ij}	t'_{ij}	T'_{ij}	t''_{ij}	T''_{ij}	x^t_{ij}
0, 1	4	0	4	0	4	0
0, 2	3	0	3	9	12	9
1, 2	8	4	12	4	12	0
1, 3	7	4	11	15	22	11
2, 3	9	12	21	13	22	1
2, 4	12	12	24	12	24	0
3, 4	2	21	23	22	24	1
3, 5	5	21	26	25	30	4
4, 5	6	24	30	24	30	0

Example 8-8. Find the earliest activity start and finish times, and the latest activity start and finish time for each of the activities in the project network for Example 8-7. Also determine the total slack, free slack, dependent slack, and independent slack for each of the activities in the project.

The activity times have been computed from the relations in (8-12), and the total slacks for these activities have also been computed from (8-13). These results are shown in Table 8-8. It can be seen that the

Table 8-9

EVENT TIMES AND ACTIVITY SLACKS

Activity	Duration	Earliest event time		Latest event time		Slack			
						Total	Free	Depend.	Indep.
(i, j)	D_{ij}	x'_i	x'_j	x''_i	x''_j	x^t_{ij}	x^f_{ij}	x^d_{ij}	x^i_{ij}
0, 1	4	0	4	0	4	0	0	0	0
0, 2	3	0	12	0	12	9	9	0	9
1, 2	8	4	12	4	12	0	0	0	0
1, 3	7	4	21	4	22	11	10	1	10
2, 3	9	12	21	12	22	1	0	1	0
2, 4	12	12	24	12	24	0	0	0	0
3, 4	2	21	24	22	24	1	1	0	0
3, 5	5	21	30	22	30	4	4	0	3
4, 5	6	24	30	24	30	0	0	0	0

critical activities have zero total slack. The total slack, free slack, dependent slack,and independent slack have also been computed from (8-14) and the results are shown in Table 8-9.

8-7 RESOURCES LEVELING

Slacks in the noncritical activities of a project make it possible to schedule activities for leveling resources, such as manpower and equipment required in the execution of the project. Basically, the problem is to schedule activities such that the requirements of key resources are leveled to relatively constant rates. If the availability of key resources is unrestricted, the process of leveling involves only the adjustment of the schedule of the noncritical activities within the project completion time determined by the critical-path method. On the other hand, if the access to key resources is limited, the process may require the adjustment of activity durations and/or project completion, if necessary.

Since many types of resources may be involved, they must be treated sequentially in the order of their importance. Unfortunately, there is no established procedure for finding an optimal solution for the problem, but once a trial solution is obtained, its relative merit can be measured against an established criterion. We shall confine our discussion to the leveling of one type of resources at a time. Let q_{ij} be the quantity of the resources required per day for activity (i, j). Let x^k_{ij} be the occurrence of activity (i, j) on the kth day of the project such that

$$x^k_{ij} = \begin{cases} 1\,, & \text{if } (i, j) \text{ occurs on } k\text{th day}\,; \\ 0\,, & \text{if it does not}\,. \end{cases} \tag{8-15}$$

Let Q_k be the total requirement of the resources on the kth day for all activities in the project, and let m be the number of time units (days, for instance) required to complete the project. Then

$$Q_k = \sum_{ij} q_{ij} x^k_{ij}\,, \qquad k = 1, 2, \ldots, m\,. \tag{8-16}$$

The average daily requirement of the resources is given by

$$\bar{Q} = \frac{1}{m} \sum_{k=1}^{m} Q_k\,. \tag{8-17}$$

The uniformity of Q_k $(k = 1, 2, \ldots, m)$ can be measured by the sum of the squares of the differences $(\bar{Q} - Q_k)$. That is, the best solution should have a minimum of the sum S, or

$$\text{Min } S = \sum_{k=1}^{m} (\bar{Q} - Q_k)^2\,. \tag{8-18}$$

Thus, when min $S = 0$, it means that the total daily requirement is constant. In view of (8-17), in which $\bar{Q}$ is constant, we have

$$\begin{aligned}\text{Min } S &= \sum_{k=1}^{m} \bar{Q}^2 - 2\bar{Q} \sum_{k=1}^{m} Q_k + \sum_{k=1}^{m} Q_k^2 \\ &= m\bar{Q}^2 - 2\bar{Q}(m\bar{Q}) + \sum_{k=1}^{m} Q_k^2 \\ &= \sum_{k=1}^{m} Q_k^2 - m\bar{Q}^2 .\end{aligned}$$

Since $m\bar{Q}^2$ is a constant, we can replace the minimization of S by the minimization of S', which represents the variable term in the above equation, as the criterion for testing the relative merit of a solution, or

$$\text{Min } S' = \sum_{k=1}^{m} Q_k^2 . \tag{8-19}$$

Thus, after a number of trial solutions are obtained, they may be compared by means of (8-19).

One method for finding trial solutions of resources leveling problems is to list all activities in a project together with their daily resources requirements, and to compare the values of S' for all trials produced by scheduling noncritical acivities for different positions of slacks. *Inasmuch as the earliest time schedule allows the removal of uncertainties about the completion time of various activities as soon as possible, it is used as the starting point of trial solutions.* The free slack in any activity can be used without considering the scheduling of any other activity, but the dependent slack can be used only if it does not interfere with its predecessors or successors. Furthermore, the activities can be listed in any order for the purpose of speeding up the trial-and-error computation. For example, if the activities are listed in the ascending order of j, which is the number at the arrowhead of each activity (i, j), then all the predecessors of event x_j will be above the listing of the activity (i, j), and all the successors of event x_j will be below the listing of that activity. The order of the number i is immaterial, although all activities with the same value of j sometimes are arranged in the order of increasing total slack. With this arrangement, the dependent slack can be used systematically without interference if we start from the bottom of the listing and move upward. The applications of this procedure will be illustrated by examples.

Example 8-9. For the project network shown in Fig. 8-9, the activities (i, j) are listed in the order of precedence in Table 8-10, together with the daily resource requirements q_{ij}. (Note that j is in the ascending order, and i is increasing for each j.) The earliest start time t'_{ij} and finish time T'_{ij}, as well as the total slack x^t_{ij} and the free slack x^f_{ij}, for each activity have been computed and recorded in the table. The activity duration D_{ij}

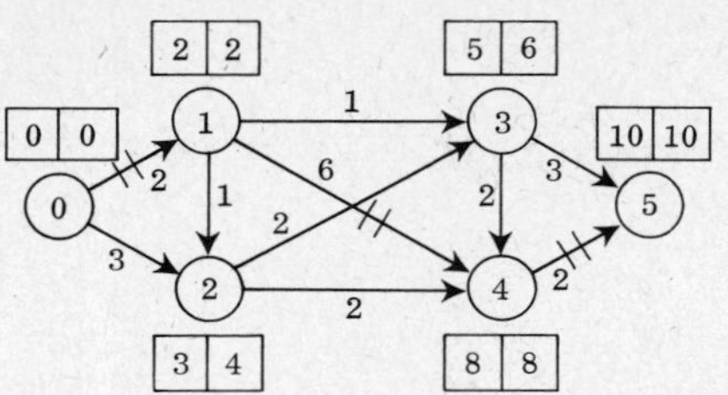

Figure 8-9

and the total slack of each activity corresponding to the earliest time schedule are shown by solid bars and hollow bars, respectively, on the bar chart at the right-hand side of the table. The shaded portion of a hollow bar represents the dependent slack while the blank portion represents the free slack. Determine a schedule which demands a relatively constant total daily resource requirement.

At the bottom of the table are the results of three trial solutions The first solution is based on the earliest time schedule, as indicated by the bar chart. Hence Q_k $(k = 1, 2, \ldots, m)$ may be obtained according to (8-16) by inspecting the occurrence of x_{ij}^k on the bar chart. For example,

$$Q_1 = q_{01} + q_{02} = 2 + 3 = 5\,,$$
$$Q_2 = q_{01} + q_{02} = 2 + 3 = 5\,,$$
$$Q_3 = q_{02} + q_{12} + q_{13} + q_{14} = 3 + 3 + 6 + 1 = 13\,,$$
$$Q_4 = Q_5 = q_{23} + q_{14} + q_{24} = 0 + 1 + 4 = 5\,,$$
$$Q_6 = Q_7 = q_{14} + q_{34} + q_{35} = 1 + 1 + 5 = 7\,,$$
$$Q_8 = q_{14} + q_{35} = 1 + 5 = 6\,,$$
$$Q_9 = Q_{10} = q_{45} = 2\,.$$

These values of Q_k are listed for $k = 1, 2, \ldots, 10$, corresponding to the days elapsed in the project. Thus for the first trial solution,

$$S' = 5^2 + 5^2 + 13^2 + 5^2 + 5^2 + 7^2 + 7^2 + 6^2 + 2^2 + 2^2 = 411\,.$$

In the second trial, we move the solid portions of the bars back to the positions for the slacks in the table, starting with the last activity listed in the table and moving upward. There are many combinations of positions in the slacks that the activities can be placed, but there is no established procedure for finding the combination of positions which leads to an optimal solution. Hence only a trial-and-error method is used here. In the second solution, for example, we move activity (3, 5) back to days 8 through 10, activity (2, 4) back to days 6 and 7, and activity (1, 3) back to the fifth day. Without drawing a new bar chart, we simply recompute Q_k by noting these changes, and the results are shown in the columns for $k = 1, 2, \ldots, 10$ corresponding to the days elapsed in the project. Thus

Table 8-10

RESOURCE LEVELING FOR EXAMPLE 8-9

Activity	Duration	Earliest		Slack			Days elapsed (k)											
(i, j)	D_{ij}	t'_{ij}	T'_{ij}	x^t_{ij}	x^f_{ij}	q_{ij}	1	2	3	4	5	6	7	8	9	10		
0, 1	2	0	2	0	0	2												
0, 2	3	0	3	1	0	3												
1, 2	1	2	3	1	0	3												
1, 3	1	2	3	3	2	6												
2, 3	2	3	5	1	0	0												
1, 4	6	2	8	0	0	1												
2, 4	2	3	5	3	3	4												
3, 4	2	5	7	1	1	1												
3, 5	3	5	8	2	2	5												
4, 5	2	8	10	0	0	2											S'	Trial
							5	5	13	5	5	7	7	6	2	2	411	1
							5	5	7	1	7	6	6	6	7	7	355	2
							5	5	7	7	5	6	7	6	7	2	347	3

for the second trial solution,

$$S' = 5^2 + 5^2 + 7^2 + 1^2 + 7^2 + 6^2 + 6^2 + 6^2 + 7^2 + 7^2 = 355\,.$$

This result indicates that the second solution is distinctly better than the first. We may also note that the maximum number of the total resource requirement per day is seven in the second solution, while that quantity is 13 in the first solution.

In the third trial, we move activity (3, 5) back to days 7 through 9, activity (2, 4) back to days 5 and 6, and activity (1, 3) back to the fourth day. The results of Q_k and S' for this solution are given the last line of the table. Note that $S' = 347$ in the third solution, compared with $S' = 355$ in the second solution. However, this improvement is more apparent than real, since the maximum number of the total resource requirement per day is seven in both cases. Hence it takes only a relatively few trials to find a reasonably good solution even though the optimal solution cannot be obtained.

Example 8–10. For the project network shown in Example 8–5, the activities (i, j) are listed in the order of precedence in Table 8–11. Note that j is listed in the ascending order, and the activities with the same j are listed in the increasing order of the slack. The daily manpower requirements q_{ij} (in terms of teams of one type of tradesmen in a construction project) for all activities are also given. The earliest start time t'_{ij} and finish time T'_{ij} as well as the total slack x^t_{ij} and the free slack x^f_{ij}, for each activity have been computed and recorded in the table. The activity duration D_{ij} and the total slack of each activity corresponding to the earliest time schedule are shown by solid bars, respectively, on the bar chart at the right-hand side of the table. All slacks are free slacks in this problem. Find a schedule which requires no more than 15 teams of tradesmen daily for all activities in the project.

At the bottom of the table are the results of three trial solutions. The first solution is based on the earliest time schedule and requires no further explanation. The second solution is obtained by moving activity (2, 4) back to days 7 through 12, and activity (1, 3) back to days 3 through 6. Although $S' = 1{,}617$ in the second solution is a considerable improvement from $S = 1{,}733$ in the first solution, the maximum total number of teams of tradesmen required daily is only reduced to 18 in the second solution, compared with 20 in the first solution. This reduction does not satisfy the manpower constraint that allows no more than 15 teams of tradesmen daily for all activities in the project.

In the solution of resources leveling problem in which the resource constraint is exceeded and cannot be satisfied by just moving the activities to the positions for slacks, the durations of some noncritical activities should

Table 8–11

RESOURCE LEVELING FOR EXAMPLE 8–10

Activity	Duration	Earliest		Slack			Days elapsed (k)													
(i, j)	D_{ij}	t'_{ij}	T'_{ij}	x^t_{ij}	x^f_{ij}	q_{ij}	1	2	3	4	5	6	7	8	9	10	11	12	S'	Trial
0, 1	1	0	1	0	0	4														
1, 2	3	1	4	0	0	6														
0, 2	2	0	4	2	2	3														
2, 3	5	4	9	0	0	8														
1, 3	4	1	9	4	4	10														
3, 4	3	9	12	0	0	1														
2, 4	6	4	12	2	2	2														
							7	19	16	16	20	10	10	10	10	3	1	1	1,733	1
							7	9	16	16	18	18	10	10	10	3	3	3	1,617	2
							7	14	11	11	13	13	15	15	15	3	3	3	1,527	3

be extended with a corresponding reduction in daily resource requirements for those activities. If the resource constraint is still exceeded even after all available slacks in the noncritical activities have been used for the extended durations, then the durations of some critical activities must be lengthened. In the latter case, each adjustment in the durations of critical activities requires a new cycle of determining the critical path and computing the slacks. Hence the selection of a schedule for activities to satisfy the resource constraint may involve many cycles of operation before an acceptable total daily resource requirement is reached.

In this example, it appears that we have reached the point where the resource constraint cannot be satisfied by just moving the activities to the positions for slacks even if we try to split some noncritical activities to two discontinuous parts, leaving the total durations unchanged. In the third trial, therefore, we extend the duration of activity (1, 3) from four days to eight days, with a corresponding reduction of daily requirement of teams of tradesmen from ten to five. With this new arrangement, the maximum total daily requirement is reduced to 15, as indicated in the last line in the table. The corresponding S' in the third trial is 1,527, which is substantially lower than $S' = 1{,}733$ in the first trial.

8-8 TIME-COST TRADEOFFS

In previous sections, the normal durations of activities in a project have been used in the application of the critical-path method. The normal durations are estimated on the basis of normal conditions of available labor and material. However, the durations of some or all of these activities can often be reduced at extra costs, and the durations of activities under an all-out crash program to expedite these activities are called *crash durations*. Obviously, if the crash durations of the activities are used in determining the critical-time path, the completion time of the project can be reduced accordingly. On the other hand, the cost of the project will increase as a result of the extra costs associated with the reduction of activity durations.

We shall consider only the *direct costs* required to perform the activities in the project. In general, the relation between the duration and the direct cost of an activity follows the general shape of the curve shown in Fig. 8-10, in which t denotes activity duration or time and c denotes activity cost. There is usually a practical limit t_1, below which the activity cannot be completed regardless of the cost we are willing to pay; and there is also a practical limit t_2, above which the cost may increase. In the range between t_1 and t_2, or at least in the range which is significant for time-cost tradeoff, the curve is usually convex and can be approximated by piecewise continuous straight line segments, as shown by the

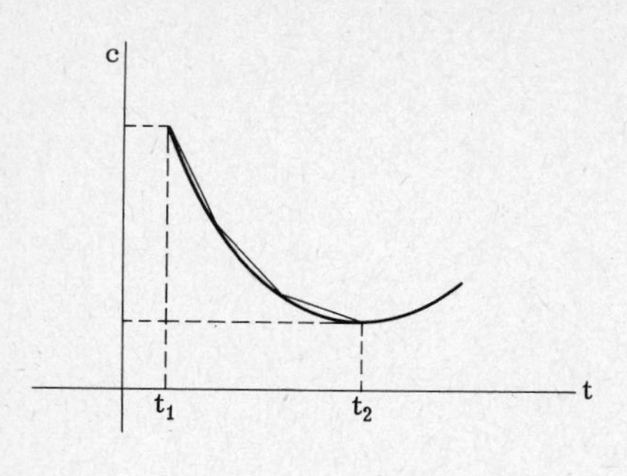

Figure 8–10

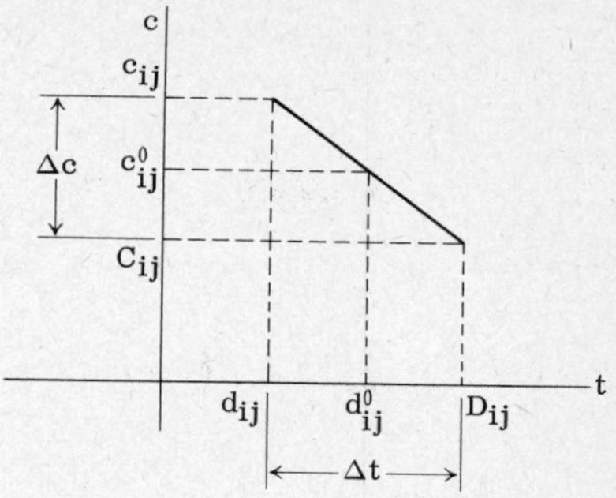

Figure 8–11

broken line in Fig. 8–10. Since this relationship between time and cost generally holds for all activities in the project, the relationship between the project completion time and project cost is expected to have the same general shape as the curve in Fig. 8–10. Very often, we encounter situations in which the project completion time based on normal duration is too long, but that based on crash durations is too costly for practical purposes. Therefore, we want to examine the intermediate positions in the range of feasible solutions. This is the essence of the *time-cost tradeoff problems*.

Let the relationship between the time and the cost of each activity in the project be approximated by a straight line (a special case of the piecewise linear function), as shown in Fig. 8–11. Let D_{ij} and d_{ij} be the normal and crash durations, respectively, for activity (i, j), and C_{ij} and c_{ij} be the activity costs associated with the normal and crash durations, respectively. Let R_{ij} be the value of the rate of change of cost with respect to change in time, i.e.,

$$R_{ij} = \left| \frac{\Delta c}{\Delta t} \right| .$$

Then, from the linear time-cost relationship in Fig. 8-11, we have

$$c_{ij} = C_{ij} + R_{ij}(D_{ij} - d_{ij}) , \tag{8–20}$$

in which R_{ij} is seen to be the absolute value of the slope of the line. Hence for any duration d^0_{ij} $(d_{ij} \leq d^0_{ij} \leq D_{ij})$, the corresponding c^0_{ij} may be obtained as follows:

$$c^0_{ij} = C_{ij} + R_{ij}(D_{ij} - d^0_{ij}) . \tag{8–21}$$

If R_{ij}, D_{ij}, and C_{ij} are known, then for any given value of d^0_{ij}, the corresponding c^0_{ij} may be computed from (8–21).

In a time-cost tradeoff problem, we seek to vary the activity durations such that the project completion time and its associated project cost for

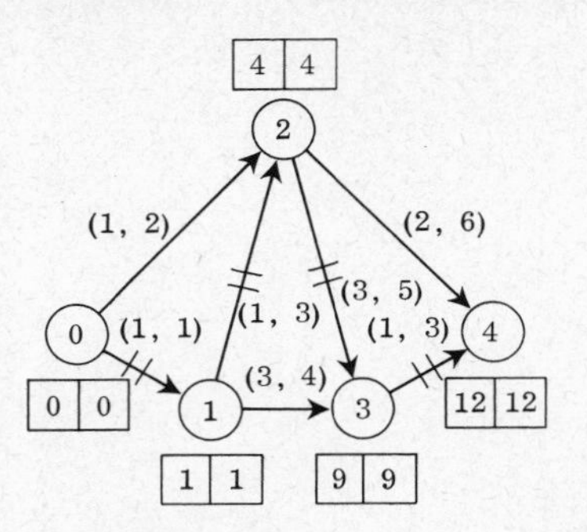

Figure 8–12

each variation can be computed and compared with those obtained for other variations. However, there are numerous possible combinations of activity durations which may lead to the same project completion time. Hence the project cost associated with a specified project completion time is not unique, and the minimum cost solution for the specified time is the desired solution.

Example 8–11. For the project network shown in Fig. 8–12, the information about R_{ij}, D_{ij}, C_{ij}, and d_{ij}, corresponding to activity (i, j), are given in the first five columns of Table 8–12. The unit for D_{ij} and d_{ij} is in days, the unit for C_{ij} is in thousands of dollars, and the unit for R_{ij} is in thousands of dollars per day. Determine the cost of completing the project on the basis of (1) all normal durations, (2) all crash durations, (3) a project crash schedule, (4) two different combinations of activity durations, each of which leads to a projection completion time of eight days, and (5) two different combinations of activity durations, each of which leads to a project completion time of 10 days.

Table 8–12

TIME-COST TRADEOFF FOR EXAMPLE 8–11

Activity	R_{ij}	12 days nomal		6 days				8 days				10 days			
	$= \left\lvert \frac{\Delta c}{\Delta t} \right\rvert$			All crash		Crash		(a)		(b)		(a)		(b)	
(i, j)		D_{ij}	C_{ij}	d_{ij}	c_{ij}	$d^{\dagger}_{ij}$	$c^{\dagger}_{ij}$	d'_{ij}	c'_{ij}	d''_{ij}	c''_{ij}	$\bar{d}_{ij}$	$\bar{c}_{ij}$	$\hat{d}_{ij}$	$\hat{c}_{ij}$
0, 1	∞	1	6	1	6	1	6	1	6	1	6	1	6	1	6
0, 2	3	2	5	1	8	2	5	2	5	2	5	2	5	2	5
1, 2	4	3	4	1	12	1	12	2	8	1	12	3	4	2	8
1, 3	1	4	5	3	6	4	5	4	5	4	5	4	5	2	5
2, 3	2	5	6	3	10	3	10	3	10	3	10	3	10	4	6
2, 4	5	6	4	2	24	4	14	5	9	6	4	6	4	6	4
3, 4	2	3	7	1	11	1	11	2	9	3	7	3	7	2	9
Total			37		77		63		52		49		41		43

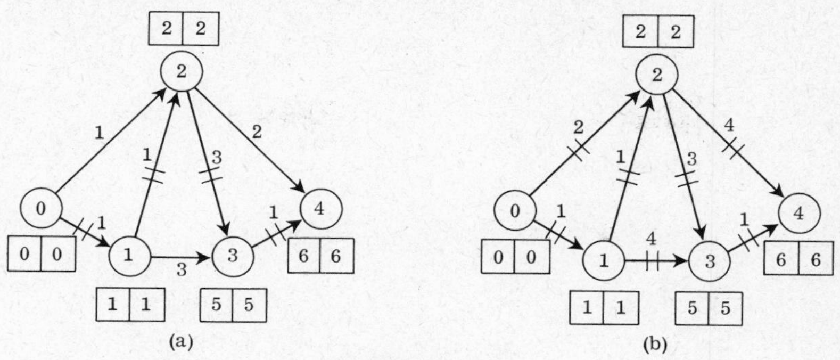

Figure 8-13

The durations shown for the activities in Fig. 8-12 are normal durations. The project completion time for all normal durations has been found to be 12 days, and each critical activity is marked by a pair of short parallel lines. The project cost associated with this project completion time is given as \$37,000.

If all activity durations in the project are changed from normal to crash, then the critical path may be determined as shown in Fig. 8-13(a), in which all the durations shown for the activities were crash. The project completion time is found to be six days on the network and the associated cost is found by summing up the activity costs determined by the relation in (8-20). Thus

$$
\begin{aligned}
c_{01} &= 6\,, \\
c_{02} &= 5 + (3)(2 - 1) = 8\,, \\
c_{12} &= 4 + (4)(3 - 1) = 12\,, \\
c_{13} &= 5 + (1)(4 - 3) = 6\,, \\
c_{23} &= 6 + (2)(5 - 3) = 10\,, \\
c_{24} &= 4 + (5)(6 - 2) = 24\,, \\
c_{34} &= 7 + (2)(3 - 1) = 11\,; \\
\text{total} &= 77\,.
\end{aligned}
$$

The project cost associated with all crash durations is seen to be \$77,000. However, it is not necessary to pay such a high cost to reduce the project completion time to six days. So long as there are noncritical activities in the network, their durations can be lengthened so that the costs for these activities can be reduced. In general, there are many ways by which the durations of noncritical activities can be lengthened without delaying the project completion time. For example, we can adopt the earliest-time schedule or the latest-time schedule, and lengthen the noncritical activi-

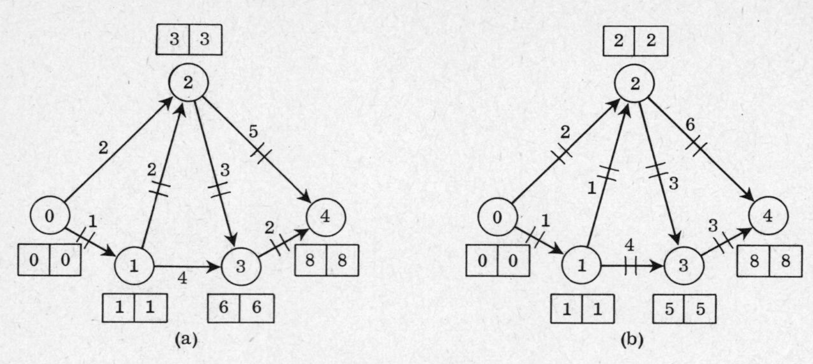

Figure 8–14

ties accordingly. In either case, the time for project completion is still the *project-crash time*, and the cost associated with each set of activity durations is called the *project-crash cost*, which is substantially below the *all-crash cost*. In this particular example, however, there is only one way by which the durations of noncritical activities can be lengthened because the project network contains very few activities. The set of activity durations $d_{ij}^{\dagger}$ which will make every activity critcal while maintaining the same project completion time is shown in Fig. 8–13(b). The cost $c_{ij}^{\dagger}$ associated with this set of activity durations is the *project-crash cost*. In the project-crash schedule, the activity durations are shortened only as necessary to achieve the project crash time, while in the all-crash program, all activity durations are automatically shortened to crash durations whether it is necessary to do so or not. The project crash cost is found to be \$63,000.

Two different combinations of activity durations, d'_{ij} and d''_{ij}, each of which leads to a project completion time of eight days, have been chosen arbitrarily, as shown in Fig. 8–14. The critical activities for the two schedules are shown in parts (a) and (b) of the figure. Note that the noncritical activities in (a) have not been lengthened because the costs for these activities cannot be further reduced below the normal costs. The costs c'_{ij} and c''_{ij} associated with these schedules are shown in Table 8–12. Similarly, two different combinations of activity durations, $\bar{d}_{ij}$ and $\hat{d}_{ij}$, each of which leads to a project completion time of 10 days, are shown in Fig. 8–15. The costs $\bar{c}_{ij}$ and $\hat{c}_{ij}$ associated with these schedules are also shown in Table 8–12.

The project costs for all schedules in Table 8–12 are plotted in Fig. 8–16. The area bounded by solid lines represents the feasible region indicated by the schedules obtained in this example, whereas the area bounded by dashed lines represents the feasible region which may exist. It is seen that if the activity durations are lengthened from the normal

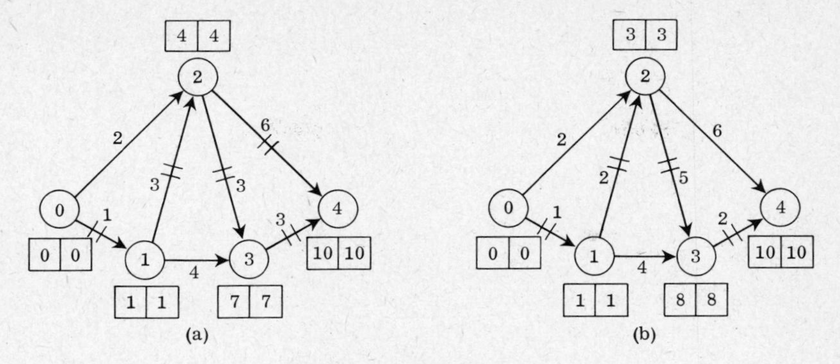

Figure 8–15

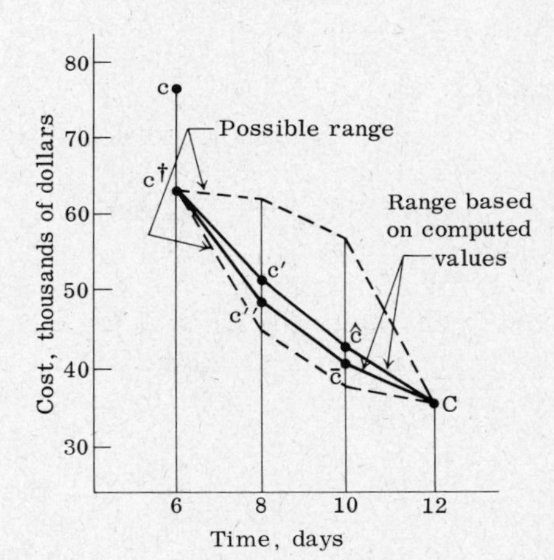

Figure 8–16

duration selectively, the project cost should not be higher than the project-crash cost if the specified project completion time is longer than the project-crash time. For project networks in which more than one project-crash schedule is possible, there will also be a number of different project-crash costs, one associated with each schedule.

8–9 ACTIVITY-ON-NODE NETWORKS

Although activity-on-node networks are not widely used, they have gained popularity for scheduling projects with activities which have complicated precedence relationships. The main advantage of activity-on-node networks is that no dummy activities are introduced except for the initial and terminal nodes, in some cases. The method of computation for activity-on-node networks is very similar to that for the activity-on-branch network.

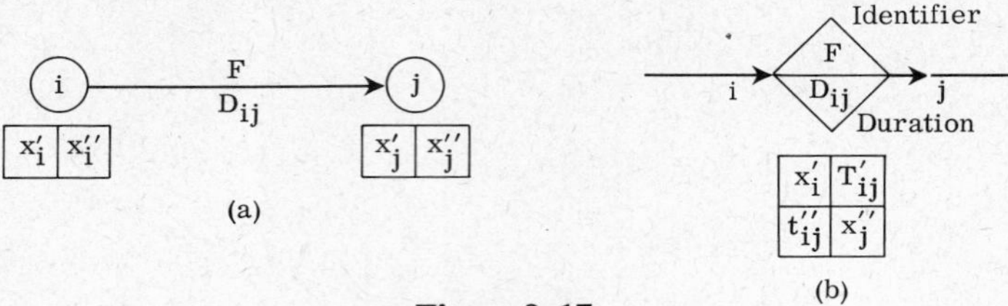

Figure 8–17

The equivalence of these two types of networks can easily be seen from the comparison of the designations common to both systems, as shown in Fig. 8–17. In part (a) of the figure, which has the conventional designations for activity-on-branch networks, the activity identifier F, as well as the activities duration D_{ij}, are placed on the branch, while the start i and the finish j of the activity (i, j) are nodes. Hence the earliest and latest event times for i (x'_i and x''_i) are placed near node i, while the corresponding times for j (x'_j and x''_j) are placed near node j. In part (b), which contains the new designations for activity-on-node networks, the events are no longer shown explicitly. Furthermore, both the activity identifier F and activity D_{ij} are placed on the node, which is shown in a diamond. The start and the finish of the activity are represented by the symbols i and j, corresponding to the instants immediately prior to and after the activity F. Hence the earliest and the latest start times t'_{ij} $(= x'_i)$ and t''_{ij} are placed on the side of the symbol i, and the corresponding finish times, T'_{ij} and T''_{ij} $(= x''_j)$ are placed on the side of symbol j. Since all the information needed for the determination of event times, critical activities, activity times, and slacks are included in these designations, the method of computation for activity-on-node networks can be carried out in a parallel manner to that for activity-on-branch networks.

Table 8–13

PRECEDENCE RELATIONSHIPS FOR EXAMPLE 8–12

Activity	Predecessors			Duration
	Case (a)	Case (b)	Case (c)	Case (a) only
A	—	—	—	2
B	*A*	*A*	*A*	3
C	*B*	(No activity)	*B*	4
D	*A*	*A*	*A*	5
E	*C*, *D*	*B*, *D*	*C*, *D*	2
F	*C*, *D*	*B*, *D*	*D*	6
G	*F*	*F*	*G*	2
H	*E*, *G*	*E*, *G*	*H*	8

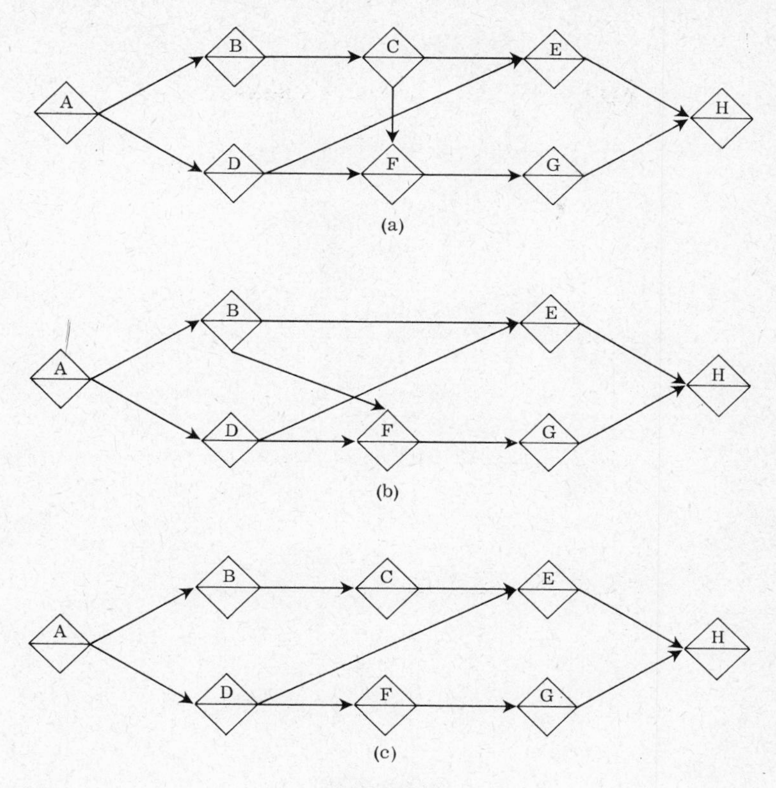

Figure 8–18

Example 8–12. Construct activity-on-node networks for project cases (a), (b), and (c), each of which has activity-precedence relationships, as shown in Table 8–13. Compare these cases with cases (a), (b), and (c) in Fig. 8–3, in which the same projects are represented by activity-on-branch networks.

The activity-on-node networks representing cases (a), (b), and (c) are shown in Fig. 8–18, in which the nodes are diamond-shaped. In part (a), the durations of activities are also shown in the network. In part (b), there is no node representing activity C because that activity has been removed. If we compare this case with a similar activity-on-branch network shown in Fig. 8–3(b), we can see that Fig. 8–18(b) shows the same precedence relationships of activities as Fig. 8–3(b), but that no dummy activity need be added on the activity-on-node network. Note especially that in Fig. 8–18(b), both B and D are uniquely identified without a dummy activity, whereas in Fig. 8–3(b), a dummy activity X must be added. Also, in Fig. 8–18(c), there is no branch between activities B and F

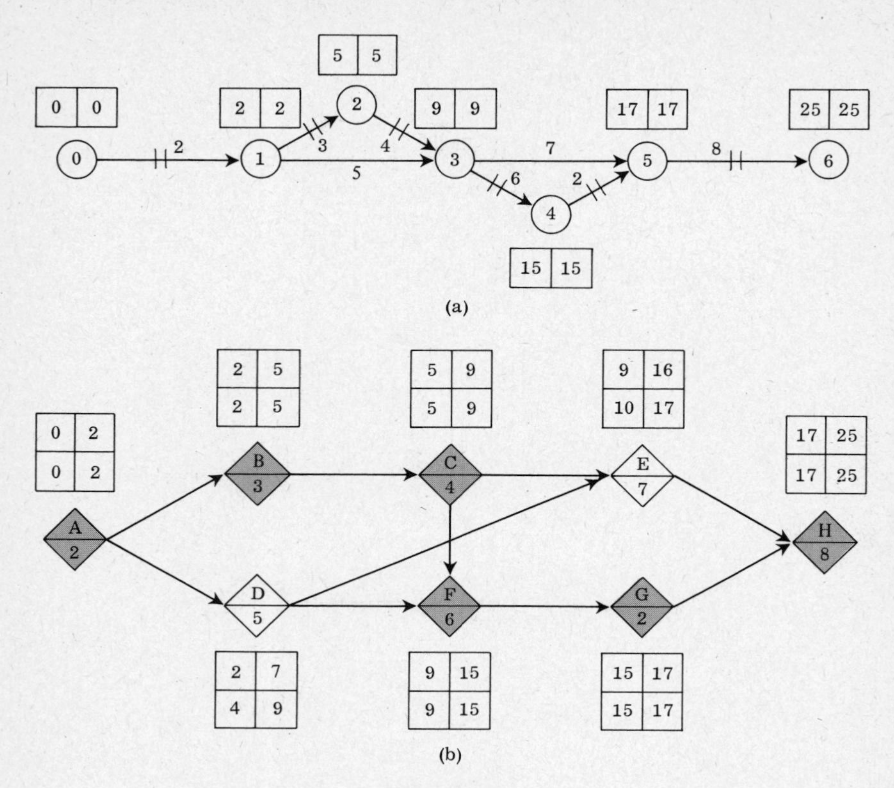

Figure 8-19

because F cannot start until its predecssor D is completed, but it is independent of B. If we compare this case with a similar activity-on-branch network shown in Fig. 8-3(c), we can see that Fig. 8-18(c) shows the same precedence relationships of activities as Fig. 8-3(c); however, no dummy activity need be added on the activity-on-node network. Note especially that in Fig. 8-18(c) activity E cannot start until both C and D are completed, but F can start after D alone is completed, whereas in Fig. 8-3(c), a dummy activity Y must be added. This example shows the advantage of the activity-on-node networks when the precedence relationships become complex, particularly when the scheduling procedure is computerized.

Example 8-13. Determine the critical path for the project in case (a) of Example 8-12 by the use of both an activity-on-branch network and an activity-on-node network.

In Fig. 8-19(a), which is an activity-on-branch network, the critical path has been determined by the method described in Sections 8-4 and

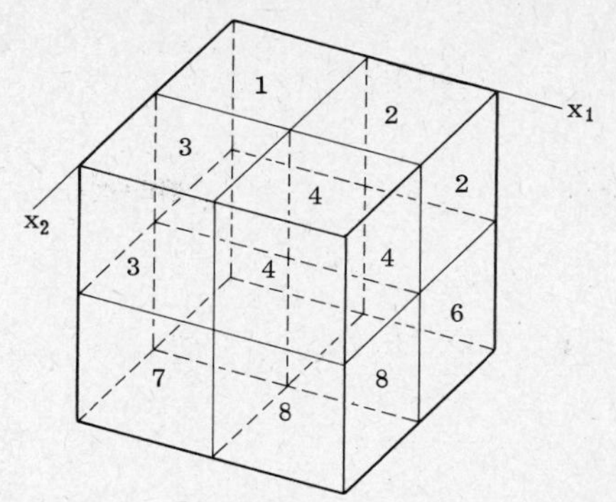

Figure 8-20

8-5. In part (b) of the figure, the same procedure can be used except that the designations are now placed in a different format. By comparing the designations on the two networks, it is not difficult to see the similarities between the two approaches. Each critical activity in part (b) is designated by a check mark next to the node.

Example 8-14. In the excavation of a building foundation, the total volume is subdivided into eight smaller zones, as shown in Fig. 8-20. Two excavators are used, with the first one assigned to zones 1, 2, 7, and 8, consecutively, and the second one assigned to zones 3, 4, 5, 6, consecutively. The excavation of each zone is considered as a separate activity. Hence the sequence of activities is governed by two factors: (1) the locations of the zones, and (2) the order of excavation specified for the excavators. Given that the excavations for zones 1, 2, . . . , 8 are designated as activities 1, 2, . . . , 8, respectively, and that the durations and precedence relationships of these activities are as shown in Table 8-14, determine the minimum completion time for the project.

In the example, we introduce dummy activities for the initial and terminal nodes (node 0 and node 9, respectively) in order to produce a

Table 8-14

PRECEDENCE RELATIONSHIPS FOR EXAMPLE 8-14

Activity	Description	Predecessors	Duration (hrs)
1	Excavate zone 1	—	23
2	Excavate zone 2	1	68
3	Excavate zone 3	—	17
4	Excavate zone 4	3	50
5	Excavate zone 5	1, 4	25
6	Excavate zone 6	2, 5	75
7	Excavate zone 7	3, 2	34
8	Excavate zone 8	4, 7	102

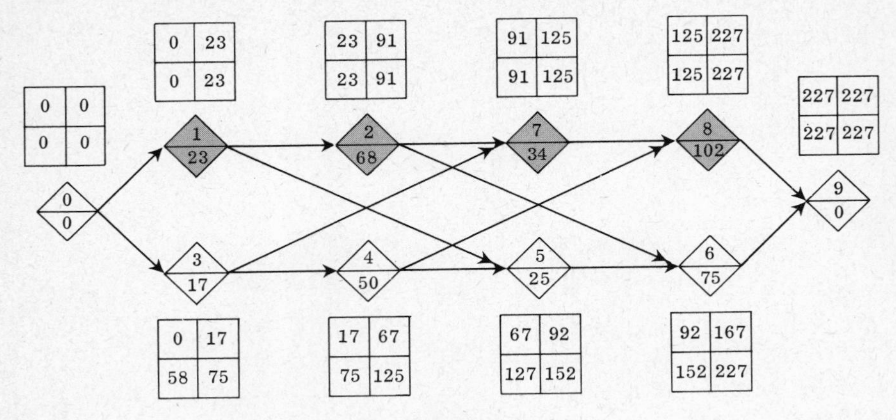

Figure 8–21

network with a single initial node and a single terminal node. The duration of each dummy activity is zero. The precedence relationships given in Table 8–14 can be checked against the locations of the zones in Fig. 8–20 and the order of excavation specified for the operators. Then an activity-on-node network may be constructed as shown in Fig. 8–21.

The eariliest event times and the latest event times are determined by using the forward pass and the backward pass, respectively. The critical activities are determined by the relations in (8–11). The results of computations of start and finish times of activities are recorded on the network according to the specified format, and each critical activity is designated by a check mark next to the node.

REFERENCES

8–1. MODER, JOSEPH J., and C. R. PHILLIPS, *Project Management with CPM and PERT*, Reinhold, New York, 1964.

8–2. SHAFFER, L. R., J. B. RITTER, and W. L. MEYER, *The Critical-Path Method*, McGraw Hill, New York, 1965.

8–3. KELLEY, J. E., JR., "Critical-Path Planning and Scheduling: Mathematical Basis," *Journal of Operations Research Society of America*, **9** (1961), 296–320.

8–4. KELLEY, J. E., JR., and M. R. WALKER, "Critical-Path Planning and Scheduling," *Proceedings of the Eastern Joint Computer Conference, 1959*, 160–173.

8–5. *DOD and NASA Guide, PERT Cost Systems Design*, Office of the Secretary of Defense and the National Aeronautics and Space Administration, U. S. Government Printing Office, Washington, D. C., 1962.

8–6. CHARNES, A., and W. W. COOPER, "A Network Interpretation and a Directed Subdual Algorithm for Critical-Path Scheduling," *Journal of Industrial Engineering*, **13**, (1962), 213–218.

PROBLEMS

P8–1 through P8–4. Construct an activity-on-branch network from the precedence relationships of activities in the project given in the table for the problem.

Table P8–1

Activity	Predecessors	Duration
A	—	6
B	*A*	7
C	*A*	1
D	—	14
E	*B*	5
F	*C, D*	8
G	*C, D*	9
H	*D*	3
I	*H*	5
J	*F*	3
K	*E, J*	4
L	*F*	12
M	*G, I*	6
N	*G, I*	2
O	*L, N*	7

Table P8–2

Activity	Predecessors	Duration
A	—	5
B	*A*	6
C	*B*	3
D	*C*	4
E	*D, G*	5
F	*A*	8
G	*F, J*	3
H	—	3
I	*H*	2
J	*I*	7
K	*F, J*	2
L	*H*	7
M	*L*	4
N	*K, M*	3

Table P8–3

Activity	Predecessors	Duration
A	—	6
B	—	12
C	—	16
D	*A*	5
E	*B*	3
F	*C*	10
G	*B, D*	9
H	*C, E*	4
I	*F*	5
J	*F*	3
K	*E, G, I*	10
L	*H, J*	6

Table P8–4

Activity	Predecessors	Duration
A	—	3
B	—	6
C	—	2
D	*C*	3
E	*C*	8
F	*B, E*	5
G	*A, F*	7
H	*B, E*	10
I	*B, E*	6
J	*B, E*	6
K	*D, J*	8
L	*G, H*	3
M	*I, K, L*	4

P8–5 through P8–8. Determine the critical path and all types of slacks for the project in Problems P8–1 through P8–4, respectively.

P8–9 and P8–10. From the network shown in the figures for the problems and the resource requirements of activities given, suggest three solutions for leveling the total daily resource requirements, and determine which in each case is the most effective.

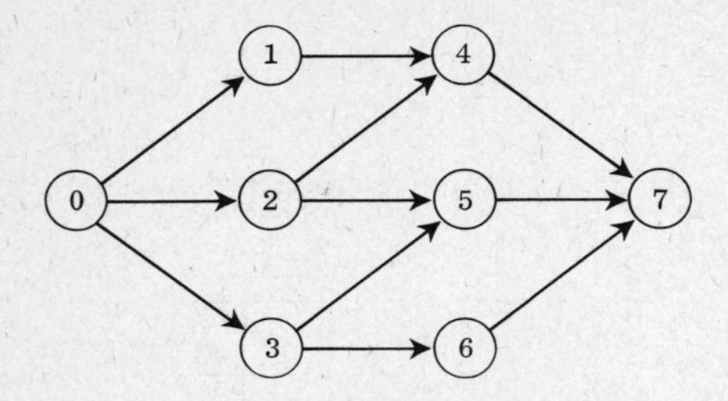

Figure P8–9

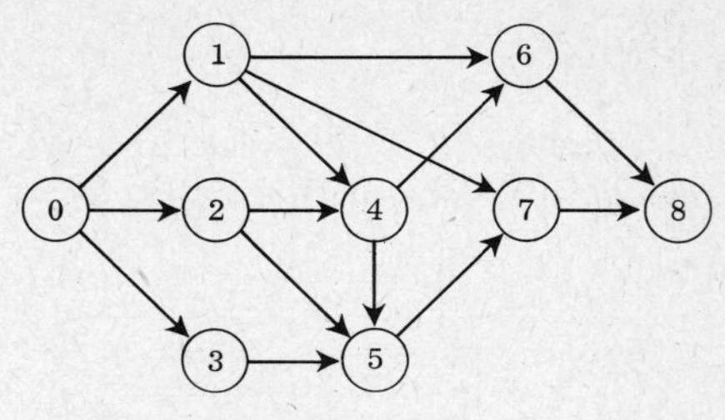

Figure P8–10

Table P8–9

Activity	Duration	Resources per day
0, 1	3	9
0, 2	5	6
0, 3	1	4
1, 4	1	10
2, 4	7	16
2, 5	6	9
3, 5	4	5
3, 6	3	8
4, 6	6	2
5, 6	4	3
6, 6	3	7

Table P8–10

Activity	Duration	Resources per day
0, 1	5	0
0, 2	1	3
0, 3	7	0
1, 4	2	9
2, 4	6	5
2, 5	4	4
3, 5	3	2
4, 5	2	14
1, 6	4	10
4, 6	3	4
1, 7	5	1
5, 7	1	2
6, 8	4	7
7, 8	5	3

Table P8–11

Activity	Predecessors	Successors	Duration (days)	Resources per day
A	*J*	*H, L*	5	1
B	—	*F, K*	5	4
C	*F*	—	2	3
D	*J*	*E*	4	2
E	*D, L*	—	2	6
F	*B*	*C, M*	6	7
G	—	*I, J*	6	5
H	*A, I, M*	—	4	3
I	*G, K*	*H, L*	7	4
J	*G, K*	*A, D*	1	5
K	*B*	*I, J*	3	4
L	*A, I, M*	*E*	3	2
M	*F*	*H, L*	4	2

P8–11. Determine the total slack and free slack in each of the activities for the project given in Table P8–11. If the project is scheduled on the basis of the earliest event times, what are the resource requirements per day for each day of the project duration?

P8–12. For the information on various activities for a project in Table 8–12, determine (a) the project-crash cost based on the earliest time schedule, and (b) the project-crash cost based on the latest time schedule.

Table P8–12

Activity	Predecessors	Successors	Normal Duration, days	Crash Duration, days	Normal Cost, $	Crash Cost, $
A	*I, M*	*K*	3	2	4	6
B	—	*I, N*	3	1	4	6
C	*F, N*	*J, M*	4	2	5	9
D	*E*	*J, M*	5	3	8	10
E	—	*D, G*	3	1	5	7
F	—	*C*	2	1	3	6
G	*E*	*H*	9	6	9	12
H	*G, L*	—	3	2	4	7
I	*B*	*A, L*	2	2	3	3
J	*C, D*	*K*	4	3	5	7
K	*A, J*	—	1	1	2	2
L	*I, M*	*H*	4	1	6	9
M	*C, D*	*A, L*	2	1	5	8
N	*B*	*C*	1	1	3	3

Table P8–13

Activity	d_{ij}	C_{ij}	R_{ij}
A	3	150	20
B	5	250	30
C	1	80	∞
D	10	400	15
E	4	220	20
F	6	300	25
G	6	260	10
H	2	120	35
I	4	200	20
J	3	180	∞
K	3	220	25
L	9	500	15
M	2	100	30
N	2	120	∞
O	5	240	10

Table P8–14

Activity	d_{ij}	C_{ij}	c_{ij}
A	2	400	460
B	4	450	510
C	1	200	250
D	3	300	350
E	3	350	430
F	5	550	640
G	2	250	300
H	1	180	250
I	2	150	150
J	6	480	520
K	1	120	150
L	4	500	560
M	3	280	320
N	2	220	260

P8–13 through P8–16. The time-cost data corresponding to each of Problems P8–1 through P8–4, respectively, are given in the table for the problem. Determine the all-crash cost and the project-crash cost based on the early time schedule for the problem. Also, suggest two different combinations of activity durations, each of which will lead to a project completion time equal to three days longer than the project crash time, and determine the costs associated with such combinations of activity durations.

Table P8–15

Activity	d_{ij}	C_{ij}	c_{ij}
A	4	70	90
B	8	150	210
C	11	200	250
D	4	60	80
E	1	40	60
F	9	120	140
G	6	100	130
H	2	50	70
I	3	70	90
J	2	60	80
K	7	120	150
L	3	70	100

Table P8–16

Activity	d_{ij}	C_{ij}	R_{ij}
A	3	50	∞
B	5	150	50
C	2	90	∞
D	2	125	40
E	5	300	30
F	3	240	20
G	5	80	15
H	6	270	30
I	6	120	∞
J	4	600	40
K	5	300	50
L	2	80	40
M	2	140	40

P8–17 through P8–20. Construct an activity-on-node network for each of Problems P8–1 through P8–4, respectively.

CHAPTER 9

DYNAMIC PROGRAMMING

9-1 CONCEPT OF MULTISTAGE DECISION PROCESS

In the optimization of a system in which the variables representing physical quantities change with time, the decision process involves a sequence of choices at various stages of time. For the sake of simplicity, let us consider a problem with only one physical quantity $x(t)$, the state of which at time $t_1, t_2, \ldots, t_n$ may be described by $x_1 = x(t_1)$, $x_2 = x(t_2), \ldots, x_n = x(t_n)$, respectively. In contrast to the optimization of a system under a static stage, which is a single-stage process, the optimization of a system under varying stages is time-dependent and can best be carried out by a multistage process in which one variable is chosen at each stage in sequence. Since the latter approach implies the notion of variation with respect to time, it is called *dynamic programming*.

In a broader sense, however, single-stage optimization problems involving a large number of variables under static stage may be artificially formulated as dynamic programming problems by considering the system in stages. For example, the choice of an optimal solution $(x_1^*, x_2^*, \ldots, x_n^*, A^*)$ in a single-stage optimization problem may be regarded as a choice of x_n^* first, x_{n-1}^* next, ..., and x_1^* last in order to obtain $A^* = \max A$. The advantage of the dynamic approach is that, at each stage, a smaller set of variables, possibly only one variable, is to be optimized. On the other hand, its disadvantage lies in the complexity of its solution *when a large number of constraints is involved.* This chapter is confined to the reinterpretation of the single-stage processes discussed earlier in the text in terms of dynamic programming. With this in mind, we shall discuss only the simplest type of multistage decision process which involves deterministic transformations at a finite number of times. This discrete deterministic process is sometimes referred to as an *n-stage process.*

In dynamic programming, the different states of a system at various stages are characterized by one or more *parameters*. At each stage, a *choice* or *decision* is made, and a *transformation* of the input state to the output state takes place as a result of the decision. After any number of decisions, the resulting state of the system and the subsequent decisions will determine the remaining stages in the process. Each sequence of decisions for all stages constitutes a *policy*. The purpose of the process is to select a policy which satisfies all constraint conditions and optimizes the objective function for the given problem. This function, which is equivalent to the criterion or merit function in linear programming, is called the *n-stage expected return*. A policy which satisfies all constraint conditions is called a *feasible policy*, and a feasible policy which optimizes the expected return is an *optimal policy*.

The mathematical theory of dynamic programming is based on the *principle of optimality* which has been postulated by Bellman as follows:† "*An optimal policy has the property that whatever the initial state and initial decision are, the remaining decisions must constitute an optimal policy with regard to the state resulting from the first decision.*" Hence a multistage process is formulated on the basis of the intrinsic structure of the optimal policy and the recurrence relation for the system. We shall illustrate these basic concepts with simple examples.

Example 9-1. Divide a straight line of length b into n divisions so that the product of these divisions will be maximum.

We formulated this problem in Chapter 2 as a simultaneous optimization problem and solved it by using Lagrange multipliers. We shall now reformulate the problem in terms of a multistage process, as indicated in Fig. 9-1. Let us define $f_n(b_n)$ as the expected return of an n-stage process of dividing b into n-divisions, and consider several related problems with the same objective of maximizing the product of divisions.

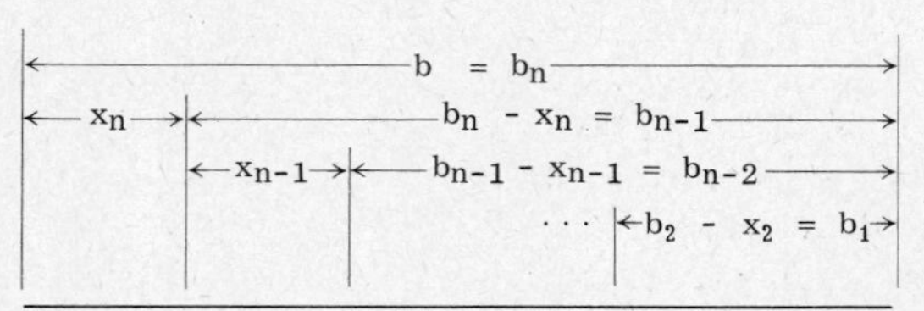

Figure 9-1

(1) Divide a line of length b_1 into one division. This is obviously a trivial problem, but it will lead to other related problems later. Hence we shall formulate it formally in the same context. Let x_1 be the first decision on the length of the division. Then, by definition, the expected return of this one-stage process is

$$f_1(b_1) = \max_{0 \leq x_1 \leq b_1} [x_1] \,. \tag{a}$$

Since the choice of x_1 can be only the full length b_1, the maximum value of x_1 is given by

$$x_1^*(b_1) = b_1 \,, \tag{b}$$

† See Reference 9-1.

which refers to the fact that the value of x_1 is a function of b_1, the given length in the one-stage process. Then, from Eq. (a), the optimal expected return is

$$f_1(b_1) = b_1 . \tag{c}$$

Since the optimal policy consists of only one decision x_1^*, it is obvious that

$$x_1^* = x_1^*(b_1) = b_1 .$$

(2) Divide a line of length b_2 into two divisions. Let x_2 be the first decision on the length of a division, and $b_2 - x_2$ be the remaining length of the line. Then, by definition, the expected return of this two-stage process is

$$f_2(b_2) = \underset{0 \leq x_2 \leq b_2}{\text{Max}} [x_2(b_2 - x_2)] . \tag{d}$$

We know how to divide a line of length $(b_2 - x_2)$ into one division optimally, once the choice of x_2 is made. In view of Eq. (c), we have

$$f_1(b_2 - x_2) = b_2 - x_2 . \tag{e}$$

Hence Eq. (d) may be rewritten in terms of the expected return of stage 1 as follows:

$$f_2(b_2) = \underset{0 \leq x_2 \leq b_2}{\text{Max}} [x_2 f_1(b_2 - x_2)] . \tag{f}$$

Referring to Eq. (d), we see that the maximization may be carried out directly by differentiation:

$$\frac{d}{dx_2} [x_2(b_2 - x_2)] = b_2 - 2x_2 = 0 .$$

The value of x_2 which maximizes the function is given by

$$x_2^*(b_2) = \frac{b_2}{2} , \tag{g}$$

which refers to the fact that the optimal choice of x_2^* is $b_2/2$ when the given length in the two-stage process is b_2. The maximization is verified by checking the second derivative of the function and the boundary points for the interval $0 \leq x_2 \leq b_2$. Hence

$$f_2(b_2) = \left(\frac{b_2}{2}\right)^2 . \tag{h}$$

The optimal policy consisting of (x_2^*, x_1^*) may be obtained from Eq. (g) for the problem and Eq. (b) for the previous problem. Thus

$$x_2^* = x_2^*(b_2) = \frac{b_2}{2} ,$$

$$x_1^* = x_1^*(b_2 - x_2^*) = b_2 - \frac{b_2}{2} = \frac{b_2}{2} .$$

(3) Divide a line of length b_3 into three divisions. Let x_3 be the first decision on the length of a division, and $b_3 - x_3$ be the remaining length of the line. Then, by induction, the expected return of this three-stage process is found to be

$$f_3(b_3) = \operatorname*{Max}_{0 \leq x_3 \leq b_3} [x_3 f_2(b_3 - x_3)] . \tag{i}$$

We can appreciate the notation $f_2(b_3 - x_3)$ at this point because, unlike the two-division problem in which the notation $f_1(b_2 - x_2)$ is almost trivial, this problem is simplified by seeking the maximization of x_3 and the expected return of stage 2, which treats the remaining length $b_3 - x_3$ as a two-division problem. On the basis of Eq. (h), we conclude that

$$f_2(b_3 - x_3) = \left(\frac{b_3 - x_3}{2}\right)^2 . \tag{j}$$

Hence from Eq. (i)

$$f_3(b_3) = \operatorname*{Max}_{0 \leq x_3 \leq b_3} \left[x_3 \left(\frac{b_3 - x_3}{2}\right)^2 \right] . \tag{k}$$

Again, the maximization is obtained by differentiating the expression in Eq. (k) and setting the derivative equal to zero. Thus

$$(b_3 - 3x_3)(b_3 - x_3) = 0 .$$

Since $x_3 = b_3$ is inadmissible for a three-division problem, we get

$$x_3^*(b_3) = \frac{b_3}{3} , \tag{l}$$

and

$$f_3(b_3) = \left(\frac{b_3}{3}\right)^3 . \tag{m}$$

The optimal policy consisting of (x_1^*, x_2^*, x_3^*) is obtained in sequence from Eqs. (l), (g) and (b), respectively:

$$x_3^* = x_3^*(b_3) = \frac{b_3}{3} ,$$

$$x_2^* = x_2^*(b_3 - x_3^*) = \frac{1}{2}\left(b_3 - \frac{b_3}{3}\right) = \frac{b_3}{3},$$

$$x_1^* = x_1^*(b_3 - x_3^* - x_2^*) = \frac{b_3}{3} ,$$

(4) Divide a line of length b_n into n divisions. Let x_n be the first decision on the length of a division, and $b_n - x_n$ be the remaining

length of the line. Then, by induction, the expected return of this n-stage process is given by

$$f_n(b_n) = \underset{0 \le x_n \le b_n}{\text{Max}} \left[x_n f_{n-1}(b_n - x_n)\right] . \tag{n}$$

By mathematical induction, we can also demonstrate that

$$f_{n-1}(b_n - x_n) = \left(\frac{b_n - x_n}{n - 1}\right)^{n-1} . \tag{o}$$

Then by substituting Eq. (o) in Eq. (n) and by carrying out maximization through differentiation we obtain, after setting the derivative equal to zero, the following relation:

$$(b_n - nx_n)(b_n - x_n)^{n-2} = 0 .$$

Since $x_n = b_n$ is inadmissible for an n-division problem, we get

$$x_n^*(b_n) = \frac{b_n}{n} , \tag{p}$$

and

$$f_n(b_n) = \left(\frac{b_n}{n}\right)^n . \tag{q}$$

The optimal policy $(x_n^*, x_{n-1}^*, \ldots, x_1^*)$ may be obtained by backward mathematical induction, starting with Eq. (p); it is found to be $(b_n/n, b_n/n, \ldots, b_n/n)$.

We have indirectly solved the original problem of dividing a line of length b into n divisions so that the product of the divisions will be a maximum. We shall now recapitulate the solution in terms of a multistage process. Given the problem

$$\text{Max } A = \underset{x_k}{\text{Max}}\, x_1 x_2 \cdots x_n , \tag{9–1}$$

subject to

$$x_k \ge 0 , \qquad k = 1, 2, \ldots, n ,$$

$$x_1 + x_2 + \cdots + x_n = b .$$

We can make an initial decision of choosing x_n from the initial state of the line of length b. In order to obtain an optimal policy, the remaining decisions must constitute an optimal policy with regard to the remaining state $(b - x_n)$. Hence the relation in (9–1) may be reformulated as follows:

$$\text{Max } A = \underset{x_n}{\text{Max}} \left[x_n \underset{x_k}{\text{Max}}\, (x_1 x_2 \cdots x_{n-1}) \right] , \tag{9–2}$$

subject to

$$x_k \ge 0 , \qquad k = 1, 2, \ldots, (n - 1) ,$$

$$x_1 + x_2 + \cdots + x_{n-1} = b - x_n .$$

Using the notation $b_n = b$, $b_{n-1} = b_n - x_n, \ldots, b_1 = b_2 - x_2$ for various states of the process, as indicated in Fig. 9-1, and the notation $f_n(b_n)$, $f_{n-1}(b_{n-1}), \ldots, f_1(b_1)$ for the expected returns for $n, (n-1), \ldots,$ one-stage, respectively, we have from (9-2)

$$f_n(b_n) = \operatorname*{Max}_{x_n} [x_n f_{n-1}(b_{n-1})] . \tag{9-3}$$

This is the recurrence relation for the given multiplicative process in (9-1), since the remaining states in the process can be obtained by the same relation upon subsequent decisions. Thus

$$\begin{aligned} f_{n-1}(b_{n-1}) &= \operatorname*{Max}_{x_{n-1}} [x_{n-1} f_{n-2}(b_{n-2})] , \\ &\vdots \\ f_2(b_2) &= \operatorname*{Max}_{x_2} [x_2 f_1(b_1)] , \\ f_1(b_1) &= \operatorname*{Max}_{x_1} [x_1] . \end{aligned}$$

By substituting $b_{n-1} = b_n - x_n$ into (9-3), the recurrence relation may also be expressed in the following form:

$$f_n(b_n) = \operatorname*{Max}_{x_n} [x_n f_{n-1}(b_n - x_n)] . \tag{9-4}$$

Thus the solution of the n-stage process in (9-1) is obtained by considering first a one-stage process, then a two-stage process, etc. until the expected return for the n-stage process is optimized.

It is interesting to note that, while the decisions for decomposing the problem are to be made in the decreasing order of $x_n, x_{n-1}, \ldots, x_1$, the problem is solved in the increasing order of $x_1, x_2, \ldots, x_n$. Therefore stage n refers to the beginning of the time sequence when there are n-stages left in the process, and stage 1 is the end of the time sequence when we move backward to solve the problem.

The optimal solution of this problem is therefore given by

$$[(x_n^*, x_{n-1}^*, \ldots, x_2^*, x_1^*), f_n(x_n)] ,$$

which represents the optimal policy and the optimal expected return of the n-stage process.

Example 9-2. Repeat the previous example by considering it as an additive process as follows:

$$\operatorname{Max} A = \operatorname*{Max}_{x_k} [\ln x_1 + \ln x_2 + \cdots + \ln x_n] , \tag{9-5}$$

subject to

$$x_k \geq 0 , \qquad k = 1, 2, \ldots, n ,$$
$$x_1 + x_2 + \cdots + x_n = b .$$

Let $R_k(x_k) = \ln x_k$, $(k = 1, 2, \ldots, n)$. Then (9-5) can be restated as

$$\text{Max } A = \underset{x_k}{\text{Max}} \sum_{k=1}^{n} R_k(x_k) . \tag{9-6}$$

The recurrence relation of the additive process in (9-6) is given by

$$f_n(b_n) = \underset{x_n}{\text{Max}} \, [R_n(x_n) + f_{n-1}(b_{n-1})] , \tag{9-7}$$

or

$$f_n(b_n) = \underset{x_k}{\text{Max}} \, [R_n(x_n) + f_{n-1}(b_n - x_n)] , \tag{9-8}$$

since the transformation from the state at stage n to the state at stage $(n - 1)$ is given by the relation

$$b_{n-1} = b_n - x_n . \tag{9-9}$$

Starting with stage 1, we have

$$f_1(b_1) = \underset{x_1}{\text{Max}} \, [R_1(x_1)] = \underset{0 \leq x_1 \leq b_1}{\text{Max}} \, [\ln x_1] .$$

Note that the boundary point $x_1 = 0$ is not significant for maximization because $\ln x_1 \to -\infty$ as $x_1 \to 0$, but the boundary point $x_1 = b_1$ is important since the maximum value lies at that point. Hence

$$x_1^*(b_1) = b_1 ,$$

and

$$f_1(b_1) = \ln b_1 .$$

At stage 2,

$$\begin{aligned} f_2(b_2) &= \underset{x_2}{\text{Max}} \, [R_2(x_2) + f_1(b_2 - x_2)] \\ &= \underset{0 \leq x_2 \leq b_2}{\text{Max}} \, [\ln x_2 + \ln (b_2 - x_2)] . \end{aligned}$$

The maximization of the interior of the interval $0 < x_2 < b_2$ is obtained by differentiation. Thus

$$\frac{d}{dx_2} [\ln x_2 + \ln (b_2 - x_2)] = \frac{1}{x_2} - \frac{1}{b_2 - x_2} = 0 ,$$

from which

$$\frac{b_2 - 2x_2}{x_2(b_2 - x_2)} = 0 , \qquad x_2^*(b_2) = \frac{b_2}{2} .$$

Note that the boundary points of the interval $0 < x_2 < b_2$ are not significant for maximization because $\ln x_2 \rightarrow -\infty$ as $x_2 \rightarrow 0$ or $x_2 \rightarrow b_2$. Hence

$$f_2(b_2) = 2 \log \frac{b_2}{2} = \log \left(\frac{b_2}{2}\right)^2.$$

Using mathematical induction, we can surmise that the optimal solution of the problem is given by

$$\left[\left(\frac{b}{n}, \frac{b}{n}, \ldots, \frac{b}{n}\right), \log \left(\frac{b_n}{n}\right)^n\right].$$

Thus the optimal policy is the same as in the previous problem, but the optimal expected return is different, as it should be.

9-2 GENERAL APPROACH TO RECURSIVE OPTIMIZATION

Dynamic programming may be used to the greatest advantage as a general procedure for recursive optimization of complex problems. Basically, any problem to be solved by dynamic programming is decomposed into n subproblems which are to be optimized recursively and are thus put together again. If a problem can be solved conveniently by the single-stage operation, then there is no need to decompose it for solution by the multistage process. However, for certain types of problems, the amount of computation in a single-stage process increases exponentially with the increase of variables, while that in a multistage process for solving the corresponding problem increases linearly with the increase of stages.

The stagewise or sequential process permits the optimization of decision variables one at a time for as many times as needed. Hence the recursive optimization procedure may be regarded as a mathematical artifact for accomplishing such a purpose even though the problem does not present itself as a sequential process. However, the optimization of a given problem in dynamic programming depends on the successful recursive optimization of its subproblems in a sequential order. Two classes of optimization problems most frequently encountered at each stage are: (1) optimization of a continuous function in a closed region, and (2) optimization of a finite number of allowable decisions. Although most examples for illustration in this chapter belong to the first category, they should not obscure the fact that many practical applications of dynamic programming fall into the second category.

The general approach to recursive optimization may be illustrated by considering the critical-path scheduling problem in Sections 8-4 and 8-5 (Chapter 8) as a multistage process. For example, consider an activity-on-branch network in which the events are denoted by nodes $i =$

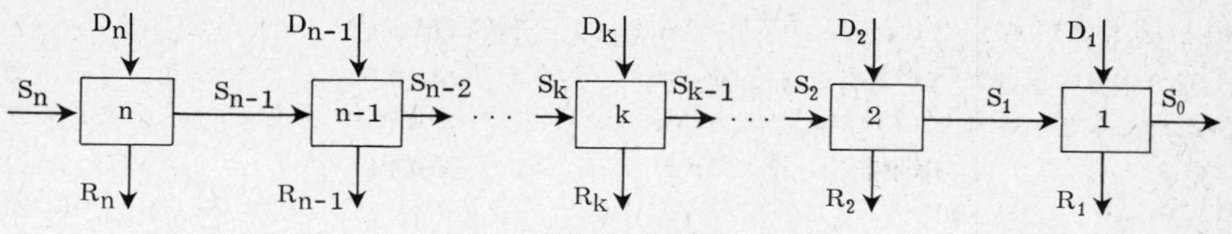

Figure 9–2

0, 1, 2, . . . , n. The event times x_i have a serial structure in that they occur in the ascending order of i. In other words, the occurrence of x_k for any node $i = k$ will influence the event times x_i for $i > k$, but it has no effect on the event times x_i for $i < k$ which have already taken place. To find the earliest event time x_i' by means of the forward pass, we in effect search for an optimal time path sequentially in the ascending order of i. At each event i, we determine the optimal time x_i' from event 0 to event i. On the other hand, to find the latest event times x_i'' by means of the backward pass, we search for an optimal time path sequentially in the descending order of i. At each event i, we determine the optimal time x_i'' from event n to event i, although the time scale of x_i'' is still based on $x_0 = 0$ as the reference point. Hence the minimum completion time of the project may be determined either by the forward pass or the backward pass if the intermediate event times are of no concern to us. In general, we can use either the *forward multistage process* or the *backward multistage process* in dynamic programming. The former starts from the initial state and proceeds through forward stepwise transformations to the final state, while the latter starts from the final state and backtracks through backward stepwise transformations to the initial state. Although the forward process seems to be more natural, the backward process leads more directly to a solution in many cases. Hence only the backward multistage process is used in this chapter.

To make the backward process to appear more natural, we denote the initial state of an n-stage process by S_n and the subsequent states by S_k ($k = 0, 1, 2, \ldots, n - 1$) in the descending order of k, ending with the final state S_0 in which the stage $k = 0$ is an artifact to be explained later. At each stage k, a decision D_k is made and a transformation T_k takes place such that the input state k is transformed to the output state $k - 1$ with a return R_k. The backward multistage process may be represented schematically by the flow diagram in Fig. 9–2, in which the initial state is the input state of stage n, and the output state from stage 1 is the final state. Hence in the decomposition of the problem, we follow in descending order of k starting from $k = n$, but in the solution of the problem, we proceed in the ascending order of k starting from $k = 0$.

9–3 FORMULATION OF MULTISTAGE PROBLEMS

The formulation of an n-stage recursive optimization problem from a single-stage optimization problem of n variables is the crucial step in dynamic programming. Whenever possible, we try to decompose a problem of n decision variables into n equivalent subproblems, each containing only one decision variable. In addition, the constraints containing these decision variables prescribe the variations of a decision variable in each stage. As shown in Example 9–1, the problem defined by (9–1) is decomposed into n-stages, each containing only one decision variable; similarly, in Example 9–2, the problem defined by (9–5) is also decomposed into n-stages. The former is called a *multiplicative process* inasmuch as the return at any stage represented by the recurrence relation in (9–3) is multiplicative to that of the previous stage; while the latter is called an *additive process*, since the return at any stage indicated by the recurrence relation in (9–7) is additive to that of the previous stage. Note also that aside from the conditions of nonnegativity, there is only one constraint in each case.

Actually, the types of problems of n decision variables that can be decomposed into n subproblems are quite restrictive. Without furnishing the proof, *it is sufficient to state that problems with additive stage returns can always be decomposed, but that problems with mutiplicative stage returns can be decomposed only if the decision variables are limited to the range of nonnegative real values*. However, the decomposition restriction on the multiplicative process is not as severe as it appears, since most decision variables in real problems can have only nonnegative real values. Other than problems with additive and multiplicative stage returns, each process must be examined to determine if it is mathematically decomposable.

The states S_k ($k = 0, 1, \ldots, n$) in an n-stage process resulting from the constraints of a single-stage problem may be defined by a set of parameters. If each constraint contains only one decision variable, as in the case of a nonnegativity condition, it restricts only the feasible region for optimization at the stage involving that variable. On the other hand, if each constraint contains more than one decision variable, it increases the *dimensionality of the space* or the *number of parameters in each state*. In general, a parameter is needed at each stage for each constraint containing more than one decision variable in the single-stage problem. Such a parameter is called a *state variable*. In problems with a single constraint containing more than one decision variable, such as Examples 9–1 and 9–2, the state S_k at stage k is defined by a single parameter, such as b_k ($k = 1, 2, \ldots, n$) in either example. In problems with m constraints, the state S_k at each stage k must be defined by its m parameters, say S_{ik} $(i = 1, 2, \ldots, m)$. For linear constraints in single-stage optimization, the decision variables are additive; hence linear constraints are referred to as additive constraints. Other constraints which occur frequently are multiplicative constraints, in which the decision variables are multipliers. The additional state variables increase the computational complexity in dynamic programming. Hence for practical applications recursive optimization is most effective when a problem of n decision variables can be decomposed to an n-stage process, with each stage containing a single decision variable and a state variable. Both Examples 9–1 and 9–2 belong to this category.

Thus we can proceed to formulate a multistage problem after dissecting the procedures of decomposing the objective function and of treating the constraints in the original one-stage problem. Since the single-stage optimization problems treated in the earlier chapters in this text can be reinterpreted as additive multistage processes in terms of dynamic programming, we shall confine our discussion of dynamic programming only to the formulation and solution of n-stage additive processes. Also, because we deal primarily with linear constraints in the single-stage optimization, we shall limit our consideration to additive constraints in the formulation of multistage problems. The decomposition of the objective function for additive processes and the treatment of positive constraints will be discussed in the next two sections.

9–4 DECOMPOSITION OF AN ADDITIVE PROCESS

We shall now examine the general structure of an additive process with particular reference to the state variables, the decision variables, the transformation functions, the return functions, and the optimal solution.

In a backward n-stage process, the initial state is the input state S_n of stage n, which represents a single-state variable at the stage if there is only one constraint, or the m parameters of the input state (m state variables) if there are m constraints in the single-stage problem. For the sake of simplicity, we shall consider the process as one with a single state variable S_n in this section. The complications caused by additional constraints (and hence additional state variables in each stage) will be discussed later.

Let D_n be the *decision variable* in stage n, which transforms the state variable S_n to that of the next state S_{n-1} through the input-output *transformation function* T_n, such that

$$S_{n-1} = T_n(S_n, D_n) \, .$$

This transformation yields a *return function* $R_n(S_n, D_n)$, representing the contribution of the stage n to the *expected n-stage return* $f_n(S_n)$. In general, at stage k ($k = n, n-1, \ldots, 2, 1$), the transformation function T_k is given by

$$S_{k-1} = T_k(S_k, D_k)\,.$$

The return function at stage k is $R_k(S_k, D_k)$, and the expected k-stage return is $f_k(S_k)$. This process can be repeated for stage $(n-1)$, and then for stage $(n-2)$, until stage 1 is reached, when there is one more decision left in the process. Symbolically, stage 0 represents the time when the n-stage process is completed; physically, stage 0 contributes nothing to the process and the expected return for that stage is zero, i.e., $f_0(S_0) = 0$.

By the definition of an additive process, the *total return* for the n-stage optimization is given by

$$A = \sum_{k=1}^{n} R_k(S_k, D_k)\,. \tag{9-10}$$

Then for the maximization of additive returns it is seen that the maximum total return is equal to the expected n-stage return, i.e.,

$$\text{Max } A = \underset{D_1 \cdots D_n}{\text{Max}} \sum_{k=1}^{n} R_k(S_k, D_k) = f_n(S_n)\,,$$

where $f_n(S_n)$ is obtained for the additive process according to the principle of optimality as follows:

$$\begin{aligned}
f_1(S_1) &= \underset{D_1}{\text{Max}}\,[R_1(S_1, D_1) + f_0(S_0)]\,,\\
f_2(S_2) &= \underset{D_2}{\text{Max}}\,[R_2(S_2, D_2) + f_1(S_1)]\,,\\
&\vdots\\
f_n(S_n) &= \underset{D_n}{\text{Max}}\,[R_n(S_n, D_n) + f_{n-1}(S_{n-1})]\,.
\end{aligned}$$

Since this relation between expected return at a stage $f_k(S_k)$ and the return function $R_k(S_k, D_k)$ at that stage recurs at every stage ($k = 1, 2, \ldots, n$), it is called the *recurrence relation* or *recursive equation*. Hence the maximization of an n-stage additive process can be stated in terms of the recurrence relation as follows:

$$f_k(S_k) = \underset{D_k}{\text{Max}}\,[R_k(S_k, D_k) + f_{k-1}(S_{k-1})]\,, \tag{9-11}$$

subject to

$$S_{k-1} = T_k(S_k, D_k)\,, \qquad k = 1, 2, \ldots, n\,, \tag{9-12}$$

where $f_0(S_0) = 0$, and D_k is applied to a specified interval satisfying the constraints. By substituting (9–12) into (9–11), the recursive equation can also be expressed in the form

$$f_k(S_k) = \operatorname*{Max}_{D_k} \{R_k(S_k, D_k) + f_{k-1}[T_k(S_k, D_k)]\} . \tag{9–13}$$

A sequence of feasible decisions $(D_n, D_{n-1}, \ldots, D_1)$ satisfying the bounds or constraints constitutes an n-stage policy. The policy $(D_n^*, D_{n-1}^*, \ldots, D_1^*)$ which leads to an optimal expected n-stage return $f_n(S_n)$ is the optimal policy. Hence the optimal solution of the problem is completely described by

$$[D_n^*, D_{n-1}^*, \ldots, D_1^*, f_n(S_n)] .$$

Conceptually, the solution of this problem will be obtained through a backward induction procedure starting from stage 1 and continuing to stage n. We wish to emphasize again that the formulation of the problem starts with stage n, and repeats itself through the recursive equation. However, the solution starts with a one-stage problem and finds its way backward to stage n. The degree of difficulty of the actual solution depends on the maximization relation of the single-stage solutions during backward induction.

It may be observed that, although the optimal expected return is unique if it is bounded, the policy leading to the optimal expected return need not be unique. In other words, there may be a number of optimal policies which yield the same optimal expected return. Furthermore, a minimization problem, as well as a maximization problem, can be reduced to a multistage problem by the use of a recursive equation in minimization.

As an illustration of the additive process, let us return to Example 9–2, in which x_k and b_k are, respectively, the decision variable and the state variable in stage k $(k = 1, 2, \ldots, n)$. Hence from the constraint

$$x_1 + x_2 + \cdots + x_n = b ,$$

we note that

$$\begin{aligned} b_n &= b , \\ b_{n-1} &= b_n - x_n , \qquad 0 \le x_n \le b_n , \\ &\vdots \\ b_1 &= b_2 - x_2 , \qquad 0 \le x_2 \le b_2 , \\ b_0 &= b_1 - x_1 , \qquad 0 \le x_1 = b_1 . \end{aligned}$$

Table 9-1

COMPARISON OF GENERAL PROBLEM AND EXAMPLE 9-2

Description	General problem	Example 9-2
State variable at stage *n*	S_n	$b_n = b$
Decision variable at stage *n*	D_n	x_n
Transformation function at *n*	$S_{n-1} = T_n(S_n, D_n)$	$b_{n-1} = b_n - x_n$
Return function at *n*	$R_n(S_n, D_n)$	$R_n(x_n) = \ln x_n$
Expected total *n*-stage return	$f_n(S_n)$	$f_n(x_n)$

By adding the above equations relating the state variable and the decision variable in each stage, it is seen that $b_0 = 0$. Hence, $x_1 = b_1$. Also, the bounds of the decision variables x_k ($k = 1, 2, \ldots, n$) are given directly by the corresponding relations and the nonnegative constraints. Hence the recurrence relation is given by

$$f_k(b_k) = \operatorname*{Max}_{x_k} [R_k(x_k) + f_{k-1}(b_{k-1})] ,$$

subject to the transformation

$$b_{k-1} = b_k - x_k .$$

A complete analogy between the general problem and Example 9-2 is shown in Table 9-1.

9-5 ADDITIVE CONSTRAINTS

When more than one constraint is given for the single-stage optimization problem, a state variable is usually required for each constraint in the corresponding multistage problem, unless the dimensionality of space can be reduced because some constraints do not contain all decision variables. If a constraint contains only some decision variables, it restricts only the feasible region for optimization at stages corresponding to those variables. When several constraints contain different sets of decision variables, the feasible regions for optimization at various stages represented by these sets may not overlap. In the case of nonnegative constraints, for example, each nonnegativity condition restricts only the feasible region for optimization at one stage corresponding to a decision variable, and does not affect other stages corresponding to other decision variables. However, except for nonnegative constraints, the increase in the number of constraints generally increases the dimensionality of the space, i.e., the number of state variables. We shall examine briefly the effects of two or more additive constraints on the decomposition of the additive process.

For example, consider a single-stage maximization problem with m additive constraints:

$$\text{Max } A = \sum_{k=1}^{n} R_k(x_k) , \tag{9–14}$$

subject to

$$\sum_{k=1}^{n} g_{ik}(x_k) \leq b_i , \qquad i = 1, 2, \ldots, m ;$$

$$x_k \geq 0 , \qquad k = 1, 2, \ldots, n ;$$

in which x_k ($k = 1, 2, \ldots, n$) are decision variables. Each additive constraint, say, the ith constraint, consists of x_k terms in the form of functions $g_{ik}(x_k)$ and a constant term b_i. For example, if $g_{ik}(x_k) = a_{ik}x_k$, where a_{ik} ($k = 1, 2, \ldots, n$) are constants, then the constraints become linear inequalities:

$$\sum_{k=1}^{n} a_{ik}x_k \leq b_i , \qquad i = 1, 2, \ldots, m .$$

Let us introduce a state variable b_{ik} at stage k for the ith constraint, which has a constant b_i. Hence the transformation at any stage k involves m input state variables b_{ik} and m output state variables $b_{i,k-1}$ as follows:

$$b_{i,k-1} = T_{ik}(b_{ik}, x_k) , \qquad i = 1, 2, \ldots, m .$$

Expressing the transformation at stage k by the relations between state variables and decision variables, we have

$$b_{i,k-1} = x_{ik} - g_{ik}(x_k) , \qquad i = 1, 2, \ldots, m . \tag{9–15}$$

Then the recurrence relation for the expected n-stage return is given by

$$\begin{aligned} &f_k(x_{1k}, \ldots, x_{mk}) \\ &\quad = \underset{x_k}{\text{Max}} \, \{R_k(x_k) + f_{k-1}[b_{1k} - g_{1k}(x_k), \ldots, b_{mk} - g_{mk}(x_k)]\} . \end{aligned} \tag{9–16}$$

It may be noted that the difference in the number of calculations between the problems with one state variable and those with two or more state variables is very great. Also, the amount of computation increases exponentially with the number of state variables. Hence special methods for simplifying the computational procedure must be introduced if dynamic programming is used to solve problems having more than one state variable in each stage. We shall not consider such special methods in this elementary text; hence most examples in our subsequent discussion will be confined to problems with only one state variable at each stage. However, one example of problems with two state variables at each stage will be given (Example 9–5) to illustrate computational difficulty in the solution of problems with more than one state variable.

9-6 LINEAR PROGRAMMING PROBLEMS

Consider the general linear programming problem in inequality form as follows (assume that $a_{ij} > 0$):

$$x_j \geq 0\,, \qquad j = 1, 2, \ldots, n\,;$$

$$\sum_{j=1}^{n} a_{ij}x_j \leq b_i\,, \qquad i = 1, 2, \ldots, m\,;$$

$$\sum_{j=1}^{n} c_j x_j = \text{Max}\, A\,.$$

The formulation of this problem in dynamic programming may be regarded as the choice of variables $x_n, x_{n-1}, \ldots, x_1$ in sequence for an additive process. The transformations at various stages $k = n, n-1, \ldots, 1$ are respectively given by

$$\begin{aligned} b_{in} &\leq b_i\,, \\ b_{i,n-1} &= b_{in} - a_{in}x_n\,, \\ &\vdots \\ b_{i1} &= b_{i2} - a_{i2}x_2\,, \\ 0 &= b_{i1} - a_{i1}x_1\,, \end{aligned}$$

where $i = 1, 2, \ldots, m$ for all states of the system $b_{in}, b_{i,n-1}, \ldots, b_{i2}, b_{i1}$. The bounds of the decisions $x_n, x_{n-1}, \ldots, x_1$ are given by

$$\theta_k = \frac{b_{ik}}{a_{ik}}\,, \qquad k = n, n-1, \ldots, 1\,.$$

Since $0 \leq x_1 = \theta_1$, we have

$$0 \leq x_k \leq \theta_k\,, \qquad k = n, n-1, \ldots, 2\,.$$

The recurrence relation is obtained by the use of principle of optimality as follows:

$$f_k[b_{1k}, b_{2k}, \ldots, b_{mk}] = \underset{0 \leq x_k \leq \theta_k}{\text{Max}} [c_k x_k + f_{k-1}(b_{1,k-1}, b_{2,k-1}, \ldots, b_{m,k-1})]\,,$$

with $f_0(b_{11}, b_{21}, \ldots, b_{m1}) = 0$. Then the expected return is given by

$$\text{Max} \sum_{j=1}^{n} c_j x_j = f_n[b_{1n}, b_{2n}, \ldots, b_{mn}]\,,$$

and the optimal policy is given by $(x_n^*, x_{n-1}^*, \ldots, x_1^*)$.

In general, the actual solution of the general problem is not at all simple. We shall therefore consider only specific cases which offer simple interpretations of linear programming problems as multistage processes.

Example 9-3. Formulate the following problem (for $a_j > 0$, $c_j > 0$):

$$x_1 \geq 0\,, \qquad x_2 \geq 0\,,$$
$$a_1x_1 + a_2x_2 \leq b\,,$$
$$c_1x_1 + c_2x_2 = \text{Max}\, A\,;$$

and specifically

$$x_1 + 2x_2 \leq 8\,,$$
$$x_1 + 3x_2 = \text{Max}\, A\,.$$

Since there is only one constraint condition in which b has no subscript, it is more convenient to use a single subscript notation for b in describing the stage number. Thus for the two-stage process, let $b_2 \leq b$, and

$$b_1 = b_2 - a_2x_2\,, \qquad \theta_2 = b_2/a_2\,, \qquad 0 \leq x_2 \leq \theta_2\,;$$
$$0 = b_1 - a_1x_1\,, \qquad \theta_1 = b_1/a_1\,, \qquad 0 \leq x_1 = \theta_1\,.$$

Hence the recurrence relation can be expressed by:

$$f_2(b_2) = \underset{0 \leq x_2 \leq \theta_2}{\text{Max}}\,[c_2x_2 + f_1(b_2 - a_2x_2)]\,,$$
$$f_1(b_1) = \underset{0 \leq x_1 = \theta_1}{\text{Max}}\,[c_1x_1]\,.$$

The solution begins with stage 1 in which x_1 is seen to take the maximum value

$$x_1^*(b_1) = \theta_1 = \frac{b_1}{a_1}\,,$$

and

$$f_1(b_1) = \frac{c_1b_1}{a_1}\,.$$

In stage 2,

$$f_1(b_2 - a_2x_2) = \frac{c_1}{a_1}(b_2 - a_2x_2)\,.$$

Then

$$f_2(b_2) = \underset{0 \leq x_2 \leq \theta_2}{\text{Max}}\left[c_2x_2 + \frac{c_1}{a_1}(b_2 - a_2x_2)\right]$$
$$= \underset{0 \leq x_2 \leq \theta_2}{\text{Max}}\left[\left(c_2 - \frac{c_1a_2}{a_1}\right)x_2 + \frac{c_1b_2}{a_1}\right].$$

For $c_2 > c_1a_2/a_1$,

$$x_2^*(b_2) = \theta_2 = \frac{b_2}{a_2}, \qquad f_2(b_2) = \frac{c_2b_2}{a_2},$$

$$x_2^* = x_2^*(b_2) = \frac{b_2}{a_2}, \qquad x_1^* = x_1^*(b_1) = \frac{b_2 - a_2x_2^*}{a_1} = 0.$$

For $c_2 < c_1a_2/a_1$,

$$x_2^*(b_2) = 0, \qquad f_2(b_2) = \frac{c_1b_2}{a_1},$$

$$x_2^* = x_2^*(b_2) = 0, \qquad x_1^* = x_1^*(b_1) = \frac{b_2 - a_2x_2^*}{a_1} = \frac{b_2}{a_1}.$$

Hence for $b_2 \leq b$, the solution is

$$[(0, b/a_2), c_2b/a_2] \quad \text{for} \quad c_2 - c_1a_2/a_1 > 0,$$
$$[(b/a_1, 0), c_1b/a_1] \quad \text{for} \quad c_2 - c_1a_2/a_1 < 0.$$

The graphical representation of this linear programming problem is shown in Fig. 9-3. For $c_2 = c_1a_2/a_1$, the objective function and the equation for the constraint coincide with each other, and any point (x_1, x_2) on the line joining $(b/a_1, 0)$ and $(0, b/a_2)$ leads to the same optimal solution for A. In dynamic programming, this is equivalent to having an infinite number of policies leading to the same optimal expected return. Note also that, $\theta_1 = (b_2 - a_2x_2)/a_1$ can be determined after x_2 is found.

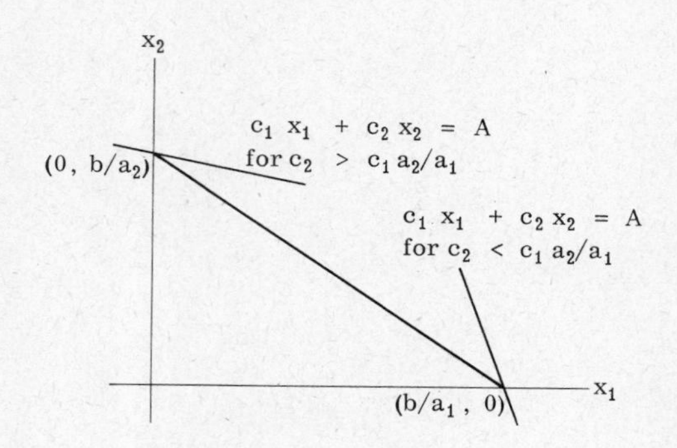

Figure 9-3

For the given numerical example, $c_2 - c_1a_2/a_1 = 3 - (1)(2)/(1) = 1(>0)$. Hence

$$x_2^* = b/a_2 = 8/2 = 4, \qquad x_1^* = 0,$$

$$\max A = c_2b/a_2 = (3)(8)/2 = 12.$$

Example 9–4. Formulate the problem (for $a_j > 0$, $c_j > 0$):

$$x_j \geq 0\,, \qquad j = 1, 2, 3\,,$$
$$a_1x_1 + a_2x_2 + a_3x_3 \leq b\,,$$
$$c_1x_1 + c_2x_2 + c_3x_3 = \text{Max}\, A\,.$$

Solve specifically for the following numerical cases:

a) $5x_1 + 3x_2 + x_3 \leq 3\,,$
$3x_1 + 2x_2 + x_3 = \text{Max}\, A\,;$

b) $4x_1 + 2x_2 + 5x_3 \leq 3\,,$
$3x_1 + 2x_2 + x_3 = \text{Max}\, A\,;$

c) $4x_1 + 4x_2 + x_3 \leq 3\,,$
$3x_1 + 2x_2 + x_3 = \text{Max}\, A\,;$

d) $2x_1 + 2x_2 + 3x_3 \leq 3\,,$
$3x_1 + 2x_2 + x_3 = \text{Max}\, A\,.$

This problem is formulated as a three-stage process. Let $b_3 \leq b$, and

$$b_2 = b_3 - a_3x_3\,, \qquad \theta_3 = b_3/a_3\,, \qquad 0 \leq x_3 \leq \theta_3\,;$$
$$b_1 = b_2 - a_2x_2\,, \qquad \theta_2 = b_2/a_2\,, \qquad 0 \leq x_2 \leq \theta_2\,;$$
$$0 = b_1 - a_1x_1\,, \qquad \theta_1 = b_1/a_1\,, \qquad 0 \leq x_1 = \theta_1\,.$$

Hence the recurrence relation becomes

$$f_3(b_3) = \underset{0 \leq x_3 \leq \theta_3}{\text{Max}}\, [c_3x_3 + f_2(b_3 - a_3x_3)]\,,$$
$$f_2(b_2) = \underset{0 \leq x_2 \leq \theta_2}{\text{Max}}\, [c_2x_2 + f_1(b_2 - a_2x_2)]\,,$$
$$f_1(b_1) = \underset{0 \leq x_1 \leq \theta_1}{\text{Max}}\, [c_1x_1]\,.$$

From stage 1, the maximization can be carried out by inspection. Thus

$$x_1^*(b_1) = \theta_1 = \frac{b_1}{a_1}\,, \qquad f_1(b_1) = \frac{c_1b_1}{a_1}\,.$$

In stage 2,

$$f_1(b_2 - a_2x_2) = \frac{c_1}{a_1}(b_2 - a_2x_2)\,.$$

Then

$$f_2(b_2) = \underset{0 \leq x_2 \leq \theta_2}{\text{Max}} \left[c_2x_2 + \frac{c_1}{a_1}(b_2 - a_2x_2)\right] = \underset{0 \leq x_2 \leq \theta_2}{\text{Max}} \left[\left(c_2 - \frac{c_1a_2}{a_1}\right)x_2 + \frac{c_1b_2}{a_1}\right].$$

For $c_2 > c_1a_2/a_1$

$$x_2^*(b_2) = \theta_2 = \frac{b_2}{a_2}, \qquad f_2(b_2) = \frac{c_2b_2}{a_2}.$$

For $c_2 < c_1a_2/a_1$,

$$x_2^*(b_2) = 0, \qquad f_2(b_2) = \frac{c_1b_2}{a_1}.$$

In stage 3, if $c_2 > c_1a_2/a_1$, then

$$f_2(b_3 - a_3x_3) = \frac{c_2}{a_2}(b_3 - a_3x_3),$$

$$f_3(b_3) = \operatorname*{Max}_{0\le x_3\le\theta_3}\left[c_3x_3 + \frac{c_2}{a_2}(b_3 - a_3x_3)\right]$$

$$= \operatorname*{Max}_{0\le x_3\le\theta_3}\left[\left(c_3 - \frac{c_2}{a_2}a_3\right)x_3 + \frac{c_2}{a_2}b_3\right].$$

For $c_3 > c_2a_3/a_2$,

$$x_3^*(b_3) = \theta_3 = \frac{b_3}{a_3}, \qquad f_3(b_3) = \frac{c_3b_3}{a_3},$$

$$x_3^* = x_3^*(b_3) = \frac{b_3}{a_3}, \qquad x_2^* = x_2^*(b_2) = \frac{b_3 - a_3x_3^*}{a_2} = 0,$$

$$x_1^* = x_1^*(b_1) = \frac{b_2 - a_2x_2^*}{a_1} = \frac{b_2}{a_1} = \frac{b_3 - a_3x_3^*}{a_1} = 0.$$

For $c_3 < c_2a_3/a_2$

$$x_3^*(b_3) = 0, \qquad f_3(b_3) = \frac{c_2b_3}{a_2},$$

$$x_3^* = x_3^*(b_3) = 0, \qquad x_2^* = x_2^*(b_2) = \frac{b_3 - a_3x_3^*}{a_2} = \frac{b_3}{a_2},$$

$$x_1^* = x_1^*(b_1) = \frac{b_2 - a_2x_2^*}{a_1} = \frac{b_2 - b_3}{a_1} = \frac{(b_3 - a_3x_3^*) - b_3}{a_1} = 0.$$

On the other hand, if $c_2 < c_1a_2/a_1$, then

$$f_2(b_3 - a_3x_3) = \frac{c_1}{a_1}(b_3 - a_3x_3),$$

$$f_3(b_3) = \operatorname*{Max}_{0\le x_3\le\theta_3}\left[\left(c_3 - \frac{c_1}{a_1}a_3\right)x_3 + \frac{c_1}{a_1}b_3\right].$$

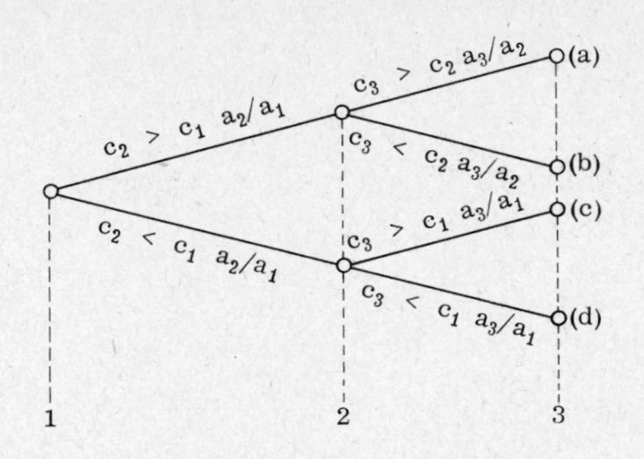

Figure 9-4

Then, for $c_3 > c_1 a_3/a_1$,

$$x_3^*(b_3) = \theta_3 = \frac{b_3}{a_3}, \qquad f_3(b_3) = \frac{c_3 b_3}{a_3},$$

$$x_3^* = x_3^*(b_3) = \frac{b_3}{a_3}, \qquad x_2^* = x_2^*(b_2) = 0,$$

$$x_1^* = x_1^*(b_1) = \frac{b_2 - a_2 x_2^*}{a_1} = \frac{b_2}{a_1} = \frac{b_3 - a_3 x_3^*}{a_1} = 0.$$

Similarly, for $c_3 < c_1 a_3/a_1$

$$x_3^*(b_3) = 0, \qquad f_3(b_3) = \frac{c_1 b_3}{a_1},$$

$$x_3^* = x_3^*(b_3) = 0, \qquad x_2^* = x_2^*(b_2) = 0,$$

$$x_1^* = x_1^*(b_1) = \frac{b_2 - a_2 x_2^*}{a_1} = \frac{b_2}{a_1} = \frac{b_3 - a_3 x_3^*}{a_1} = \frac{b_3}{a_1}.$$

The conditions for maximization at various stages are summarized in Fig. 9-4. The following numerical examples illustrate the four cases shown in the figure.

a) $c_2 - c_1 a_2/a_1 = 2 - (3)(3)/5 = 1/5 \quad (>0)$,
$c_3 - c_2 a_3/a_2 = 1 - (2)(1)/3 = 1/3 \quad (>0)$,
$x_3^* = b/a_3 = 3, \qquad x_2^* = 0, \qquad x_1^* = 0$,
$\max A = c_3 b/a_3 = (1)(3)/1 = 3$.

b) $c_2 - c_1 a_2/a_1 = 2 - (3)(2)/4 = 1/2 \quad (>0)$,
$c_3 - c_2 a_3/a_2 = 1 - (2)(5)/2 = -4 \quad (<0)$,
$x_3^* = 0, \qquad x_2^* = b/a_2 = 3/2, \qquad x_1^* = 0$,
$\max A = c_2 b/a_2 = (2)(3)/2 = 3$.

c) $c_2 - c_1a_2/a_1 = 2 - (3)(4)/4 = -1 \quad (<0)\,,$
$c_3 - c_1a_3/a_1 = 1 - (3)(1)/4 = 1/4 \quad (>0)\,,$
$x_3^* = b/a_3 = 3/1 = 3\,, \qquad x_2^* = 0\,, \qquad x_1^* = 0\,,$
$\max A = c_3b/a_3 = (1)(3)/1 = 3\,.$

d) $c_2 - c_1a_2/a_1 = 2 - (3)(2)/2 = -1 \quad (<0)\,,$
$c_3 - c_1a_3/a_1 = 1 - (3)(3)/2 = -7/2 \quad (<0)\,,$
$x_3^* = 0\,, \qquad x_2^* = 0\,, \qquad x_1^* = b/a_1 = 3/2\,,$
$\max A = c_1b/a_1 = (3)(3)/2 = 9/2\,.$

Example 9–5. Consider the following linear programming problem ($a_{ij} > 0$, $b_i > 0$, and $c_j > 0$):

$$x_j \geq 0\,, \qquad j = 1, 2\,,$$
$$a_{11}x_1 + a_{12}x_2 \leq b_1\,,$$
$$a_{21}x_1 + a_{22}x_2 \leq b_2\,,$$
$$c_1x_1 + c_2x_2 = \text{Max}\,A\,.$$

Specifically, solve the following numerical example:

$$x_1 \geq 0\,, \qquad x_2 \geq 0\,,$$
$$x_1 + 2x_2 \leq 8\,,$$
$$x_1 + 4x_2 \leq 12\,,$$
$$x_1 + 3x_2 = \text{Max}\,A\,.$$

This is a two-stage maximization problem which has two constraints instead of one. The additional constraint introduces additional bounds to the decisions. Let the stage number be represented by the second subscript of b, that is $b_{ik} = b_i$, in which k is the stage number and i is the number of the constraint. Thus

$$b_{i2} = b_i\,, \qquad b_{i1} = b_{i2} - a_{i2}x_2\,, \qquad i = 1, 2\,;$$

and the upper bounds of x_i at various stages are given by

$$\theta_{i2} = b_{i2}/a_{i2}\,, \qquad \theta_{i1} = b_{i1}/a_{i1}\,, \qquad i = 1, 2\,.$$

Then the recurrence relation for the two-stage process is obtained according to the principle of optimality as follows:

$$f_2(b_{12}, b_{22}) = \underset{0 \leq x_2 \leq \theta_{i2}}{\text{Max}} [c_2x_2 + f(b_{12} - a_{12}x_2, b_{22} - a_{22}x_2)]\,,$$
$$f_1(b_{11}, b_{21}) = \underset{0 \leq x_1 = \theta_{i1}}{\text{Max}} [c_1x_1]\,.$$

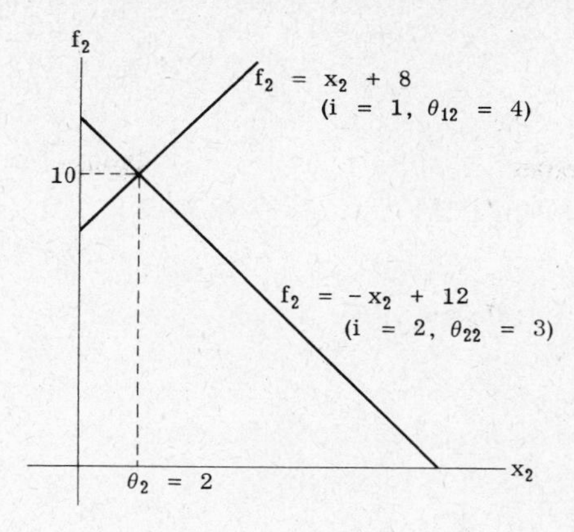

Figure 9–5

For stage 1, it is seen that the maximization is dependent on all constraint conditions involving x_1. In this particular case, it turns out that the maximization is realized when

$$x_1^*(b_{11}, b_{21}) = \underset{i}{\text{Min}}\,[\theta_{i1}] = \underset{i}{\text{Min}}\left[\frac{b_{i1}}{a_{i1}}\right].$$

The main difficulty of the solution lies in the fact that the minimum value of θ_{i1} is conditioned on the value of x_2 which is to be determined in stage 2. Hence

$$f_1(b_{11}, b_{21}) = c_1\theta_{i1} = c_1\frac{b_{i1}}{a_{i1}},$$

in which a particular i cannot be selected at this stage. For stage 2, therefore, the maximization must satisfy the requirements imposed by both constraints of $i = 1$ and $i = 2$ in stage 1. That is, for $i = 1$ and 2,

$$\begin{aligned} f_2(b_{12}, b_{22}) &= \underset{0\le x_2\le\theta_{i2}}{\text{Max}}\left[c_2x_2 + \frac{c_1}{a_{i1}}(b_{i2} - a_{i2}x_2)\right] \\ &= \underset{0\le x_2\le\theta_{i2}}{\text{Max}}\left[\left(c_2 - \frac{c_1a_{i2}}{a_{i1}}\right)x_2 + \frac{c_1}{a_{i1}}b_{i2}\right]. \end{aligned}$$

The conditions for maximization are shown schematically in Fig. 9–5. For the given numerical example, it is seen that for $i = 1$, $\theta_{12} = b_{12}/a_{12} = 8/2 = 4$,

$$\begin{aligned} &\left(c_2 - \frac{c_1a_{12}}{a_{11}}\right)x_2 + \frac{c_1}{a_{11}}b_{12} \\ &\quad = [3 - (1)(2)/1]x_2 + (1)(8)/1 = x_2 + 8\,; \end{aligned}$$

and for $i = 2$, $\theta_{22} = b_{22}/a_{22} = 12/4 = 3$,

$$\left(c_2 - \frac{c_1 a_{22}}{a_{21}}\right) x_2 + \frac{c_1}{a_{21}} b_{22}$$
$$= [3 - (1)(4)/1]x_2 + (1)(12)/1 = -x_2 + 12 .$$

Thus the maximum value for $f_2(b_{12}, b_{22})$ occurs at $x_2 = \theta_2$, where both conditions for maximization are satisfied, as shown in Fig. 9-5. For the numerical example, this value is obtained from the solution of the simultaneous equations representing the conditions for maximization. From the solution of the system of equations

$$f_2 = x_2 + 8 ,$$
$$f_2 = -x_2 + 12 ,$$

we have $x_2 = 2$ and $f_2 = 10$. Hence

$$x_2^* = x_2^*(b_{12}, b_{22}) = 2 ,$$
$$f_2(b_{12}, b_{22}) = 10 .$$

Then

$$x_1^* = x_1^*(b_{11}, b_{21}) = \operatorname*{Min}_i \left[\frac{b_{i2} - a_{i2}x_2^*}{a_{i1}}\right] .$$

Since

$$\frac{b_{12} - a_{12}x_2^*}{a_{11}} = \frac{8 - (2)(2)}{1} = 4 ,$$
$$\frac{b_{22} - a_{22}x_2^*}{a_{21}} = \frac{12 - (3)(2)}{1} = 6 ,$$

we get

$$x_1^* = \text{Min}\,(4, 6) = 4 .$$

The optimal solution obtained from dynamic programming is $x_1^* = 4$, $x_2^* = 2$ and $A^* = f_2 = 10$. This optimal solution can easily be verified by the direct solution of the linear programming problem.

9-7 NONLINEAR PROGRAMMING PROBLEMS

The dynamic programming approach may also be used to solve nonlinear programming problems although the same difficulties encountered in the solution of linear programming problems are also present here. We shall therefore confine our discussion to a simple nonlinear problem with only one constraint.

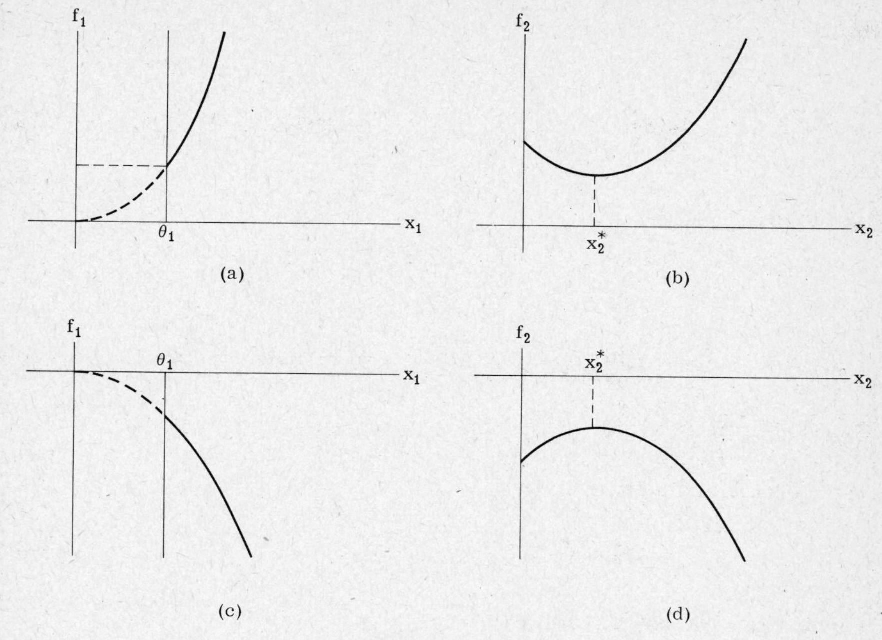

Figure 9–6

Example 9–6. Formulate the following problem (for $a_j > 0$, $b > 0$ and $c_j > 0$):

$$\text{Min } z = c_1x_1^2 + c_2x_2^2 ,$$
$$a_1x_1 + a_2x_2 \geq b ,$$
$$x_1 \geq 0 , \qquad x_2 \geq 0 ,$$

and solve the numerical example when $c_1 = c_2 = a_1 = a_2 = b = 1$.

The problem can be formulated as a two-stage process in which $b_2 \geq b$ and

$$b_1 = b_2 - a_2x_2 , \qquad \theta_2 = b_2/a_2 , \qquad 0 \leq x_2 \leq \theta_2 ,$$
$$0 = b_1 - a_1x_1 , \qquad \theta_1 = b_1/a_1 , \qquad 0 \leq x_1 = \theta_1 .$$

The recurrence relation is given by

$$f_2(b_2) = \underset{0 \leq x_2 \leq \theta_2}{\text{Min}} [c_2x_2^2 + f_1(b_2 - a_2x_2)] ,$$
$$f_1(b_1) = \underset{0 \leq x_1 = \theta_1}{\text{Min}} [c_1x_1^2] .$$

For stage 1, the minimization is obtained by observing that the global minimum occurs at the boundary point $x_1 = \theta_1$, as shown in Fig. 9–6(a).

Hence

$$x_1^*(b_1) = \theta_1 = \frac{b_1}{a_1},$$

$$f_1(b_1) = c_1\left(\frac{b_1}{a_1}\right)^2 = \frac{c_1}{a_1^2} b_1^2 .$$

For stage 2,

$$\begin{aligned} f_2(b_2) &= \min_{0 \le x_2 \le \theta_2} \left[c_2 x_2^2 + \frac{c_1}{a_1^2}(b_2 - a_2 x_2)^2 \right] \\ &= \min_{0 \le x_2 \le \theta_2} \left[\left(c_2 + \frac{c_1 a_2^2}{a_1^2}\right) x_2^2 - \frac{2 c_1 a_2 b_2}{a_1^2} x_2 + \frac{c_1 b_2^2}{a_1^2} \right]. \end{aligned}$$

The global minimum of this function is seen to occur in the interior, as indicated in Fig. 9–6(b). Taking the first derivative of function f_2 with respect to x_2 and setting it equal to zero, we have

$$2c_2 x_2 + \frac{2c_1}{a_1^2}(b_2 - a_2 x_2)(-a_2) = 0 .$$

Hence a critical point is located at

$$x_2 = \frac{c_1 a_2 b_2}{c_2 a_1^2 + c_1 a_2^2} .$$

The second derivative of function f_2 with respect to x_2 shows that

$$2\left(c_2 + \frac{c_1 a_2^2}{a_1^2}\right) > 0 .$$

The critical point is therefore a minimum. Finally, for $b_2 = b$, we get

$$f_2(b_2) = \left[\frac{c_1 c_2 b_2}{c_2 a_1^2 + c_1 a_2^2}\right]^2 \left(\frac{a_2^2}{c_2} + \frac{a_1^2}{c_1}\right) = \frac{c_1 c_2 b^2}{c_2 a_1^2 + c_1 a_2^2},$$

$$x_2^* = x_2^*(b_2) = \frac{c_1 a_2 b}{c_2 a_1^2 + c_1 a_2^2},$$

$$x_1^* = x_1^*(b_1) = \frac{c_2 a_1 b}{c_2 a_1^2 + c_1 a_2^2}.$$

For the numerical example $c_1 = c_2 = a_1 = a_2 = b = 1$, we have

$$\begin{aligned} &\text{Min } z = x_1^2 + x_2^2 , \\ &x_1 + x_2 \ge 1 , \\ &x_1 \ge 0 , \qquad x_2 \ge 0 . \end{aligned}$$

Then,

$$f_2(b_2) = \left(\frac{1}{1+1}\right)^2(1+1) = \frac{1}{2},$$

$$x_2^* = \frac{1}{1+1} = \frac{1}{2}, \qquad x_1^* = \frac{1}{1+1} = \frac{1}{2}.$$

If we had attempted to solve this problem by first changing the minimization to maximization by multiplying each relation by -1, we would have started from

$$\text{Max } A = -c_1x_1^2 - c_2x_2^2,$$
$$-a_1x_1 - a_2x_2 \leq -b,$$
$$x_1 \geq 0, \qquad x_2 \geq 0.$$

However, the bounds of x_1 and x_2 would still be the same. Hence the recurrence relation becomes

$$f_2(b_2) = \underset{0 \leq x_2 \leq \theta_2}{\text{Max}} \left[-c_2x_2^2 + f_1(b_2 - a_2x_2)\right],$$
$$f_1(b_1) = \underset{0 \leq x_1 = \theta_1}{\text{Max}} \left[-c_1x_1^2\right].$$

Then the maximization for stages 1 and 2 will simply be as shown in Figs. 9–6(c) and (d), respectively.

9-8 NETWORK PROBLEMS

The dynamic programming approach may also be applied to network analysis. Since the stage-wise optimization in a network problem generally involves a choice from a finite number of allowable decisions, a general procedure for the solution of network problems will be applicable to many practical sequential decision problems. However, we shall confine our discussion to a special procedure developed for a more restrictive type of network problems. Hence only a simple example of determining the shortest route in a network is given as an illustration.

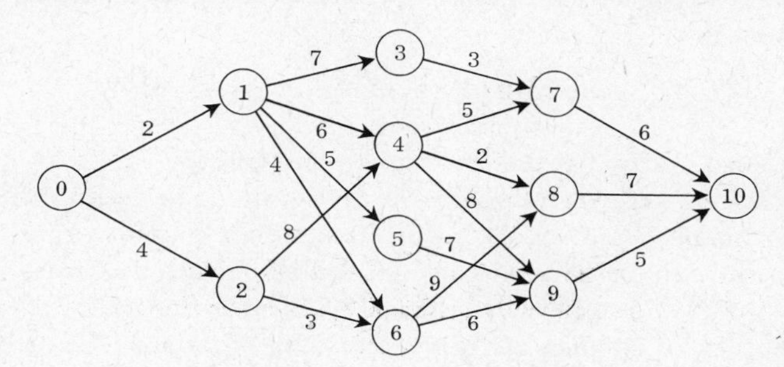

Figure 9–7

Example 9–7. For the network shown in Fig. 9–7, the distance between each pair of nodes is given on the branch linking the nodes. If two nodes are not accessible to each other directly, there will be no branch connecting them and their distance may be regarded as infinite. Determine the shortest route between the origin at node 0 and the destination at node 10.

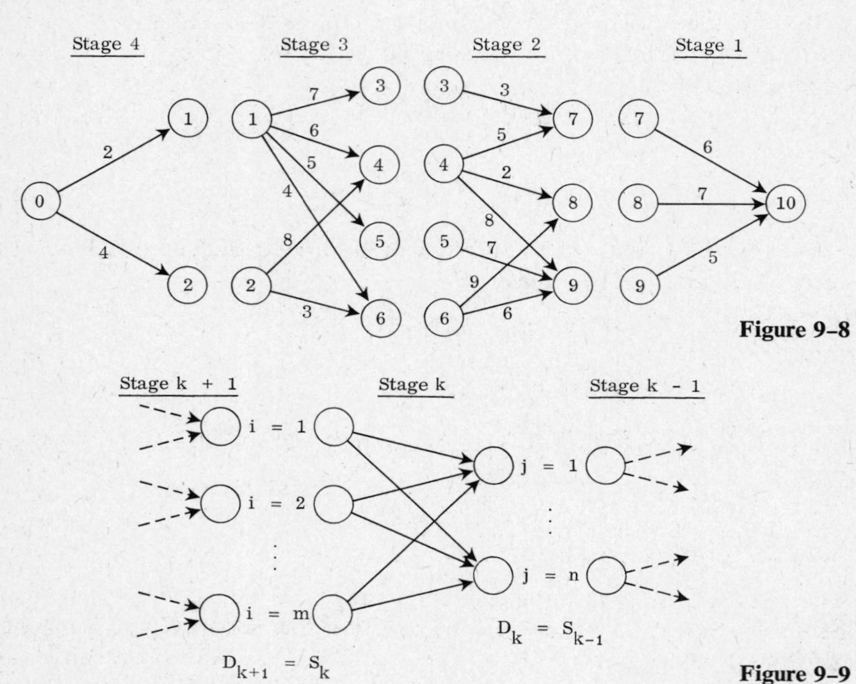

Figure 9–8

Figure 9–9

This problem has previously been solved in Chapter 7. In using the dynamic programming approach, we separate the routes into stages, as shown in Fig. 9–8, and note that it is a four-stage process. For a particular stage k, as shown in Fig. 9–9, the recurrence relation is similar to that given in (9–11), or

$$f_k(S_k) = \underset{D_k}{\text{Min}}\,[R_k(S_k, D_k) + f_{k-1}(S_{k-1})]\,,$$

with $f_0(S_0) = 0$. This expression can be interpreted in terms of the relations of nodes and branches in the network. Since the minimization process in this problem involves a finite number of allowable decisions, the operation can be carried out in a matrix form. A decision at stage k dictates the distance between node i at the source and the node j at the destination of stage k, as represented by S_k^i and D_k^j, respectively, in Fig. 9–9. In other words, the input-output transformation may be regarded as a process of making decision D_k by selecting a particular destination D_k at stage k according to the input information S_k^j, thus generating the k-stage return, R_k^{ij}. Furthermore, it is seen that the destination D_k^j at stage k is identical to the source S_{k-1}^j at stage $(k-1)$. Hence the return function $R_k(S_k, D_k)$ at stage k is seen to be as represented by the matrix in Table 9–2(a). However, the choice of S_{k-1}^j

Table 9–2

SOLUTION OF A NETWORK PROBLEM

(a) Return Function at Stage k

		$D_k = S_{k-1}$			
		S_{k-1}^1	S_{k-1}^2	...	S_{k-1}^n
S_k	S_k^1	R_k^{11}	R_k^{12}	...	R_k^{1n}
	S_k^2	R_k^{21}	R_k^{22}	...	R_k^{2n}
	...	...	...	...	...
	S_k^m	R_k^{m1}	R_k^{m2}	...	R_k^{mn}

(b) Expected Total $(k-1)$ Stage Return

		S_{k-1}			
		S_{k-1}^1	S_{k-1}^2	...	S_{k-1}^n
S_k	S_k^1	f_{k-1}^1	f_{k-1}^2	...	f_{k-1}^n
	S_k^2	f_{k-1}^1	f_{k-1}^2	...	f_{k-1}^n
	...	...	...	...	...
	S_k^m	f_{k-1}^1	f_{k-1}^2	...	f_{k-1}^n

(c) Determination of Total k-Stage Return

		D_k			
		D_k^1	D_k^2	...	D_k^n
S_k	S_k^1	F_k^{11}	F_k^{12}	...	F_k^{1n}
	S_k^2	F_k^{21}	F_k^{22}	...	F_k^{2n}
	...	...	...	...	...
	S_k^m	F_k^{m1}	F_k^{m2}	...	F_k^{mn}

Table 9-3

SOLUTION FOR EXAMPLE 9-7

(a) Stage 1

		R_1	f_0	F_1
		#10	#10	#10
S_1	#7	6	0	(6)
	#8	7	0	(7)
	#9	5	0	(5)

(b) Stage 2

		R_2			f_1			F_2		
		#7	#8	#9	#7	#8	#9	#7	#8	#9
S_2	#3	3	∞	∞	6	7	5	(9)	∞	∞
	#4	5	2	8	6	7	5	11	(9)	13
	#5	∞	∞	7	6	7	5	∞	∞	(12)
	#6	∞	9	6	6	7	5	∞	16	(11)

(c) Stage 3

		R_3				f_2				F_3			
		#3	#4	#5	#6	#3	#4	#5	#6	#3	#4	#5	#6
S_3	#1	7	6	5	4	9	9	12	11	16	(15)	17	(15)
	#2	∞	8	∞	3	9	9	12	11	∞	17	∞	(14)

(d) Stage 4

		R_4		f_3		F_4	
		#1	#2	#1	#2	#1	#2
S_4	#0	2	4	15	14	(17)	18

depends on the total return of the previous $(k - 1)$ stages, which has been determined at stage $(k - 1)$. Since the procedure of obtaining the total k-stage return will be explained later, it is sufficient to note at this time that the total $(k - 1)$ stage return may be obtained in a similar manner. The result of the total $(k - 1)$ stage return f_{k-1}, for each j in stage $(k-1)$, is given by f_{k-1}^j in part (b) of the table. Then the optimization process in stage k can be obtained by using the recurrence relation:

$$f_k^i = \operatorname*{Min}_{D_k}\,[R_k^{ij} + f_{k-1}^j] = \operatorname*{Min}_{D_k}\,[F_k^{ij}]\,,$$

in which R_k^{ij} and f_{k-1}^j are given in parts (a) and (b) of the table respectively, and $F_k^{ij} = R_k^{ij} + f_{k-1}^j$ (for all i and j) are shown in part (c). Thus the minimum value of F_k^{ij} for each i can be selected from the rows of the matrix in this table. These minimum values f_k^i $(i = 1, 2, \ldots, m)$ are to be used for the optimization process in stage $(k+1)$.

Since a network must be separated into n stages in a solution by dynamic programming, it is sometimes necessary to introduce fictitious nodes and branches in order to make the separation possible. If the distance in each of the fictitious branches is assigned a value of zero, the addition of fictitious nodes and branches will not change the physical situation of the problem.

Referring to Fig. 9–8, we see that the multistage optimization process can be carried out as shown in Table 9–3, in which ♯ designates the node point. Note that at each stage, the table combines the matrices for R_k^{ij}, F_{k-1}^j, and f_k^i, in that order. At stage 1, there is only one destination, i.e., node 10. Hence R_1 is a 3×1 matrix with coefficients equal to given distances, and $f_0 = 0$, since the network terminates at node 10. Then the expected total one-stage return f_1 is determined by selecting the minimum value of each row of matrix F_1 and, in this case, the minimum value happens to be the only value of each row. These values are circled in the table. In stage 2, R_2 is a 3×3 matrix with coefficients equal to given distances. The minimum value in each row at the end of stage 1 becomes the coefficients in each column for the matrix f_1 in stage 2. The sum of matrices R_2 and f_1 gives the matrix F_2, from which the minimum value of each row is selected for entering into the next stage. In stage 3, the same procedure is repeated. It is only necessary to point out that, in the matrix F_3, two coefficients are tied for minimum in the first row, and both values are circled. In entering stage 4, this minimum value is shown in the matrix for f_3 like any other untied minimum value. Since there is only one origin or source in the last stage, i.e., node 0, the minimization process is a choice for the minimum value in only one row. Hence the expected total return for the four-stage process is $f_4 = 17$.

As denoted in Table 9–2(c), $D_k = S_{k-1}$. Hence the node S_{k-1} corresponding to the minimum value in the row S_k of the matrix F_k that corresponds to an optimal decision D_{k+1}^* in stage $(k+1)$ is the optimal decision D_k^* for stage k. Thus, in Table 9–3, the optimal decisions for stages 4, 3, 2, and 1 are determined, respectively, in the following order:

$$D_4^* = \#1 \,,$$
$$D_3^* = \#4 \quad \text{or} \quad \#6 \,,$$
$$D_2^* = \#8 \quad \text{from} \quad \#4 \qquad \text{or} \qquad \#9 \quad \text{from} \quad \#6 \,,$$
$$D_1^* = \#10 \quad \text{from either} \quad \#4 \quad \text{or} \quad \#6 \,.$$

Hence the optimal policies for the solution of the network problem are the routes linking (#0, #1, #4, #8, #10) or (#0, #1, #6, #9, #10), and the minimum distance in either case is 17.

9-9 ADDITIONAL EXAMPLES OF APPLICATION

In this chapter, we have presented the basic concept of dynamic programming and its application to the solution of multistage decision problems. The primary purpose of this brief discussion is to illustrate the versatility of the dynamic programming approach. Thus, instead of dealing with physical problems which involve serial operations, the emphasis has been placed on the use of dynamic programming for the reformulation of single-stage decision problems. However, the broad range of applicability of dynamic programming can be seen from additional examples of practical interest. Two such examples are described here, but the detailed solutions will not be included.†

Example 9-8. Select the most economical structural system for a 40-ft-high platform consisting of a rectangular deck of 30 ft × 20 ft, four supporting columns with adequate bracing, and a foundation which transfers the column loads safely to the ground. The components have been designed for a total imposed vertical load of 120,000 lb, uniformly distributed on the platform, the dead load and cost data of alternative designs for each of the three components are given in Table 9-4. Interpolation and extrapolation of data are permissible.

The problem may be formulated as a three-stage backward optimization process, using (a) the deck as stage 3, (b) the columns with bracing as stage 2, and (c) the foundation as stage 1. While the design of components follows the order of deck, columns, and foundation because the loads, including dead loads, are transmitted in this manner, the construction of the components is carried out in the reversed order. Hence it is not unnatural to optimize the cost by using the backward process. However, for this problem, the optimization of the subproblem at stage

† See References 9-4 and 9-5 in connection with Examples 9-8 and 9-9, respectively.

1 is interwoven with the selection of components at stages 2 and 3. Hence the subproblem is solved by considering a range of possible values for the state variable. In other words, we can recast a problem involving an exhaustive search in the framework of dynamic programming. This example illustrates the versatility of the dynamic programming approach. However, the detailed solution will not be discussed here.

Table 9–4

LOAD AND COST DATA FOR EXAMPLE 9–8

(a) Deck Data

Type of deck	Load, lb	Weight, lb	Load + Wt, lb	Cost, dollars
a. One-way concrete slab and beams	120,000	75,000	195,000	1100
b. One-way concrete pan joist and beams	120,000	37,000	157,000	1650
c. Two-way concrete slab and edge beams	120,000	49,000	169,000	1300
d. Concrete waffle slab and edge beams	120,000	28,000	148,000	1800
e. Steel beams and steel deck	120,000	12,000	132,000	3000
f. Steel bar joists and steel deck	120,000	14,000	134,000	2400
g. Steel beams and composite concrete deck	120,000	30,000	150,000	2000

(b) Columns and Bracing Data

Type of columns and bracing	Load, lb	Weight, lb	Load + Wt, lb	Cost, dollars
a. Reinforced concrete tied columns	200,000	135,000	335,000	3300
	170,000	100,000	270,000	2500
	150,000	75,000	225,000	1900
	130,000	60,000	190,000	1400
b. Reinforced concrete spiral columns	200,000	115,000	315,000	3900
	170,000	85,000	255,000	3000
	150,000	64,000	214,000	2200
	130,000	50,000	180,000	1400
c. Structural steel	200,000	27,000	227,000	6000
	170,000	23,000	193,000	5200
	150,000	20,000	170,000	4500
	130,000	18,000	148,000	4000

(Cont.)

Table 9-4 (Continued)

(c) Foundation Data

Type of foundation	Load, lb	Weight, lb	Load + Wt, lb	Cost, dollars
a. Spread foundation	330,000	120,000	450,000	1800
	270,000	90,000	360,000	1400
	220,000	66,000	286,000	1000
	190,000	50,000	240,000	750
	150,000	34,000	184,000	500
b. Drilled concrete piles	330,000	105,000	435,000	1650
	270,000	88,000	358,000	1350
	220,000	72,000	292,000	1100
	190,000	62,000	252,000	950
	150,000	49,000	199,000	750
c. Driven steel piles	330,000	10,000	340,000	1400
	270,000	8,200	278,200	1150
	220,000	6,800	226,800	960
	190,000	6,200	196,200	880
	150,000	5,500	155,500	780

Example 9-9. In a 24-hr period, the arrival rate of passengers per hour heading downtown from an express bus station is known. (The total number of arrivals is 1000 passengers, with peaks at 8:00 A.M. and 5:00 P.M.) During this period, sequential decisions with regard to dispatching vehicles are to be made at equal increments of passenger arrivals. If the vehicle capacity (for c passengers) is reached, a dispatch will automatically be scheduled; otherwise, the decision must be made to dispatch one vehicle or no vehicle. The cost of dispatching a vehicle is V dollars, but the cost of passenger delays for not dispatching a vehicle is equal to the product of the number of waiting passengers and the waiting time (also in dollars if a rate of one dollar per passenger-hour is used for computing the delay cost). Given that 100 decisions are to be made in the 24-hr period, determine an optimal schedule that will minimize the average delay cost, assuming that the delay costs are additive.

For this problem, it is more natural to reverse the order in designating the states of the backward multi-stage optimization process, since the number of decisions chronologically increases with the passage of time. There are 100 stages for the 100 decisions, which are spaced at equal increments of passenger arrivals. The number of passengers p arriving in any time increment is constant, and is equal to the total number of arrivals divided by the total number of decisions, or $p = 1000/100 = 10$.

Furthermore, the distribution of decisions in the 24-hr period is proportional to the passenger arrival rate. Since the criterion for optimization is the average delay cost, some passengers may have long delays.

In the new order of designation for the multistage optimization process, let the initial time t_0 be the initial state, and the time t_k be the input state at stage k ($k = 0, 1, \ldots, n - 1$). Referring to Fig. 9–10, the recurrence relation of the n-stage optimization process may be obtained as follows:

$$f_k(S_k) = \operatorname*{Min}_{D_k} [R_k(S_k, D_k) + f_{k+1}(S_{k+1})] ,$$

subject to

$$S_{k+1} = T_k(S_k, D_k) , \qquad k = 0, 1, \ldots, n - 1 ,$$

with $f_n(S_n) = 0$, and D_k applied to a given range. Thus in this problem the number of waiting passengers is used as the state variable. At stage k, the decision variable D_k is limited to the choice of

$$D_k = \begin{cases} 1 , & \text{if a vehicle is dispatched,} \\ 0 , & \text{if no vehicle is dispatched.} \end{cases}$$

The state variable S_k ($\leq c$) corresponds to the number of waiting passengers at time t_k. If D_k^* denotes the optimal choice of D_k at stage k, then by considering the number of passengers arriving in the time increment $t_{k+1} - t_k$, the transformation becomes

$$S_{k+1} = S_k[1 - D_k^*] + p .$$

The return at stage k is the cost corresponding to the decision D_k, that is,

$$R_k(S_k, D_k) = \begin{cases} V , & \text{for} \quad D_k = 1 , \\ S_k(t_{k+1} - t_k) , & \text{for} \quad D_k = 0 . \end{cases}$$

The expected k-stage return $f_k(S_k)$ is the optimal cost at time t_k for all stages from the current stage to the final stage. At the final stage $n - 1$, $f_n(S_n) = 0$, since no cost is involved at the output state of the final stage. By working backward for the full range of k, we obtain the expected k-stage return $f_k(S_k)$ for the given values of S_k. The detailed solution of this problem will not be shown here.

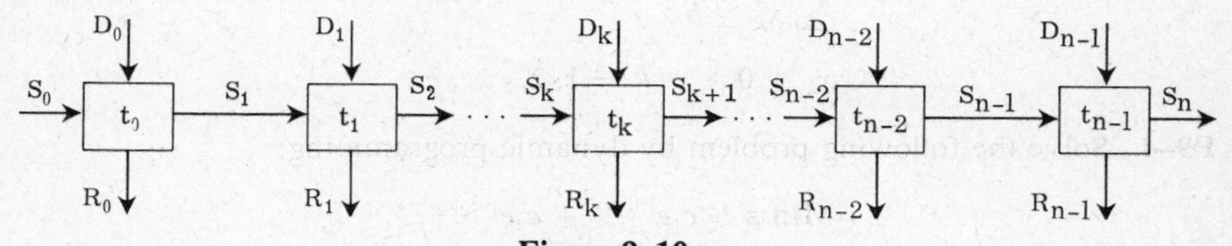

Figure 9–10

REFERENCES

9–1. Bellman, R., *Dynamic Programming*, Princeton University Press, Princeton, N. J., 1957.

9–2. Bellman, R., and Dreyfus, S., *Applied Dynamic Programming*, Princeton University Press, Princeton, N. J., 1962.

9–3. Nemhauser, G. L., *Introduction to Dynamic Programming*, Wiley, New York, 1966.

9–4. Aguilar, R. J., *An Application of Dynamic Programming to Structural Optimization*, Division of Engineering Research, Bulletin No. 91, Louisiana State University, Baton Rouge, La., 1967.

9–5. Ward, D. L., *Optimal Dispatching Policies by Dynamic Programming*, Dept. of Civil Engineering, MIT, Cambridge, Mass., 1966.

PROBLEMS

P9–1. Solve the following problem by dynamic programming (for $a_j > 0$, $c_j > 0$):

$$\text{Min } z = c_1 x_1 + c_2 x_2 ,$$

subject to

$$a_1 x_1 + a_2 x_2 \geq b ,$$
$$x_1 \geq 0 , \qquad x_2 \geq 0 .$$

P9–2. Solve the following problem by dynamic programming (for $a_j > 0$, $c_j > 0$):

$$\text{Max } A = c_1 x_1^2 + c_2 x_2^2 ,$$

subject to

$$a_1 x_1 + a_2 x_2 \leq b ,$$
$$x_1 \geq 0 , \qquad x_2 \geq 0 .$$

P9–3. Solve the following problem by dynamic programming:

$$\text{Max } A = \sum_{k=1}^{n} x_k^2$$

subject to

$$\sum_{k=1}^{n} x_k \geq b ,$$
$$x_k \geq 0 , \qquad k = 1, 2, \ldots, n .$$

P9–4. Solve the following problem by dynamic programming:

$$\text{Min } z = c_1 e^{-d_1 x_1} + c_2 e^{-d_2 x_2} ,$$

subject to

$$a_1x_1 + a_2x_2 = b\,,$$
$$x_1 \geq 0\,, \qquad x_2 \geq 0\,,$$

in which e is the base of natural logarithms, and all coefficients are positive.

P9–5 through P9–8. Determine the shortest route by dynamic programming for problems P7–13 through P7–16, respectively.

CHAPTER 10

OPTIMIZATION WITH COMPETITION

10–1 COMPETITIVE GAMES

In some decision problems, conflicting situations may arise so that the decision maker has to consider not only his own preference to the solution but also those of his competitors, since the outcome is determined by the decisions of all participants. Such situations can often be cast as *competitive games*, in which competitors have different preferences over the outcome. Obviously, a competitive situation does not exist if each participant desires a solution which is not in conflict with those desired by others. A competitive game, therefore, has the following characteristics:

1. There are a finite number of competitors called *players*.

2. Each of the players has a plan listing possible courses of action, called *strategy*, which is, in general, different from those of other players.

3. A *play* of the game takes place when each of the players chooses a course of action independent of the choices of others.

4. The *outcome* of a play is the consequence of a particular set of courses of action undertaken by the players, and determines the *payoff* to each player.

Each player of a competitive game may be an individual or a team of individuals. However, to avoid complications of conflicting situations among different individuals on the same team, we shall consider players as persons. Hence, competitive games with a finite number of competitors are referred to as *n-person games*. To simplify our discussion further, we shall consider only *two-person games*, in which only two players are involved. Fortunately two-person games not only are useful in their own right, but also preserve the principal notion of the *theory of games*, which is a powerful mathematical approach to the optimization of competitive problems developed by von Newmann.

In general, in a two-person game, a player does not necessarily gain what the other player loses. However, if the players have strictly opposite preferences for the outcomes, then one's gain is the other's loss. Hence strictly competitive games are also called zero-sum games, in which the algebraic sum of the payoffs to all players in a game is zero. Thus a payoff may either be positive or negative.

The decision in a competitive game involves the choice of a rational strategy among a number of possible actions. Under conditions of certainty, each course of action leads invariably to a specific outcome, and the plan for a definite course of action is called a *pure strategy*. Under conditions of uncertainty, each action leads to a set of possible outcomes, depending on the probabilities of occurrence, and the plan for choosing a course of action according to certain probability distribution for the possible outcomes is termed a *mixed strategy*.

In a two-person zero-sum game, each player knows the strategies, pure or mixed, which are available to him and his opponent; he knows that the outcome depends upon his own and his opponent's choices for these alternative strategies; and he knows his own and his opponent's preferences for the outcomes of the game. The purpose of the theory of games is to establish a *criterion* for selecting an *optimal strategy* such that the *security level of each player may be guaranteed*. A game is said to have a *solution* if the set of optimal strategies for both players leads to a payoff expected by both of them. The set of optimal strategies is said to be *in equilibrium*, and the corresponding payoff is called the *value* of the game.

Although the theory of games has a wide range of applications, we shall confine our discussion only to an introduction to two-person, zero-sum games with pure and mixed strategies. The primary purpose of this chapter is to show the relationship between a two-person zero-sum game with mixed strategies and its equivalent in linear programming. We shall begin with an example of a conflicting situation which leads to a two-person zero-sum game.

Example 10–1. Two aircraft companies, A and B, are competing for a contract in the production of a special-purpose airplane. Company A has three preliminary designs, A_1, A_2, and A_3; company B also has three preliminary designs, B_1, B_2, and B_3. Although the designs of each company are well-guarded secrets, some experts have reasonably good estimates on the strength and weakness of these designs. As a matter of fact, an article appears in a respectable trade magazine rating the desirability of each pair of designs, as shown in Table 10–1. It is well-known that the opinion expressed in this magazine usually receives considerable attention

Table 10–1

PREFERENCE RATIOS FOR VARIOUS DESIGNS BY PAIRS

Company A	A_1	A_1	A_1	A_2	A_2	A_2	A_3	A_3	A_3
Company B	B_1	B_2	B_3	B_1	B_2	B_3	B_1	B_2	B_3
Preference ratio A/B	$\frac{4}{5}$	$\frac{5}{4}$	$\frac{4}{3}$	$\frac{1}{3}$	$\frac{1}{4}$	$\frac{5}{3}$	$\frac{3}{5}$	$\frac{3}{2}$	$\frac{1}{2}$

of persons knowledgeable in the field, including officials who will decide on the award of the contract. If each company is permitted to submit only one proposed design in soliciting the contract, how should each company select its design?

Let us assume that the companies believe that this article will indeed influence the opinion of the officials who will decide on the award of the contract, and wish to make their decisions accordingly. We shall first analyze the standing of the companies in matching various designs on a fraction or percentage basis, as shown in Table 10-2. For example, in pairing A_1 and B_1, $4/(4 + 5) = 0.444$ or 44.4% and $5/(4 + 5) = 0.556$ or 55.6%, as shown in parts (a) and (b) of the table, respectively.

Table 10-2

COMPANY STANDINGS BASED ON PREFERENCE COEFFICIENTS

(a) Company A				(b) Company B			
	B_1	B_2	B_3		B_1	B_2	B_3
A_1	0.444	0.556	0.571	A_1	0.556	0.444	0.429
A_2	0.250	0.200	0.625	A_2	0.750	0.800	0.375
A_3	0.375	0.600	0.333	A_3	0.625	0.400	0.667

The fraction or percentage that each company is ahead of the other may be computed from Table 10-2. Thus the coefficients in (a) minus the corresponding coefficients in (b) represent the advantage of A over B; and those in (b) minus the corresponding ones in (a) denote the advantage of B over A. These results may be regarded as payoffs to A and B, respectively, as shown in Table 10-3.

Table 10-3

PAYOFF MATRICES

(a) Payoff to A				(b) Payoff to B			
	B_1	B_2	B_3		B_1	B_2	B_3
A_1	−0.112	+0.112	+0.142	A_1	+0.112	−0.112	−0.142
A_2	−0.500	−0.600	+0.250	A_2	+0.500	+0.600	−0.250
A_3	−0.250	+0.200	−0.334	A_3	+0.250	−0.200	+0.334

Note that the coefficients in the payoff matrix to B are opposite in sign to those in the payoff matrix to A, as expected for a two-person zero-sum game. Hence we need to consider only one of them, say Table 10-3(a).

Let us first put ourselves in the position of company A. It wants to have as high a percentage over company B as it can, but is well aware that B will try to prevent that from happening. If A chooses A_2 in the hope that B may choose B_3 for a maximum payoff of $+0.250$, it may end up with a minimum payoff of -0.600 if B chooses B_2 instead of B_3. Therefore, A reasons that, to play safe, it can expect only the *minimum* payoff corresponding to every course of his own action. For A_1, the minimum gain is -0.112; for A_2, it is -0.600; and for A_3, it is -0.334. However, A is free to choose the maximum of these minimum gains; and A's strategy is to choose A_1, which guarantees a gain of -0.112 over B. Although this is not a happy choice for A, since it stands to lose, there is no other alternative which can guarantee a higher level of security. This is clearly a pessimistic approach.

Now, let us look at the situation from the position of company B. It wants A to have a minimum payoff from Table 10–3(a). If B chooses B_2 in the hope that A may choose A_2 for a minimum payoff of -0.600, it may end up with a payoff to A equal to of $+0.200$ if A chooses A_3 instead of A_2. Hence B reasons that, to play safe, it must be prepared for the *maximum* payoff to A corresponding to every course of his own action. For B_1, the maximum gain for A is -0.112; for B_2, it is $+0.200$; and for B_3, it is $+0.250$. However, B is free to choose the minimum of these maximum values; and B's strategy is to choose B_1, which guarantees that A can gain no more than -0.112.

In this problem, it so happens that the set of strategies selected by A and B leads to the same value -0.112 for the pair of choices A_1 and B_1, which is the *value* of the game. When this situation happens, *the value of a two-person zero-sum game is obtained from the minimum of the rows and the maximum of the columns in the payoff matrix.*

Example 10–2. Repeat the previous example given that the table of preference ratios for the designs has been changed to that of Table 10–4 in a second evaluation by the magazine.

Table 10–4

NEW PREFERENCE RATIOS

Company A	A_1	A_1	A_1	A_2	A_2	A_2	A_3	A_3	A_3
Company B	B_1	B_2	B_3	B_1	B_2	B_3	B_1	B_2	B_3
Preference ratio A/B	$\frac{5}{3}$	$\frac{4}{5}$	$\frac{4}{3}$	$\frac{3}{5}$	$\frac{3}{2}$	$\frac{3}{2}$	$\frac{1}{3}$	$\frac{3}{7}$	$\frac{5}{4}$

The new payoff matrix to A has been obtained in Table 10–5(a). We can proceed to reason in the positions of A and B as before. The minima of rows 1, 2, and 3 are -0.112, -0.250, and -0.500, respectively; and the maxima of columns 1, 2, and 3 are $+0.250$, $+0.200$, and $+0.200$,

Table 10-5

NEW PAYOFF MATRIX TO A

(a)

	B_1	B_2	B_3
A_1	$+0.250$	-0.112	$+0.142$
A_2	-0.250	$+0.200$	$+0.200$
A_3	-0.500	-0.400	$+0.112$

(b)

	B_1	B_2
A_1	$+0.250$	-0.112
A_2	-0.250	$+0.200$

respectively. Thus the maximum of the minimum gain by A is -0.112, and the minimum of the maximum loss by B is $+0.200$. Since these two values are not equal, the game is said to have no solution.

It is interesting to note in Table 10-5(a) that every element in column 3 is greater than or equal to the corresponding element in column 2. Since only the smaller of the two is pertinent in determining the minimum in each row, column 3 need not be considered. Similarly, we note that every element in row 2 is greater than the corresponding element in row 3. Since only the larger of the two is pertinent in determining the maximum in each column, row 3 need not be considered. By eliminating column 3 and row 3 in Table 10-5(a), a matrix of smaller dimension as shown in Table 10-5(b) is obtained. Although the size of the problem has been reduced by such eliminations, the fact that this game has no solution remains as long as we insist on finding a set of equilibrium or optimal strategies which is simultaneously the minimum of its rows and the maximum of its columns.

Out of desperation, both A and B decide to leave the selection of strategies to chance. For example, A may select the strategy by throwing a dice. If a one or a two turns up, he will select A_1; otherwise he will select A_2. On the other hand, B may select the strategies by flipping a coin. If a head turns up, he will select B_1; otherwise he will select B_2. Then the probability distribution of selecting A_1 and A_2 by A is $(\frac{1}{3}, \frac{2}{3})$, and the probability distribution of selecting B_1 and B_2 by B is $(\frac{1}{2}, \frac{1}{2})$. When we know the probabilities that each player selects as his strategy, we can make some kind of prediction, which is called the *expected payoff* of the game. For the payoff matrix in Table 10-5(b), if A chooses his strategy according to probability and B chooses a pure strategy, the expected payoff for a column may be regarded as the weighted average of the coefficients in the column using the row probabilities as weights. Thus the expected payoffs for columns 1 and 2 are, respectively,

$$(\tfrac{1}{3})(+0.250) + (\tfrac{2}{3})(-0.250) = -0.083\,, \tag{a}$$

$$(\tfrac{1}{3})(-0.112) + (\tfrac{2}{3})(+0.200) = +0.096\,. \tag{b}$$

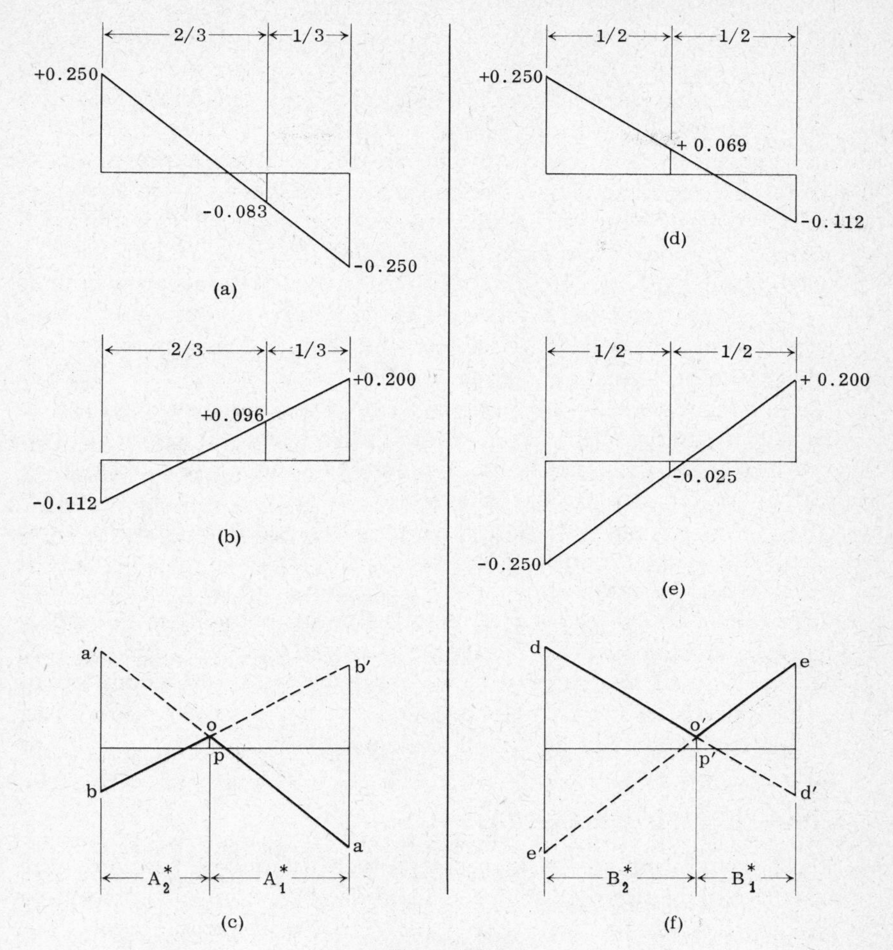

Figure 10–1

Similarly, if B chooses his strategy according to probability and A chooses a pure strategy, the expected payoffs for rows 1 and 2 are, respectively,

$$(\tfrac{1}{2})(+0.250) + (\tfrac{1}{2})(-0.112) = +0.069\,, \tag{c}$$

$$(\tfrac{1}{2})(-0.250) + (\tfrac{1}{2})(+0.200) = -0.025\,. \tag{d}$$

Equations (a), (b), (c), and (d) give the expected payoffs for matching the pairs (A, B_1), (A, B_2), (A_1, B), and (A_2, B), respectively, for the given probability distributions of A and B for choosing respective strategies. The relations of payoff vs. probability in these equations are shown graphically by (a), (b), (d), and (e), respectively, in Fig. 10–1.

If each player is looking for the guaranteed level of security, A can select his probability distribution for the rows in such a way as to maximize the smallest expected payoff in a column because this level of security can be guaranteed. Similarly, B can select his probability distribution for the columns in such a way as to minimize the largest expected payoff in a row. Thus a game which does not have a value for pure strategies may still be solved by the introduction of mixed strategies. This approach is illustrated graphically in parts (c) and (f) of the figure. Note that the ordinate in each figure denotes the payoff to A and that the abscissa denotes the probability of selecting a mixed strategy. Part (c) is a summary of A's expectation for each of his mixed strategies if B uses pure strategy B_1 and B_2, as represented by lines aa' and bb', respectively. The solid line aob, therefore, represents the least A can get for various strategies of B; thus A will naturally select the maximum payoff represented by point o. Figure 10-1(f), on the other hand, is a summary of B's expectation for each of his mixed strategies if A uses pure strategy A_1 and A_2, as represented by lines dd' and ee', respectively. The solid line $do'e$ therefore represents the most B can lose for various strategies of A; thus B will naturally select the minimum payoff, represented by point o'. If op in part (c) is equal to $o'p'$ in part (f), then the points p and p' in the respective figures give the pair of optimal mixed strategies (A_1^*, A_2^*) and (B_1^*, B_2^*) for the solution of the game. As we shall see later, the condition of $op = o'p'$ indeed holds true. Hence the game has a solution according to the modified rule of using mixed strategies.

10-2 GAMES WITH PURE STRATEGIES

We shall now formulate the general problem of two-person zero-sum games with a pure strategy, which is identified by a number representing the selected course of action. Let A and B be the players. Player A has m courses of action in the order of $i = 1, 2, \ldots, m$; and B has n courses of action in the order of $j = 1, 2, \ldots, n$. If A chooses the action i, and B chooses action j, the payoff to A is a_{ij} and payoff to B is $b_{ij} = -a_{ij}$. The payoff to A for different combinations of actions played by A and B can be represented by a matrix as follows:

$$[a_{ij}] = \begin{bmatrix} a_{11} & a_{12} & \cdots & a_{1j} & \cdots & a_{1n} \\ a_{21} & a_{22} & \cdots & a_{2j} & \cdots & a_{2n} \\ \vdots & & & & & \\ a_{i1} & a_{i2} & \cdots & a_{ij} & \cdots & a_{in} \\ \vdots & & & & & \\ a_{m1} & a_{m2} & \cdots & a_{mj} & \cdots & a_{mn} \end{bmatrix}.$$

Thus each action of A is given by a row, and each action of B is given by a column in the matrix. What is the criterion of optimality and what is the expected outcome of the game?

It is obvious that A wants a_{ij} to be as large as possible, and B wants a_{ij} as small as possible. However, A controls only the choice of i, while B controls only the choice of j. Thus, if A is to make a choice first, B will permit A to choose only the minimum a_{ij} for a given i, but A is free to choose the i which gives a maximum among these minimum values, i.e.,

$$\underset{i \leq m}{\text{Max}} \, (\underset{j \leq n}{\text{Min}} \, a_{ij}) \, ,$$

On the other hand, if B is to make a choice first, A gets no more than the maximum a_{ij} for any j that B may choose. Since B is free to choose j, A can expect to get only the minimum of these maximum values, i.e.,

$$\underset{j \leq n}{\text{Min}} \, (\underset{i \leq m}{\text{Max}} \, a_{ij}) \, ,$$

Because this approach is basically pessimistic, A wants only the guaranteed level of minimum payoff from B, while he may possibly get more; similarly, B wants the guaranteed level of maximum payoff to A, while he may possibly pay less. Hence, in general,

$$\underset{i \leq m}{\text{Max}} \, (\underset{j \leq n}{\text{Min}} \, a_{ij}) \leq \underset{j \leq n}{\text{Min}} \, (\underset{i \leq m}{\text{Max}} \, a_{ij}) \, ; \tag{10–1}$$

If it happens that the relationship is an equality,

$$\underset{i \leq m}{\text{Max}} \, (\underset{j \leq n}{\text{Min}} \, a_{ij}) = \underset{j \leq n}{\text{Min}} \, (\underset{i \leq m}{\text{Max}} \, a_{ij}) = a_{i^*j^*} \, ; \tag{10–2}$$

the matrix $[a_{ij}]$ is said to have a *saddle point* $a_{i^*j^*}$ where i^* and j^* are optimal strategies whereby A cannot do better than to choose i^* and B cannot do better than to choose j^*. The saddle point, which is simultaneously the minimum of the rows and the maximum of the columns of the matrix $[a_{ij}]$, is called the *value of the game*.

The significance of a saddle point or a set of optimal strategies i^* and j^* is that it has the following properties:

1. If A chooses i^*, then no matter which action among j that B chooses, A can get at least $a_{i^*j^*}$, that is,

$$a_{i^*j^*} \leq a_{i^*j} \, .$$

2. If B chooses j^*, then no matter which action among i that A chooses, B can hold A to at most $a_{i^*j^*}$, that is,

$$a_{ij^*} \leq a_{i^*j^*} \, .$$

3. Neither A nor B can have a higher level of security than by choosing i^* and j^*, respectively, since

$$a_{ij^*} \leq a_{i^*j^*} \leq a_{i^*j} \tag{10–3}$$

and

$$a_{i^*j^*} = v^* , \tag{10–4}$$

which is the value of the game. The pair of optimal pure strategies i^* and j^* is called the *solution of the game.*

It may be noted that even if (10–2) is satisfied and the game has a value, the set of optimal strategies i^* and j^* is not necessarily unique. However, any pair of optimal or equilibrium strategies will lead to the same payoff. Hence a game may have several saddle points. Furthermore, the structure of the game is not altered, in so far as the optimal strategies are concerned, if the same positive quantity is added to all elements of the payoff matrix.

The dimension of a payoff matrix may be reduced if every coefficient in a row (or column) of the matrix is greater than or equal to the corresponding coefficient in another row (or column). These rows (or columns) are said to be in a *dominant-recessive* relationship, because one of them can be eliminated from the payoff matrix without affecting the choice of optimal strategies. The one which is eliminated is called a *recessive* strategy, and the one which remains is called a *dominant* strategy.

Example 10–3. Determine the saddle point(s) and the value of the game represented by the payoff matrix in Table 10–6(a). Verify that the same saddle point(s) will remain when all elements of the payoff matrix are increased by one, as indicated in Part (b) of the table.

We note from Table 10–6(a) that the pair $i = 2$ and $j = 1$ is a saddle point because $a_{21} = 3$ is simultaneously the minimum of its row and the maximum of its column. Similarly, the pair $i = 3$ and $j = 1$ is also a saddle point because $a_{31} = 3$ also satisfies this condition. Hence the value of the game is 3. However, the pair $i = 3$ and $j = 3$ is not a saddle point, even though $a_{33} = 3$.

In Table 10–6(b), the saddle points are also located at $i = 2$ and $j = 1$,

Table 10–6

PAYOFF MATRICES FOR PURE STRATEGIES

(a)

i \ j	1	2	3
1	2	−2	−3
2	3	4	6
3	3	5	3

(b)

i \ j	1	2	3
1	3	−1	−2
2	4	5	7
3	4	6	4

Table 10–7

PAYOFF MATRIX FOR PURE STRATEGY

(a)

i \ j	1	2	3
1	3	−1	8
2	5	7	6
3	4	6	4

(b)

i \ j	1	2
1	3	−1
2	5	7

and at $i = 3$ and $j = 1$ because an addition of the same positive quantity to all elements of the payoff matrix does not alter the *relative* magnitudes of the elements. However, the value of the game is now increased by one, which is the number added to every element of the matrix.

Example 10–4. Determine the saddle point and the value of the game in Table 10–7(a).

From Table 10–7(a) it can be seen that the pair $i = 2$ and $j = 1$ is a saddle point because $a_{21} = 5$ is simultaneously the minimum of its row and the maximum of its column, and the value of the game is 5. The problem can be simplified by noting the dominant-recessive relationships in the rows and columns. Since every element in row 2 is greater than that in row 3, and since only the larger of the two numbers in a column need be considered in the determination of a maximum in a column, row 3 can be eliminated. Hence row 2 is said to be dominant and row 3 recessive; that is, row 3 is dominated by row 2. Similarly, every element in column 3 is greater than that in column 1, but only the smaller of the two numbers in a row need be considered in the determination of a minimum in a row. Hence column 3, which has consistently larger elements than column 1, is eliminated, and column 3 is said to be dominated by column 1. After the elimination of row 3 and column 3, the remaining elements of the payoff matrix are as shown in Table 10–7(b), from which the saddle point is found to be located at ($i = 2$, $j = 1$), and the value of the game is $a_{21} = 5$. Thus the elimination of recessive rows or columns does not alter the pair of optimal strategies, but does simplify the selection when the dimension of a given payoff matrix is large.

10–3 GAMES WITH MIXED STRATEGIES

If a game played with pure strategies has no value, we may attempt to solve it by the introduction of mixed strategies for the players. Let x_i ($i = 1, 2, \ldots, m$) be the probability of selecting action i by A, and y_j ($j = 1, 2, \ldots, n$) be the probability of selecting action j by B. Then, for

$x_i \geq 0$, and $y_j \geq 0$,

$$\sum_{i=1}^{m} x_i = 1 , \qquad \sum_{j=1}^{n} y_j = 1 ,$$

because the sum of each probability distribution is a certainty.

Suppose that A uses a mixed strategy x_i ($i = 1, 2, \ldots, m$) and B uses a pure strategy j. Then the expected payoff to A is seen to be the expected payoff based on column j of the matrix

$$[a_{ij}] = \begin{bmatrix} a_{11} & a_{12} & \cdots & a_{1j} & \cdots & a_{1n} \\ a_{21} & a_{22} & \cdots & a_{2j} & \cdots & a_{2n} \\ \vdots & & & & & \\ a_{i1} & a_{i2} & \cdots & a_{ij} & \cdots & a_{in} \\ a_{m1} & a_{m2} & \cdots & a_{mj} & \cdots & a_{mn} \end{bmatrix} .$$

Thus the expected payoff k_j corresponding to x_i is defined by

$$a_{ij}x_1 + a_{2j}x_2 + \cdots + a_{mj}x_m = k_j ,$$

or

$$\sum_{i=1}^{m} a_{ij}x_i = k_j . \tag{10-5}$$

For every j ($j = 1, 2, \ldots, n$) there is an expected payoff k_j for column j. Player A concedes that he will get only the minimum of all possible k_j, that is,

$$k = \operatorname*{Min}_{1 \leq j \leq n} k_j ,$$

since B controls the choice of j. However, A can select a mixed strategy $x_i \geq 0$ ($i = 1, 2, \ldots, m$) to *maximize* k, satisfying the following system of inequalities:

$$\begin{aligned} a_{11}x_1 + a_{21}x_2 + \cdots + a_{m1}x_m &= k_1 \geq k , \\ a_{12}x_1 + a_{22}x_2 + \cdots + a_{m2}x_m &= k_2 \geq k , \\ &\vdots \\ a_{1n}x_1 + a_{2n}x_2 + \cdots + a_{mn}x_m &= k_n \geq k , \end{aligned} \tag{10-6}$$

and

$$x_1 + x_2 + \cdots + x_m = 1 . \tag{10-7}$$

Now, suppose that B uses a mixed strategy y_j ($j = 1, 2, \ldots, n$) and that A uses a pure strategy i, then the expected payoff to A is the expected payoff of row i in matrix a_{ij}. Thus the expected payoff h_i corresponding to y_j is defined by

$$\sum_{j=1}^{n} a_{ij}y_j = h_i . \tag{10-8}$$

For every i ($i = 1, 2, \ldots, m$) there is an expected payoff h_i for row i. Player B realizes that he can only hold A to the maximum of all possible h_i, that is,

$$h = \operatorname*{Max}_{1 \le i \le m} h_i \,,$$

since A controls the choice of i. However, B can select a mixed strategy $y_j = 0$ ($j = 1, 2, \ldots, n$) to *minimize* h, satisfying the following system of inequalities:

$$\begin{aligned}
a_{11}y_1 &+ a_{12}y_2 + \cdots + a_{1n}y_n = h_1 \le h \,, \\
a_{21}y_1 &+ a_{22}y_2 + \cdots + a_{2n}y_n = h_2 \le h \,, \\
&\vdots \\
a_{m1}y_1 &+ a_{m2}y_2 + \cdots + a_{mn}y_n = h_m \le h \,,
\end{aligned} \tag{10-9}$$

and

$$y_1 + y_2 + \cdots + y_n = 1 \,. \tag{10-10}$$

If both A and B use mixed strategies x_i and y_j respectively, the expected payoff to A is obtained by averaging the expected payoffs k_j in (10–5) for all j, or by averaging the expected payoffs h_i in (10–8) for all i. Thus the expected payoff to A is either given by

$$v = \sum_{j=1}^{n} y_j \left(\sum_{i=1}^{m} a_{ij} x_i \right) \qquad \text{or by} \qquad v = \sum_{i=1}^{m} x_i \left(\sum_{j=1}^{n} a_{ij} y_j \right).$$

If there exists a pair of mixed strategies $x_i = x_i^*$ and $y_j = y_j^*$ such that both expressions for v equal v^*, then

$$v^* = \sum_{j=1}^{n} \sum_{i=1}^{m} a_{ij} x_i^* y_j^* \,. \tag{10-11}$$

10–4 THE MINIMAX THEOREM

Although each player chooses his own mixed strategy, i.e., the probability distribution, the particular pair of x_i^* and y_j^* which is eventually played in the game is selected according to the criterion of a guaranteed level of security for each player. For the convenience of subsequent discussion, let us restate our problem in matrix form. (See the matrix notation in Section 4–2.) Let $\mathbf{X}$ ($x_1, x_2, \ldots, x_m$) and $\mathbf{Y}$ ($y_1, y_2, \ldots, y_n$) be column vectors representing the probability distributions for the strategies of A and B, respectively. Let $\mathbf{A} = [a_{ij}]$ be an $m \times n$ matrix representing the payoff to A. Let $\mathbf{K}$ ($k_1, k_2, \ldots, k_n$) be a row vector, and $\mathbf{H}$ ($h_1, h_2, \ldots, h_m$) be a column matrix such that, from (10–5),

$$\mathbf{K}^T = \mathbf{X}^T \mathbf{A}$$

and from (10-8),

$$\mathbf{H} = \mathbf{AY},$$

where **K** is the set of expected payoffs to A, when A uses mixed strategy **X** and B uses pure strategy j ($j = 1, 2, \ldots, n$); and H is the set of expected payoffs to A, when B uses mixed strategy **Y** and A uses pure strategy i ($i = 1, 2, \ldots, m$). If both A and B use mixed strategies **X** and **Y**, respectively, the expected payoff to A is

$$v = \mathbf{K}^T\mathbf{Y} = \mathbf{X}^T\mathbf{H} = \mathbf{X}^T\mathbf{AY}. \tag{10-12}$$

Suppose that A chooses his strategy first. He expects to get only the minimum for all possible $\mathbf{Y}$ ($y_1, y_2, \ldots, y_n$) controlled by B. However, he can select $\mathbf{X}$ ($x_1, x_2, \ldots, x_m$) to maximize the minimum, i.e.,

$$\operatorname*{Max}_{\mathbf{X}} \left(\operatorname*{Min}_{\mathbf{Y}} \mathbf{X}^T\mathbf{AY} \right).$$

Similarly, B expects to pay the maximum for all possible $\mathbf{X}$ ($x_1, x_2, \ldots, x_m$) controlled by A, but he can select $\mathbf{Y}$ ($y_1, y_2, \ldots, y_n$) to minimize the maximum, i.e.,

$$\operatorname*{Min}_{\mathbf{Y}} \left(\operatorname*{Max}_{\mathbf{X}} \mathbf{X}^T\mathbf{AY} \right).$$

Since the approaches of both players are conservative, it is clear that

$$\operatorname*{Max}_{\mathbf{X}} \operatorname*{Min}_{\mathbf{Y}} \mathbf{X}^T\mathbf{AY} \leq \operatorname*{Min}_{\mathbf{Y}} \operatorname*{Max}_{\mathbf{X}} \mathbf{X}^T\mathbf{AY}. \tag{10-13}$$

The minimax theorem of von Neumann states that a pair of mixed strategies $\mathbf{X}^*$ and $\mathbf{Y}^*$ always exists such that

$$\operatorname*{Max}_{\mathbf{X}} \operatorname*{Min}_{\mathbf{Y}} \mathbf{X}^T\mathbf{AY} = \operatorname*{Min}_{\mathbf{Y}} \operatorname*{Max}_{\mathbf{X}} \mathbf{X}^T\mathbf{AY} = \mathbf{X}^{*T}\mathbf{AY}^*, \tag{10-14}$$

where the pair of mixed strategies $\mathbf{X}^*$ and $\mathbf{Y}^*$ is called the optimal mixed strategies. The expected payoff to A corresponding to $\mathbf{X}^*$ and $\mathbf{Y}^*$ is given by

$$v^* = \mathbf{X}^{*T}\mathbf{AY}^* = \sum_{j=1}^{n} \sum_{i=1}^{m} a_{ij} x_i^* y_j^*, \tag{10-15}$$

which is called the value of the game. The proof of this theorem will not be given here.

The pair of optimal mixed strategies has the following properties:

1. If A chooses $\mathbf{X}^*$, then no matter which strategy B chooses, A can get at least v^*, since

$$v^* = \sum_{j=1}^{n} \sum_{i=1}^{m} a_{ij} x_i^* y_j^* \leq \sum_{j=1}^{n} \sum_{i=1}^{m} a_{ij} x_i^* y_j. \tag{10-16}$$

2. If B chooses $\mathbf{Y}^*$, then no matter which strategy A chooses, B can hold A to at most v^*, since

$$\sum_{j=1}^{n}\sum_{i=1}^{m} a_{ij}x_i y_j^* \leq \sum_{j=1}^{n}\sum_{i=1}^{m} a_{ij}x_i^* y_j^* = v^* . \tag{10–17}$$

3. Neither A nor B can have a higher level of guarantee than by choosing x_i^* and y_j^*, respectively, since

$$\sum_{j=1}^{n}\sum_{i=1}^{m} a_{ij}x_i y_j^* \leq v^* \leq \sum_{j=1}^{n}\sum_{i=1}^{m} a_{ij}x_i^* y_j . \tag{10–18}$$

Hence the pair of optimal mixed strategies $\mathbf{X}_i^*$ and $\mathbf{Y}_j^*$ is called the solution of the game.

Example 10–5. Solve the game in Example 10–2 by the introduction of mixed strategies.

We can use the payoff matrix of reduced dimension in Table 10–5(b). From (10–12), we have

$$\begin{aligned} v &= [x_1 \quad x_2] \begin{bmatrix} +0.250 & -0.112 \\ -0.250 & +0.200 \end{bmatrix} \begin{bmatrix} y_1 \\ y_2 \end{bmatrix} \\ &= (+0.250x_1 - 0.250x_2)y_1 + (-0.112x_1 + 0.200x_2)y_2 \\ &= x_1(+0.25y_1 - 0.112y_2) + x_2(-0.250y_1 + 0.200y_2) . \end{aligned}$$

According to (10–16),

$$(+0.250x_1^* - 0.250x_2^*)y_1 + (-0.112x_1^* + 0.200x_2^*)y_2 \geq v^* .$$

Since this relation is true for all $y_j \geq 0$, where $\sum_j y_j = 1$, it is true for both ($y_1 = 1$, $y_2 = 0$) and ($y_1 = 0$, $y_2 = 1$). Thus

$$+0.250x_1^* - 0.250x_2^* \geq v^* , \tag{a}$$

$$-0.112x_1^* + 0.200x_2^* \geq v^* , \tag{b}$$

and

$$x_1^* + \qquad x_2^* = 1 . \tag{c}$$

Similarly, from (10–17), we get

$$+0.250y_1^* - 0.112y_2^* \leq v^* , \tag{d}$$

$$-0.250y_1^* + 0.200y_2^* \leq v^* , \tag{e}$$

and

$$y_1^* + \qquad y_2^* = 1 . \tag{f}$$

Thus the game problem is expressed by this set of inequalities and equations. As we shall see later, this set of relations can be converted into a linear programming problem. For the time being, let us suggest a trial-and-error method in assuming all inequalities to be equations. If a solution for x_1^*, x_2^*, y_1^*, and y_2^* satisfies all six equations, we have a pair of optimal mixed strategies. If not, we will restore one or more of the inequality signs until a solution is obtained. For this particular problem, the pair of optimal mixed strategies satisfying the six relations is found to be as follows:

$$x_1^* = 0.554\,, \qquad x_2^* = 0.446\,,$$
$$y_1^* = 0.384\,, \qquad y_2^* = 0.616\,.$$

The value of the game may be obtained from any of the four equations containing v^*. As a check on the assumption of changing all inequalities to equations, we compute v^* from all four of them, namely:

a) $(0.250)(0.554) - (0.250)(0.446) = +0.027\,,$

b) $-(0.112)(0.554) + (0.200)(0.446) = +0.027\,,$

d) $(0.250)(0.384) - (0.112)(0.616) = +0.027\,,$

e) $-(0.250)(0.384) + (0.200)(0.616) = +0.027\,,$

Hence, $v^* = +0.027$ is the value of the game.

10-5 REDUCTION OF A GAME TO A LINEAR PROGRAM

In the remaining sections of this chapter, we shall consider the similarity between games with mixed strategies and linear programs. Once a game is reduced to a linear program, it can be solved by the methods developed in earlier chapters for the solution of linear programming problems.

If all $a_{ij} \geq 0$ in the payoff matrix to A, the values of h and k in the relations of (10-6) and (10-9), respectively, will always be positive, since $x_i \geq 0$ and $y_j \geq 0$. If not all $a_{ij} \geq 0$, we can always add a positive number to every element of the matrix so that $a_{ij} \geq 0$. Hence we can assume that all $a_{ij} \geq 0$ without losing generality. Let $x_i'k = x_i$ and make the substitutions in the relations of (10-6) and (10-7). We have for $x_i' \geq 0$,

$$\begin{aligned} a_{11}x_1' + a_{21}x_2' + \cdots + a_{m1}x_m' &\geq 1\,, \\ a_{12}x_1' + a_{22}x_2' + \cdots + a_{m2}x_m' &\geq 1\,, \\ &\vdots \\ a_{1n}x_1' + a_{2n}x_2' + \cdots + a_{mn}x_m' &\geq 1\,, \end{aligned} \tag{10-19}$$

and

$$x_1' + x_2' + \cdots + x_m' = \text{Min}\,\frac{1}{k}\,,$$

since minimization of $1/k$ is equivalent to the maximization of k for $k \geq 0$. Similarly, let $y_j'h = y_j$ and make the substitutions in the relations of (10-8) and (10–9). We have for $y'_j \geq 0$,

$$\begin{aligned} a_{11}y'_1 + a_{12}y'_2 + \cdots + a_{1n}y'_n &\leq 1 , \\ a_{21}y'_1 + a_{22}y'_2 + \cdots + a_{2n}y'_n &\leq 1 , \\ &\vdots \\ a_{m1}y'_1 + a_{m2}y'_2 + \cdots + a_{mn}y'_n &\leq 1 , \end{aligned} \tag{10–20}$$

and

$$y'_1 + y'_2 + \cdots + y'_n = \text{Max}\,\frac{1}{h} .$$

Note that (10–19) and (10–20) represent a set of primal-dual problems in linear programming. If an optimal solution exists for the primal, it also exists for the dual. Then, for $x'_i = x^0_i$ $(i = 1, 2, \ldots, m)$ and $y'_j = y^0_j$ $(j = 1, 2, \ldots, n)$, which are the values of primal and dual variables leading to the optimal solution, we have

$$\min \frac{1}{k} = \max \frac{1}{h} = \frac{1}{v^0} , \tag{10–21}$$

or

$$\max k = \min h = v^0 , \tag{10–22}$$

where v^0 is the value of the game corresponding to the adjusted $a_{ij} \geq 0$. Hence the pair of optimal mixed strategies in a game can be obtained from the solution of the primal-dual linear programming problems.

Example 10–6. Reformulate the problem in Example 10–5 as linear programming problems.

First, we shall add a positive number, $+0.250$, to every element of the payoff matrix in Example 10–5 so that all $a_{ij} \geq 0$. Thus

$$\begin{bmatrix} +0.250 & -0.112 \\ -0.250 & +0.200 \end{bmatrix} + \begin{bmatrix} +0.250 & +0.250 \\ +0.250 & +0.250 \end{bmatrix} = \begin{bmatrix} 0.500 & 0.138 \\ 0 & 0.450 \end{bmatrix} .$$

Then from Relations (10–19) we have, for $x'_1 \geq 0$, $x'_2 \geq 0$,

$$\begin{aligned} 0.500x'_1 \qquad\qquad &\geq 1 , \\ 0.138x'_1 + 0.450x'_2 &\geq 1 , \\ x'_1 + \qquad x'_2 &= \text{Min}\,\frac{1}{k} ; \end{aligned}$$

and from the relations in (10–20), we have

$$\begin{aligned} 0.500y'_1 + 0.138y'_2 &\leq 1 , \\ 0.450y'_2 &\leq 1 , \\ y'_1 + \qquad y'_2 &= \text{Max}\,\frac{1}{h} . \end{aligned}$$

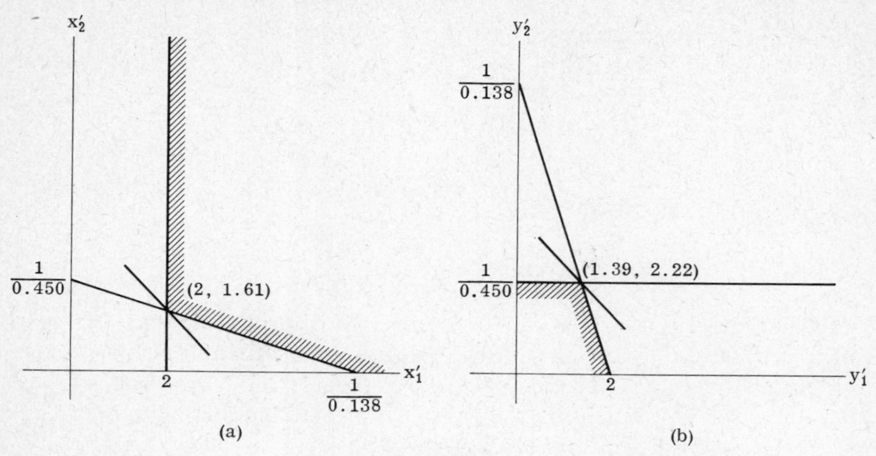

Figure 10-2

The solutions of the primal and dual problems are given graphically in Fig. 10-2, (a) and (b) respectively. Thus

$$x_1^0 = 2.00\,, \qquad x_2^0 = 1.61\,, \qquad \frac{1}{k} = 2.00 + 1.61 = 3.61\,,$$

$$y_1^0 = 1.39\,, \qquad y_2^0 = 2.22\,, \qquad \frac{1}{h} = 1.39 + 2.22 = 3.61\,,$$

Hence

$$v^0 = \max k = \min h = \frac{1}{3.61} = 0.277\,.$$

We can then convert these results to those for the original payoff matrix by noting that

$$x_i^* = x_i^0 k \qquad \text{and} \qquad y_j^* = y_j^0 h\,.$$

Hence

$$x_1^* = x_1^0 k = \frac{2.0}{3.61} = 0.554\,, \qquad x_2^* = x_2^0 k = \frac{1.61}{3.61} = 0.446\,,$$

$$y_1^* = y_1^0 h = \frac{1.39}{3.61} = 0.384\,, \qquad y_2^* = y_2^0 h = \frac{2.22}{3.61} = 0.616\,.$$

Also,

$$v^* = v^0 - 0.250 = 0.277 - 0.250 = +0.027\,,$$

in which 0.250 is the number that was added to the payoff matrix before we converted the game problem to linear programming problems.

If we had ignored the dominant-recessive relationship and started the problem with the matrix in Table 10-5(a) of Example 10-2, we would have obtained

$$\begin{bmatrix} +0.250 & -0.112 & +0.142 \\ -0.250 & +0.200 & +0.200 \\ -0.500 & -0.400 & +0.112 \end{bmatrix} + 0.500 \begin{bmatrix} 1 & 1 & 1 \\ 1 & 1 & 1 \\ 1 & 1 & 1 \end{bmatrix}$$

$$= \begin{bmatrix} 0.750 & 0.388 & 0.642 \\ 0.250 & 0.700 & 0.700 \\ 0 & 0.100 & 0.612 \end{bmatrix}.$$

Then

$$\begin{aligned} 0.750x_1' + 0.250x_2' &\geq 1\,, \\ 0.388x_1' + 0.700x_2' + 0.100x_3' &\geq 1\,, \\ 0.642x_1' + 0.700x_2' + 0.612x_3' &\geq 1\,, \\ x_1' + x_2' + x_3' &= \text{Min}\,\frac{1}{k}\,; \end{aligned}$$

and

$$\begin{aligned} 0.750y_1' + 0.388y_2' + 0.642y_3' &\leq 1\,, \\ 0.250y_1' + 0.700y_2' + 0.700y_3' &\leq 1\,, \\ 0.100y_2' + 0.612y_3' &\leq 1\,, \\ y_1' + y_2' + y_3' &= \text{Max}\,\frac{1}{h}\,. \end{aligned}$$

It can easily be shown by the direct solution of the primal-dual linear programming problem that

$$x_1^* = 0.554\,, \qquad x_2^* = 0.446\,, \qquad x_3^* = 0\,,$$
$$y_1^* = 0.384\,, \qquad y_2^* = 0.616\,, \qquad y_3^* = 0\,,$$

and

$$v^* = +0.027\,.$$

Hence the elimination of the recessive rows and columns has, in effect, eliminated actions with zero probability in the mixed strategies.

10-6 SADDLE-POINT EQUIVALENCE FOR PRIMAL-DUAL LINEAR PROGRAMS

The primal-dual linear programming problems defined by (10-19) and (10-20) can be written in the general inequality form consistent with the game with mixed strategies. The *primal problem* is given by

$$x'_i \geq 0\,, \qquad i = 1, 2, \ldots, m\,;$$
$$\sum_{i=1}^{m} a_{ij}x'_i \geq b_j\,, \qquad j = 1, 2, \ldots, n\,; \tag{10-23}$$
$$\sum_{i=1}^{m} c_j x'_j = \text{Min}\,\frac{1}{k} = \text{Min}\,z\,;$$

and the *dual problem* is given by

$$y'_j \geq 0\,, \qquad j = 1, 2, \ldots, n\,;$$
$$\sum_{j=1}^{n} a_{ij}y'_j \leq c_i\,, \qquad i = 1, 2, \ldots, m\,; \tag{10-24}$$
$$\sum_{j=1}^{n} b_j y'_j = \text{Max}\,\frac{1}{h} = \text{Max}\,A\,.$$

If feasible solutions exist to both the primal and the dual problems, an optimal solution exists to both such that max A = min z, or $A^0 = z^0$. The complete solution to each problem is given by the set $(x_1^0, x_2^0, \ldots, x_m^0, z^0)$ for the primal, and by the set $(y_1^0, y_2^0, \ldots, y_n^0, A^0)$ for the dual.

Let us multiply the jth constraint inequality in the relations (10-23) by $y'_j \geq 0$ and sum from $j = 1$ to $j = n$. Then

$$\sum_{j=1}^{n}\sum_{j=1}^{m} a_{ij}x'_i y'_j \geq \sum_{j=1}^{n} b_j y'_j\,.$$

For $y'_j = y_j^0$ $(i = 1, 2, \ldots, n)$,

$$\sum_{j=1}^{n}\sum_{i=1}^{m} a_{ij}x'_i y_j^0 \geq \sum_{j=1}^{n} b_j y_j^0 = A^0\,.$$

Similarly, multiply the ith constraint inequality in the relations (10-24) by $x'_i \geq 0$ and sum from $i = 1$ to $i = m$. Thus

$$\sum_{i=1}^{n}\sum_{i=1}^{m} a_{ij}y'_j x'_i \leq \sum_{i=1}^{m} c_i x'_i\,.$$

For $x'_i = x_i^0$ $(i = 1, 2, \ldots, m)$,

$$\sum_{j=1}^{n}\sum_{i=1}^{m} a_{ij}y'_j x_i^0 \leq \sum_{i=1}^{m} c_i x_i^0 = z^0\,.$$

From the duality theorem in linear programming we have $A^0 = z^0$. Hence

$$\sum_{j=1}^{n}\sum_{i=1}^{m} a_{ij}y'_j x_i^0 \leq z^0 = A^0 \leq \sum_{j=1}^{n}\sum_{i=1}^{m} a_{ij}x'_i y_j^0\,. \tag{10-25}$$

If $y'_j = y^0_j$ for the left-hand inequality, and $x'_i = x^0_i$ for the right-hand inequality, the relation in (10–25) becomes

$$z^0 = A^0 = \sum_{j=1}^{n} \sum_{i=1}^{m} a_{ij} x^0_i y^0_j \,. \tag{10–26}$$

Thus the sets x^0_i and y^0_j for the optimal solution of the primal-dual problems in linear programming are analogous to the optimal mixed strategies in a two-person zero-sum game. The optimal solution $z^0 = A^0$ may be regarded as the *saddle point of the primal-dual linear programming problem*.

10–7 CONCLUSION

In this chapter, we have discussed competitive games with pure and mixed strategies. The equivalence of games with mixed strategies and linear programs has been established, and the solution of such games may therefore be obtained from the optimal solution of the equivalent primal-dual linear programming problems.

More significant, perhaps, is the fact that through the use of games we have illustrated the possible conditions under which decisions are made. For decision making under certainty, such as in games with pure strategies, the objective is clear and the payoff is an adequate measure of preference. For decision making under uncertainty, such as in games with mixed strategy, the probability of occurrence associated with each outcome must be found.

These properties of decision making under conditions of certainty and uncertainty are not confined to optimization with competition, but are equally applicable to optimization without competition. In this book, we have considered primarily decision making under certainty, i.e., *deterministic decision making*. This chapter serves at least as a reminder of the vast number of nondeterministic decision problems which must be dealt with by the use of probability and statistical methods. The discussion of nondeterministic decision problems in engineering design and operation merits a separate treatise, and is beyond the scope of this volume.

REFERENCES

10–1. VON NEUMANN, J., and MORGENSTERN, O., *Theory of Games and Economic Behavior*, 3rd Ed., Princeton University Press, Princeton, N. J., 1953.

10–2. WILLIAMS, J. D., *The Compleat Strategyst*, McGraw-Hill, New York, 1954.

10–3. VAJDA, S., *The Theory of Games and Linear Programming*, Wiley, New York, 1954.

PROBLEMS

P10-1 through P10-4. Determine the values of the games for the problems with payoff matrices shown in Tables P10-1 through P10-4.

Table P10-1

		B			
		1	2	3	4
A	1	30	3	9	15
	2	27	33	15	22
	3	17	18	7	4
	4	9	13	11	26

Table P10-2

		B			
		1	2	3	4
A	1	5	15	4	13
	2	3	12	9	11
	3	7	13	8	9
	4	1	5	6	4

Table P10-3

		B		
		1	2	3
A	1	−2	8	3
	2	4	−3	1
	3	9	7	5

Table P10-4

		B		
		1	2	3
A	1	16	12	8
	2	3	5	6
	3	7	19	4

P10-5 through P10-8. For the problems with a payoff matrix shown in Tables P10-5 through P10-8, determine the values of the games by solving the equivalent linear programming problems.

Table P10-5

		B		
		1	2	3
A	1	8	5	7
	2	9	6	−1
	3	4	5	2

Table P10-6

		B		
		1	2	3
A	1	3	7	6
	2	5	11	4
	3	4	8	9

Table P10-7

		B			
		1	2	3	4
A	1	12	−1	4	5
	2	4	3	2	7
	3	6	3	−3	2
	4	9	7	3	8

Table P10-8

		B			
		1	2	3	4
A	1	4	5	3	2
	2	7	−2	6	5
	3	8	2	7	6
	4	9	8	4	3

P10–9. Given that the payoff matrix for a zero-sum two-person game corresponding to mixed strategies (x_1, x_2) and (y_1, y_2) is

$$[a_{ij}] = \begin{bmatrix} a_{11} & a_{12} \\ a_{21} & a_{22} \end{bmatrix},$$

in which there is no saddle point, show that the value of the game is

$$v^* = \frac{a_{11}a_{22} - a_{12}a_{21}}{a_{11} + a_{22} - a_{12} - a_{21}}.$$

SELECTED ANSWERS TO PROBLEMS

Chapter 1

P1–1. $M = 0$, $3PL/16$, $PL/4$, $3PL/16$, 0 for $x = 0$, $L/4$, $L/2$, $3L/4$, L, respectively.

P1–2. (a) $z^* = d$, (b) $z^* = \sqrt{3}d$

P1–6. $A^* = 16$, $x^* = 2\sqrt{3}$, $y^* = 2$

P1–8. Group I: \$45,000; Group II: \$25,000

Chapter 2

P2–4. $M = (P/2L)(L - b/2)^2$

P2–6. $(0, 0, +2)$ is a relative max, $(0, 0, -2)$ is a relative min

P2–8. $a = 0.4268$, $b = -4.1715$, $c = 13.7002$

P2–10. $x_1^* = 0.65$, $x_2^* = 1.35$, $z^* = -6.25$

Chapter 3

P3–2. $x_1^* = 1$, $x_2^* = 0$, $z^* = 1$

P3–4. $x_1^* = 2$, $x_2^* = 2$, $z^* = 12$

P3–6. $x_1^* = \frac{2}{3}$, $x_2^* = \frac{1}{3}$, $z^* = \frac{2}{3}$

P3–7. $y_1^* = 0$, $y_2^* = \frac{1}{2}$, $z^* = -1$

P3–9. $x_1^* = 9$, $x_2^* = 14$, $A^* = 37$

P3–11. $x_1^* = 0.4$, $x_2^* = 0.9$, $z^* = 9.2$

P3–13. $y_1^* = 1.2$, $y_2^* = 1.4$, $y_3^* = 0$, $z^* = -9.2$

P3–15. Infeasible; $w = 5$

P3–17. 2.67 million gal/day from water company; 1.33 million gal/day from stream. Cost = \$333/day.

Chapter 4

P4–1. $y_1^* = 0$, $y_2^* = \frac{1}{2}$, $z^* = -1$

P4–3. $x_1^* = 9$, $x_2^* = 14$, $A^* = 37$

P4–6. $x_1^* = 1$, $x_2^* = 0$, $z^* = 1$

P4–8. $x_1^* = 2$, $x_2^* = 2$, $z^* = 12$

P4–12. $x_1^* = x_2^* = x_5^* = x_6^* = x_7^* = 0$, $x_3^* = \frac{1}{2}$, $x_4^* = \frac{3}{2}$, $z^* = 24$

P4–14. (a) $\Delta z = 0$; (b) $\Delta z = 0$

Chapter 5

P5–1. $z^* = 55$

P5–3. $z^* = 61$

P5–6. (a) $z^* = 205$; (b) $z^* = 190$

P5–7. $z^* = 138$

P5–9. $z^* = 380$

P5–10. $z^* = 37$

P5–12. $z^* = 77$

P5–13. $z^* = 4$

Chapter 6

P6–5. $F^* = 18$
P6–6. $F^* = 19$
P6–9. $F^* = 18$
P6–10. $F^* = 39$
P6–11. $F^* = 19$

Chapter 7

P7–1. $z^* = 137$
P7–3. $z^* = 450$
P7–5. $z^* = 980$
P7–7. $z^* = 980$
P7–9. $z^* = 880$
P7–11. $z^* = 1080$
P7–13. $d^* = 21$
P7–15. $d^* = 14$

Chapter 8

P8–5. $z^* = 41$
P8–7. $z^* = 41$
P8–9. $z^* = 18$. For earliest start time schedule, $S' = 8090$
P8–11. $z^* = 20$. For earliest start time schedule, $S' = 2034$
P8–12. (a) 93; (b) 91.
P8–13. Normal duration, 41 days; normal cost, \$3340;
All-crash duration, 30 days; all-crash cost, \$3685
P8–15. Normal duration, 41 days; normal cost, \$1110;
All-crash duration, 30 days; all-crash cost, \$1450

Chapter 9

P9–2. $x_1^* = b/a_1,\ x_2^* = 0,\ A^* = c_1b^2/a_1^2$, if $c_1b^2/a_1^2 \geq c_2b^2a_2^2$;
$x_1^* = 0,\ x_2^* = b/a_2,\ A^* = c_2b^2/a_2^2$, if $c_1b^2/a_1^2 \leq c_2b^2a_2^2$.

P9–4.
$$x_1^* = \frac{a_2}{d_1a_2 + d_2a_1}\left[\ln\left(\frac{d_1c_1a_2}{d_2c_2a_1}\right) + \frac{d_2b}{a_2}\right]$$
$$x_2^* = \frac{a_1}{d_1a_2 + d_2a_1}\left[\ln\left(\frac{d_2c_2a_1}{d_1c_1a_2}\right) + \frac{d_1b}{a_1}\right]$$

P9–6. $d^* = 18$
P9–8. $d^* = 14$

Chapter 10

P10–1. $v^* = 15$
P10–3. $v^* = 5$
P10–5. $v^* = 5.21$
P10–7. $v^* = 3.44$

INDEX

BCDE79876543210